Lecture Notes in Energy

Volume 96

Lecture Notes in Energy (LNE) is a series that reports on new developments in the study of energy: from science and engineering to the analysis of energy policy. The series' scope includes but is not limited to, renewable and green energy, nuclear, fossil fuels and carbon capture, energy systems, energy storage and harvesting, batteries and fuel cells, power systems, energy efficiency, energy in buildings, energy policy, as well as energy-related topics in economics, management and transportation. Books published in LNE are original and timely and bridge between advanced textbooks and the forefront of research. Readers of LNE include postgraduate students and non-specialist researchers wishing to gain an accessible introduction to a field of research as well as professionals and researchers with a need for an up-to-date reference book on a well-defined topic. The series publishes single- and multi-authored volumes as well as advanced textbooks.

Indexed in Scopus and EI Compendex The Springer Energy board welcomes your book proposal. Please get in touch with the series via Anthony Doyle, Executive Editor, Springer (anthony.doyle@springer.com)

Xin-Rong Zhang · Trygve Magne Eikevik
Editors

CO_2 Refrigeration Cycle and Systems

Springer

Editors
Xin-Rong Zhang
College of Engineering
PKU
Beijing, China

Trygve Magne Eikevik
Department of Energy and Process Engineering
NTNU
Trondheim, Norway

ISSN 2195-1284 ISSN 2195-1292 (electronic)
Lecture Notes in Energy
ISBN 978-3-031-22511-6 ISBN 978-3-031-22512-3 (eBook)
https://doi.org/10.1007/978-3-031-22512-3

This Springer imprint is published by the registered company Springer Nature Switzerland AG
The registered company address is: Gewerbestrasse 11, 6330 Cham, Switzerland

Preface

Since the late 1980s, interest in utilizing CO_2 as a refrigerant has increased considerably as the usage of CFC and HCFC became a pressing issue. Several factors contributed to the rise of CO_2usage. For example, from the viewpoint of protecting the ozone layer and preventing global warming, there is now a strong demand for technology based on ecologically safe 'natural' working fluids like carbon dioxide. CO_2 is a non-flammable natural fluid with no Ozone Depletion Potential (ODP) and a negligible Global Warming Potential (GWP). Besides, CO_2 is currently responsible for over 60% of the greenhouse effect. Furthermore, HCFC and HFC exacerbate CO_2 emission tremendously and equivalently. Fortunately, this effect can be minimized by recycling CO_2 and using it as a refrigerant. The CO_2 thermodynamic and transport properties are favorable in terms of heat transfer and pressure drop, where the critical pressure and temperature of CO_2 are 7.38 MPa (73.8 bar) and 31.1 °C, respectively. Due to these advantages, CO_2 fluid has received much attention in recent years in some new energy systems, especially in the CO_2 transcritical compression refrigeration thermodynamics cycle of cooling, refrigeration, and air conditioners.

CO_2 cooling and refrigeration is an ongoing technology that is promising and will be leading the future refrigeration field and market. Therefore, this book covers both fundamentals and applications of CO_2 vapor compression refrigeration thermodynamic cycles. Specifically, new application areas will be presented in this new book, for example, the creation of snow, industrial cooling, and refrigeration. Furthermore, this system and technology are closely related to phase change flow and heat transfer, fluid compressing and expanding flow, supercritical fluid flow and heat transfer, etc., which are also focused on in this book.

CO_2 refrigeration is an ongoing eco-friendly technology that is promising and will be leading the future refrigeration and air-conditioning field and market. This book is useful for undergraduates, postgraduates, researchers, engineers, and policymakers interested in gaining academic and applicable knowledge of the most potential refrigeration technology using CO_2.

The first six chapters highlight the fundamental principles and contents of physics and thermodynamics on which the CO_2 refrigeration cycle is based. The next five chapters present the application aspects, which mainly include commercial, air-conditioning, industrial, ice rink, trans-triple-point refrigeration methods, etc.

I wish to thank Prof. T. M. Eikevik, NTNU, Norway and Prof. H. Yamaguchi, Doshisha University of Japan. Great appreciation is also expressed to all the chapter authors, the China National Petroleum Corporation-Peking University Strategic Cooperation Project of Fundamental Research, the National Key Research and Development Program of China (No. 2021YFF0306803), and Mrs. Liu Jia.

Beijing, China — Xin-Rong Zhang
Trondheim, Norway — Trygve Magne Eikevik

Contents

Chapter 1
CO_2 Refrigeration Cycle and Systems

Yi-Sai Gao, Zhao-Rui Peng, and Xin-Rong Zhang

1.1 Introduction

1.1.1 The Challenge and Trend of Refrigeration

Refrigeration plays a pivotal role in today's society in preserving food, providing thermal comfort in the built environment, industry, and a lot of fields. According to the International Institute of Refrigeration, refrigeration consumes about 15% of all electricity consumed worldwide. Modern life would be difficult to visualize without refrigeration. The most widely used current applications of refrigeration are for the air-conditioning of private homes and public buildings, and the refrigeration of foodstuffs in homes, restaurants, and large storage warehouses. In commerce and manufacturing, there are also many uses for refrigeration. In compressed air purification, it is used to condense water vapor from compressed air to reduce its moisture content. In oil refineries, chemical plants, and petrochemical plants, refrigeration is used to maintain certain processes at their required low temperatures. In transporting temperature-sensitive foodstuffs and other materials by trucks, trains, airplanes, and sea-going vessels, refrigeration is a necessity as well.

Most of the commercial refrigeration and air-conditioning systems operate on a vapor compression refrigeration cycle in which the refrigerant changes phase first from liquid to gas and then gas to liquid in a closed cycle to generate cooling in the evaporator. Today's refrigerants are predominantly from a group of compounds called halocarbons (halogenated hydrocarbons) or specifically chlorofluorocarbons (CFCs), hydro-chlorofluorocarbons (HCFCs), hydrofluorocarbons (HFCs). CFCs,

Y.-S. Gao · Z.-R. Peng · X.-R. Zhang (✉)
Department of Energy and Resources Engineering, College of Engineering, Peking University, Beijing 100871, China
e-mail: xrzhang@pku.edu.cn

Beijing Engineering Research Center of City Heat, Beijing 100871, China

X.-R. Zhang and T. M. Eikevik (eds.), *CO_2 Refrigeration Cycle and Systems*, Lecture Notes in Energy 96, https://doi.org/10.1007/978-3-031-22512-3_1

HCFCs as well as HFCs are nontoxic, non-flammable, and are used extensively as coolants for commercial and home refrigeration units, aerosol propellants, electronic cleaning solvents, and blowing agents. When securely contained in a properly operating system, refrigerants do not impact climate change; however, system leaks and improper recovery of refrigerants during repairs or at the end of life result in these harmful gases entering the atmosphere, leading to environmental problems, such as ozone depletion and global warming.

In detail, as for ozone depletion, Sherwood Rowland and Mario Molina predicted that chlorofluorocarbon refrigerant gases would reach the high stratosphere and damage the protective mantle of the oxygen allotrope, ozone in 1974. In 1985 the "ozone hole" was discovered over the Antarctic and by 1990 Rowland and Molina's prediction was proved correct. When some kinds of these molecules drift into the stratosphere (like CFCs and HCFCs), the UVB and UV-C radiation from the sun releases their chlorine atoms. Complex chemical reactions in the atmosphere result in the formation of chlorine monoxide, which reacts with the ozone molecule to form oxygen and regenerates more chlorine atoms that carry on converting the ozone molecules. Each chlorine atom can destroy as many as 100,000 ozone molecules over 100 years. Thus, even a small amount of CFC or HCFC can cause tremendous damage to the ozone layer.

Nowadays, tough environmental laws and stringent government policies have revolutionized the refrigeration sector, especially concerning the cycle fluid known as the refrigerant. After years of successful deliberations for tackling the grievous problem of ozone depletion, the United Nations' Environmental Protection Agency concluded a multinational agreement called the "Montreal Protocol" for controlling the use of gases threatening the ozone layer. Besides, the Kyoto Agreement, an international environmental treaty, was produced at the United Nations Conference on Environment and Development (UNCED). The treaty is intended to achieve stabilization of greenhouse gas concentrations in the atmosphere at a level that would prevent dangerous anthropogenic interference with the climate system. With the development of technology and society, increasingly more attention has been paid to energy saving and environmental protection based on the high efficiency of refrigeration. Because traditional refrigerants have a serious impact on the ozone layer and climate warming, searching for alternatives has been realized as issues of common concern all over the world.

Another crucial problem is the harm of traditional refrigerants to human health once they leak out. Especially, people working with these fluids are more likely to be injured during the operation. For example, upon contact with moisture, including tissue, hydrofluorocarbons (HFCs) and hydro-chlorofluorocarbons (HCFCs) immediately convert to hydrofluoric acid, which is highly corrosive and toxic, and requires immediate medical attention upon exposure. Breathing in hydrogen fluoride at high levels or in combination with skin contact can cause death from an irregular heartbeat or fluid buildup in the lungs. Therefore, due to safety considerations, more natural and non-toxic refrigerants are urgently required.

Refrigerant selection is a key design decision that influences the mechanical design of refrigeration equipment. Factors that must be considered in refrigerant selection include performance, safety, reliability, environmental acceptability, and cost. However, the primary requirements are safety, reliability and, nowadays, environmental friendliness (in terms of ozone depletion and global warming potential). As for the mentioned environmental concerns, an alternative to traditional refrigerants is to apply naturally occurring and ecologically safe substances, the so-called natural working fluids. Nowadays, the refrigeration industry has accepted the challenge and trend of traditional refrigerants phase-out, and new eco-friendly nature refrigerants are replacing traditional refrigerants in all commercial and industrial applications.

1.1.2 CO_2—A Potential Natural Refrigerant

1.1.2.1 Refrigerant Selection

The most important substances of natural working fluids are hydrocarbons (such as isobutane, propane, etc.), ammonia, carbon dioxide, and water (Pearson 2004). Among these, ammonia is suffered from its toxicity and flammability, with reluctance; hydrocarbons are generally perceived as dangerous for use in large charge industrial systems and so have not received serious consideration. Water is another natural refrigerant but it is inappropriate for low-temperature plants and air systems are far too inefficient to be considered for most applications.

When safety concerns are raised (toxicity and flammability), R744, a CO_2-based refrigerant gas, becomes one of the best substitutes. The importance of CO_2 as an attractive alternative has increased manifolds because of its environmental and economic properties. Particularly its use in the supercritical range is in active consideration in mobile air conditioners and other applications. The use of CO_2 has already started in cascade systems for temperatures down to −54 °C, and in heat pumps for hot water. Improvement of the energy efficiency of the refrigerating cycles and development of commercial refrigerating system components are priorities, especially in vehicles' air conditioning systems.

Carbon dioxide is a natural substance, which has a lot of advantages (Ge and Cropper 2009):

(1) Environmentally friendly: ODP (ozone depletion potential) = 0.
(2) It is the highest oxidation state of carbon having very stable chemical properties which will not produce harmful gases even in high-temperature decomposition.
(3) It is safe, non-toxic, and non-flammability.
(4) It does not have to take the recycling, regeneration, and other measures in the operation and maintenance.
(5) It has the thermodynamic properties suited to refrigeration cycle and equipment, like evaporation latent heat is large, unit volume of cooling capacity is high, heat

transfer performance is good, viscosity is low, specific heat capacity is large, and so on.

Along with eco-friendliness, CO_2 systems have various advantages over conventional systems such as compatibility with normal lubricants and common machine construction materials, greatly reduced compression ratio, easy availability, high volumetric refrigerant capacity, and excellent heat transfer properties. In addition, the phase transition temperature of carbon dioxide is very low, which means it is a refrigerant with great potential for refrigeration. Especially, the energy-saving capability of CO_2 refrigeration systems is significant. Due to the low critical temperature of CO_2, the gas cooler is operated above the critical pressure and the evaporator is operated below that pressure; hence the cycle is called the transcritical cycle. The main factor is compression ratio which influences the COP (Coefficient of performance) of refrigeration systems directly. Even though the compressor works at higher pressure, the compression ratio is so low that the efficiency of the compressor is relatively high. Due to the special thermal physical characters of the supercritical fluid, such as higher diffusivity and lower viscosity, irreversible energy loss result in fluid inner friction, and heat exchange will reduce. As a result, when a CO_2 refrigeration cycle works at the supercritical condition, the energy efficiency ratio (EER) will increase.

In short, natural working fluid carbon dioxide has emerged as a potential refrigerant due to its zero ODP, negligible GWP, and favorable heat transfer properties.

1.1.2.2 The Origin and Development of CO_2 Refrigeration

CO_2 is an 'old' refrigerant, and it is, therefore, natural to start the story by briefly looking back on the history of 'carbonic' systems. This section outlines the early history, including some views on why the use declined after World War II. The recent revival of CO_2 is also discussed.

During the first decades of the twentieth century, CO_2 was widely used as a refrigerant, mainly in marine systems but also in air conditioning and stationary refrigeration applications. Alexander Twining appears to be the first to propose CO_2 as a refrigerant in his 1850 British Patent (Bodinus and Will 1999), but the first CO_2 system was not built until the late 1860s by the American Thaddeus S.C. Lowe (Thevenot 1979). Lowe, who received a British Patent in 1867, did not develop his ideas further (Donaldson and Nagengast 1994). In Europe, Carl Linde built the first CO_2 machine in 1881 (Kohlendioxid 1998). Franz Windhausen of Germany advanced the technology considerably and was awarded a British Patent in 1886. The company J. & E. Hall in Britain purchased the patent rights in 1887, and after having further improved the technology, Hall commenced manufacture in about 1890 (Donaldson and Nagengast 1994). Hall made the first two-stage CO_2 machine in 1889 (Thevenot 1979). The primary application was in marine refrigeration, a field where CO_2 dominated as a refrigerant until 1950–1960.

In Europe, CO_2 machines were often the only choice due to legal restrictions on the use of toxic or flammable refrigerants like NH_3 and SO_2 (Plank 1929). In the United States, CO_2 was used in refrigerating systems from about 1890 and in comfort cooling from about 1900 (Donaldson and Nagengast 1994). The refrigeration applications included small cold storage systems, display counters, food markets, kitchen and restaurant systems, while comfort-cooling systems were installed for instance in passenger ships, hospitals, theatres, and restaurants. Most of these systems used calcium chloride solution as a secondary refrigerant. Compressors were slow-running double- or single-acting crosshead machines with atmospheric crankcase pressure, and expansion valves were usually of the manual-control type. Condensers were often water-cooled double-pipe units (Bodinus 1999).

The safety compared to refrigerants like NH_3 and SO_2 gave CO_2 a preference on boards of ships and in public buildings. The commonly reported disadvantages of CO_2 were loss of capacity and low COP at high heat rejection temperature, compared to other common refrigerants. Especially in warm climates, this gave CO_2 a disadvantage. Refrigerant containment at high pressure was difficult with the sealing technology available at that time. By operation at supercritical high-side pressure or by various two-stage arrangements, the capacity and efficiency loss could be reduced. The so-called multiple-effect compression, as devised by Voorhees in 1905 (Voorhees 1905), is one example of the improvements that were made. When a supercritical high-side pressure operation was needed, this was obtained by charging more refrigerant into the system.

As the CFC fluids were introduced in the 1930s and 1940s, these 'safety refrigerants' eventually replaced the old working fluids in most applications. Although the major argument in their favor was improved safety compared to the fluids like ammonia and sulfur dioxide, CO_2 was also displaced by this transition to CFCs. There is no single reason why the use of CO_2 declined, but several factors probably contributed. These factors included high-pressure containment problems, capacity and efficiency loss at high temperature (aggravated by the need to use air cooling instead of water), aggressive marketing of CFC products, low-cost tube assembly in competing systems, and a failure of CO_2 system manufacturers to improve and modernize the design of systems and machinery.

With the CFC problem becoming a pressing issue in the late 1980s, the whole industry was searching for viable refrigerant alternatives. In Norway, Professor Gustav Lorentzen believed that the old refrigerant CO_2 could have a renaissance. In a 1989 international patent application (Lorentzen 1990), he devised a 'transcritical' CO_2 cycle system, where the high-side pressure was controlled by the throttling valve. One of the intended applications for this system was automobile air-conditioning, a sector that dominated the global CFC refrigerant emissions, and also an application where a non-toxic and non-flammable refrigerant was needed. The potential for more compact components due to high pressure was also an interesting feature.

In 1992, Lorentzen and Pettersen (1992) published the first experimental results on a prototype CO_2 system for automobile air conditioning. A comparison was made between a state-of-the-art R-12 system and a laboratory prototype CO_2 system with equal heat exchanger dimensions and design-point capacity. Although simple cycle calculations indicated that the CO_2 system efficiency would be inferior, several practical factors made the actual efficiencies of the two systems equal.

Based on these and other results, the interest in CO_2 as a refrigerant increased considerably throughout the nineties, despite the resistance from the fluorocarbon industry (Morley and Bivens 1995) and conservative parts of the automotive industry (Bhatti 1997). Many developments and co-operation projects were initiated by the industry and the research sector, including the European industry consortium project 'RACE' on car air conditioning, the European 'COHEPS' project on CO_2 heat pumps, and the CO_2 activities within the international IEA (International Energy Agency) Annexes on Natural Working Fluids and Selected Issues in CO_2 systems. Nowadays this refrigerant has gained much more attention owing to its no toxicity, no flammability, low cost, and no hazard to the environment, and there has been a considerable increase in the interest and development activity internationally. Several novel designs are being used in the industry including cascade, transcritical, transcritical booster, secondary loop, and so on. It is clear that carbon dioxide is one of the best refrigerants and as environmental regulations become more intense, it will be the ultimate refrigerant of the future.

1.2 The Purpose and Content of this Book

Although CO_2 refrigeration has made many achievements in theory and application, it is still in the rapid development stage after all. Compared with the traditional refrigeration method, its theoretical system is not perfect, and its application in many fields is still in the exploratory stage. Accordingly, CO_2 refrigeration needs more attention. In this book, detailed thermodynamic and heat transfer-based analyses of CO_2 refrigeration have been carried out. Besides, the review on transcritical CO_2 cycle for various refrigeration applications is presented as well including both academic and industrial research to date. In recent years, we have carried out a lot of research work in both fundamental cycle and application systems of CO_2 refrigeration. In addition to introducing the basic content of CO_2 refrigeration, this book also contains some of the research results we have achieved in the CO_2 refrigeration field, as well as the experience accumulated in the relevant scientific research practice.

This chapter gives a complete introduction on why this book is focused on CO_2 refrigeration; what contents this book will cover. In addition, this chapter will present the content outlines of the fundamental part and application part, respectively, in this book.

Chapter 2 first introduces the existing refrigerant fluids used in the refrigeration system. Based on these introductions, the reader can be aware of the problems related to the existing refrigerant fluids, especially on the environmental aspects. Then this chapter presents the requirements for the refrigerant fluids in the refrigeration systems. To meet the requirements, natural working fluids are recommended for future use, in which carbon dioxide (CO_2) will be emphasized. Finally, the history of CO_2 working fluid is presented and its choice is discussed and its future is estimated.

In Chap. 3, the theories related to thermodynamic cycles in refrigeration are presented. The property of CO_2 pure substances introduced in Chap. 2 is a basis comprising CO_2 refrigeration cycles. After that, basic processes, phase change process, and compression and expansion process will be presented for the refrigeration thermodynamic cycles. This chapter will also give the reader a clear illustration of various CO_2 vapor-compression refrigeration thermodynamic cycles and their characteristics. Based on the above contents, the reader can have a strong base to further understand CO_2 refrigeration cycles and systems, also including their importances.

Chapter 4 mainly introduces the theories and engineering points related to the CO_2 expansion process in CO_2 transcritical refrigeration cycles. Various expansion processes are studied by thermodynamic methods. The types and the characteristics of the CO_2 expander will be introduced here, such as capillary tubes, electronic expansion valves, and ejectors. The analyses and designs, including theories involved in the CO_2 expander, will be presented in this chapter.

The evaporation process in the CO_2 refrigeration cycle is one of the main points improving the efficiency of a refrigeration system. In Chaps. 5 and 6, the fundamentals in the aspects of CO_2 boiling flow and heat transfer in channels will be presented, which also include the updated research progress. Based on these analyses, various CO_2 evaporator designs are introduced here. Besides, CO_2 internal heat exchanger design and fundamentals will also be stated in this chapter.

Commercial cooling is one of the most important applications for CO_2 refrigeration cycles. This application status and its system will be presented in Chap. 7. The designs and thermodynamic analyses for the main components in the system are also stated. In addition, main application areas, such as supermarket cooling, other building cooling, are given for analyzing the CO_2 refrigeration cycle and their systems. Finally, influences of various climatic conditions on the CO_2 refrigeration cycle and system are presented and the promising future is estimated.

Chapter 8 introduces the CO_2 refrigeration systems for ice rink and snow making. Background and technological status are first given for those CO_2 refrigerations. Not only theoretical but also application aspects are presented. Furthermore, a comparative study will be analyzed by thermodynamic methods and then the promising designs will be given.

Chapter 9 gives another hot topic in the CO_2 vapor compression refrigeration field, automotive air conditioning. In this chapter, the background and this application status will be introduced. Then the related CO_2 thermodynamic cycles are presented. In addition, the component design will be stated and thermodynamic methods are also utilized to analyze the automotive air conditioning using CO_2.

Chapter 10 presents industrial CO_2 cooling systems, in which cooling and combined cooling and heating will be respectively introduced. Component and system are also analyzed and designed by theoretical methods, such as thermodynamic methods. Three important application areas, power generation plant, food processing industry, and transportation are taken as examples to present how CO_2 refrigeration cycles can be useful to avoid fossil fuel energy consumption for environment protection.

Chapter 11 presents a new cryogenic refrigeration method using CO_2. This refrigeration is achieved by micro CO_2 solid particle sublimation, not by CO_2 liquid evaporation. In this chapter, new cryogenic CO_2 refrigeration cycles will be introduced here, which achieves −80 °C refrigeration by using the CO_2 solid–gas sublimation process. Basic flow dynamics and heat transfer of micro CO_2 particle sublimation are also presented.

We would like to express our deep thanks to many distinguished professors who shared their expertise and time in this book. Besides, the support from the Beijing Engineering Research Center of City Heat are gratefully acknowledged. This book provides an opportunity to cover the fundamentals and applications of CO_2 refrigeration cycles and systems. We hope that this book will influence other scientists, students, engineers, and government departments all over the world to research and develop CO_2 refrigeration. Your recommendations, comments, and criticisms are appreciated.

References

Bhatti M (1997) A critical look at R-744 and R-134a mobile air conditioning systems. SAE Paper No. 970527

Bodinus WS (1999) The rise and fall of carbon dioxide systems. In: Will HM (ed) The first century of air conditioning. ASHRAE, Atlanta, GA, pp 29–34

Donaldson B, Nagengast B. Heat and cold: mastering the great indoors. ASHRAE, Atlanta, GA

Ge YT, Cropper RT (2009) Simulation and performance evaluation of finned-tube CO_2 gas coolers for refrigeration systems. Appl Therm Eng 29:957–965

Kohlendioxid (1998) Besonderheiten und Einsatzchancen als Kältemittel. Statusbericht des Deutschen Kälte- und Klimatechnischen Vereins. Nr 20. DKV, Stuttgart

Lorentzen G (1990) Trans-critical vapour compression cycle device. International Patent Publication WO 90/07683

Lorentzen G, Pettersen J (1992) New possibilities for non-CFC refrigeration. In: Pettersen J (ed) IIR international symposium on refrigeration, energy and environment, Trondheim, Norway, pp 147–63

Morley J, Bivens D (1995) Trends in environmental issues and implications for automotive air conditioning. In: Vehicle thermal management systems conference, London, pp 405–412

Pearson SF (2004) Natural working fluids. IEA Heat Pump Newsletter 22

Plank R (1929) Amerikanische Kältetechnik. VDI-Verlag, Berlin

Thevenot R (1979) A history of refrigeration throughout the world (Fidler JC, Trans). IIR, Paris

Voorhees G (1905) Improvements relating to systems of fluid compression and to compressors thereof. British Patent 4448

Chapter 2
Natural Refrigerants and Carbon Dioxide

Jun-Ming Yin, Zhao-Rui Peng, and Xin-Rong Zhang

Global awareness of environmental conservation has trigged the update of refrigerants. Carbon dioxide (R744), on behalf of natural fluids, is universally viewed as the most promising future alternative. This chapter presents a fundamental and latest refrigerant review of natural fluid, R744 in particular. Four evolution phases of refrigerants are illustrated in the first part, as well as a summary of future promising alternatives for sole refrigerants or blend components. Requirements for working fluids in refrigeration systems are concluded based on the former phase division. To satisfy these requirements, natural working fluids, containing ammonia (R717), air (R729), Water (R718) and Hydrocarbons, are discussed in the following. In the fourth part, the thermophysical properties of supercritical and subcritical CO_2 are particularly focused on, as well as recent applications of R744-based refrigeration systems around the world. Finally, the history of CO_2 serving as the refrigerant and outlook are summarized and analyzed.

2.1 Existing Refrigerant Fluids

Cooling demand has been a concomitant of human history, along with clean water, enough food, and other fundamental physiological needs. Refrigeration technology has dated back to the ancient age, since when natural ice, water, and other evaporation process have been utilized to generate cold. Research studies into phase change made their debuts in the seventeenth century, laying solid foundations for man-made refrigeration systems. Evans firstly proposed an ice-making method by harnessing a

J.-M. Yin · Z.-R. Peng · X.-R. Zhang (✉)
Department of Energy and Resources Engineering, College of Engineering, Peking University, Beijing 100871, China
e-mail: xrzhang@pku.edu.cn

Beijing Engineering Research Center of City Heat, Beijing 100871, China

X.-R. Zhang and T. M. Eikevik (eds.), *CO_2 Refrigeration Cycle and Systems*, Lecture Notes in Energy 96, https://doi.org/10.1007/978-3-031-22512-3_2

volatile fluid in a closed cycle loop in 1805 (Evans 1805). Cold energy was extracted from the evaporation process, and the vapor was then transported to a water-cooled condenser where liquid phase was acquired for the next circulation. Regretfully, no evidence of a working machine was witnessed until 1834 when Perkins patented the first vapor-compression refrigeration apparatus, simultaneously appearing the concept of refrigerants (Perkins 1834). Since then, refrigeration technology has sprung up, along with the invention, update, and alternative of refrigerants.

The burgeoning development of civil society has spawned the surging importance of environmental conservation, which triggered the update of refrigerants. Consequently, the entire development history of refrigerants can be divided into 4 phases (Dilshad et al. 2020), shown in Fig. 2.1.

Refrigerants in phase I include common solvents and volatiles, with natural fluids ranking first. The first refrigerant put into use was sulfur ether, introduced by Perkins (Perkins 1834). Other ethers, H_2O, NH_3, CO_2, SO_2, $HCOOCH_3$, HCs, and CCl_4, were gradually brought in during the period of 1840s to 1920s (Yitai et al. 2017). Refrigeration performance was the first concern in evaluating a new refrigerant at this stage. Except for water and carbon dioxide, most of the early refrigerants were toxic or combustible, and some were reactive.

Refrigerant Gen2, mainly CFCs and HCFCs, were renowned for safety and durability. Under the guidance of refrigeration system popularization, people turned to chemically stable, nontoxic, noncombustible, and efficient refrigerants, among which CFCs and HCFCs stuck out. Commercial production of R12 commenced in 1931, and R11 in 1932. In this period, CO_2, on behalf of natural fluids, was abandoned on account of many issues in the 1950s. For example, the overall system efficiency run by CFCs and HCFCs overshadowed that of CO_2-based apparatus. Yet, NH_3 has been continually employed in large-scale industrial cryogenic refrigeration systems, especially in the production, storage of food and soft drinks. Nevertheless, severe related environmental issues were followed after the prevailing adoption of CFCs and HCFCs, namely, ozone depletion and global warming (Molina and Rowland 1974). As a result, aiming at cutting down the production and application of substances that deplete the ozone layer, say, CFCs and Halons, the world-renowned *Montreal Protocol* came out in 1987 (Protocol 1987). Consequently, refrigerants have evolved into phase III.

In view of ozone protection, refrigerants that do not contain chlorine or bromine were favored in phase III. Typical refrigerants included: R134a, near-azeotropic

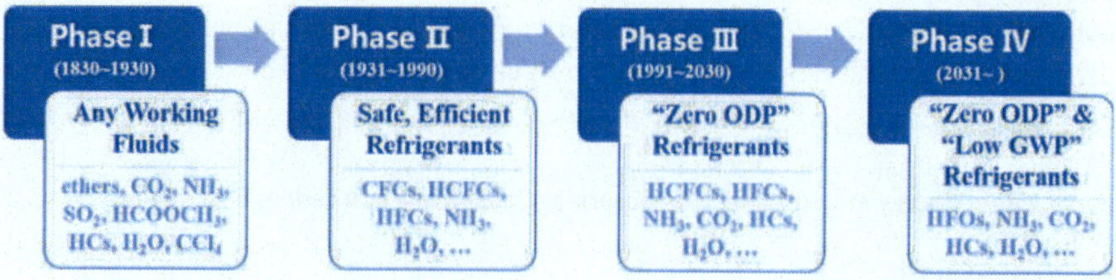

Fig. 2.1 The development path of refrigerants

working fluid R410A, and non-azeotropic working fluid R407C. The same characteristic of these refrigerants is zero ozone depletion potential (ODP). At this stage, R134a, on behalf of HFCs, serving as the substitute of R12, has been consumed on large scale and dominated in refrigerants. However, the second related environmental problem occurred, which is, global warming. It's said that the C–H bond in HFCs will take in radiation at 8–12 μm. Hence, the global warming potentials (GWP) of HFCs are commonly 1000–2000 times that of CO_2. Table 2.1 summarizes the properties of typical refrigerants. One can see that the GWP of typical HFCs, such as R134a, is 1300. Considering this, *Kyoto Protocol*, endorsed in 1997, implemented the objective to alleviate the onset of global warming by reducing six greenhouse gases: CO_2, CH_4, N_2O, HFCs, PFCs, and SF_6 (Böhringer 2003). The following *Paris agreement* in 2015 has emphasized mitigation, adaption, and finance of greenhouse-gas-emissions (Paris Agreement 2015).

As artificial refrigerants develop, related environmental issues appear. It comes to people that the more man-made substance deviates from the natural state, the more it will accumulate, and the larger potential disaster it may lead to. There currently exist mainly two pathways to future alternative refrigerants, one is new man-made zero ODP and low GWP refrigerants, and the other is natural fluids (Yitai et al. 2017). Examples of the former include HFCs (R152a) and HFOs (R1234yf, R1234ze). McLinden et al. recently conducted a comprehensive study searching for future low-GWP refrigerants via thermodynamic and environmental screening criteria. Only a few fluids possess the combination of chemical, environmental, thermodynamic, and safety properties, part of the selected fluids are shown in Table 2.2. Some argue that since the destination of those refrigerants is air, and the effects of man-made refrigerants on the ecotope are hard to estimate in the long run, revival of natural fluids which have coexisted with the eco-system for billion years will be a preferred choice. Moreover, the *Kigali Amendment*, signed in 2016, has urged to gradually reduce the consumption and production of HFCs. Natural working fluids chiefly encompass NH_3, CO_2, water, air, N_2, and HCs. Europe has initiated a "natural fluids" campaign since the 1990s (Lorentzen 1994, 1995). After that, natural fluids being one of the primary candidates of future refrigerants has reached a common consensus. And CO_2 is widely viewed as the most promising one in natural fluid refrigerants alternatives, notably in the field imposing restrictions on flammability and toxicity. It's estimated CO_2 will substitute the existing R134a in the future. Because of the relatively

Table 2.1 Properties of typical refrigerants

Category	CFCs	HCFCs	HFCs		HCs	Natural fluids (inorganic)	
Refrigerant	R12	R22	R134a	R152a	R290	R717	R744
Chemical formula	CCl_2F_2	$CHClF_2$	CH_2FCF_3	CH_3CHF_2	$CH_3CH_2CH_3$	NH_3	CO_2
ODP	1	0.05	0	0	0	0	0
GWP	10,200	1,760	1,300	138	3	0	1

low critical temperature (31.1 °C), how to design and operate CO_2-based systems efficiently calls for more research studies. Water, another natural fluid, has gained increasing momentum recently as a refrigerant. A refrigeration capacity of 1000 kW system employing water as a working medium has been constructed in Denmark (Kilicarslan and Müller 2005). In terms of ammonia, safety issues have hindered its wide applications, so does hydrocarbons. Table 2.3 summarizes existing typical applications related to existing refrigerants and their potential future alternatives.

To sum up, cold demand grows as human society proceeds, which promotes the appearance of safe and efficient refrigerants. However, man-made alternatives are found to be inducement of severe environmental issues. People finally turn their eyes on inorganic natural fluids, which have coexisted with the ecosystem for billions of years.

2.2 Requirements for Working Fluids

In different phases, requirements for refrigerants vary as practical issues occur. Whether to work or not was the first concern at early times. The leakage of refrigerants exposed safety issues to the public, say, flammability and toxicity. Prevailing adoptions of CFCs and HCFCs drew world attention to ozone depletion and global warming. Consequently, man must append and update new criteria in choosing refrigerants. To date, requirements for ideal working fluids in refrigeration systems can be summarized in the following six aspects:

(a) Safety. On account of the potential leakage and waste disposal, refrigerants may mix into the surrounding air, hence, refrigerants that are non-toxic and non-combustible, or at least less deleterious to people are required.
(b) Environmental-friendly nature. High ODP and GWP refrigerants are not allowed.
(c) Compatibility. Ideal working fluids for the next generation should be less corrosive to metals or metalloids. Compatibility with lubricating oil is also wanted.
(d) Thermophysical parameters. Low viscosity, high thermal conductivity, low surface tension, large C_p, stable thermochemical property, and large latent heat are favored in choosing new refrigerants.
(e) Cycle performance. Similar or better COP, cold capacity per unit volume, compression ratio, evaporating pressure, condensing pressure, and exhaust temperature of a new medium in comparison with existing refrigerants are plus items. Specifically, high-pressure systems require compact components and occupy less footprint, but demand stringent design. Volumetric efficiency is related to the compression ratio. An over-high compression ratio results in a decreased volumetric efficiency and thus energy loss. A larger cold capacity per unit volume translates into a more compact compressor and a lower capital investment. The larger COP, the better energy conservation.

Table 2.2 Future promising alternatives for sole refrigerants or blend components and their properties. Part from ref. McLinden et al. (2017)

IUPAC name	ASHRAE designation	Structure	GWP_{100}	ODP	Flammability	Toxicity	Critical pressure (MPa)	Critical temperature (°C)
Hydrocarbons and dimethylether								
Ethane	R170	$CH_3–CH_3$	6	0	High	Low	4.87	32.17
Propylene	R1270	$CH_2=CH–CH_3$	2	0	High	Low	4.55	91.06
Propane	R290	$CH_3–CH_2–CH_3$	3	0	High	Low	4.25	96.74
Methoxymethane	R-E170	$CH_3–O–CH_3$	1	0	High	Medium	5.33	127.23
Cyclopropane	R-C270	$– CH_2–CH_2–CH_2–$	86	0	High	Low	5.58	125.15
Fluorinated alkanes (HFCs)								
Fluoromethane	R41	CH_3F	116	0	High	Low	5.897	44.13
Difluoromethane	R32	CH_2F_2	677	0	Low	Low	5.78	78.11
Fluoroethane	R161	$CH_2F–CH_3$	4	0	High	Low	5.01	102.1
1,1-Difluoroethane	R152a	$CHF_2–CH_3$	138	0	High	Low	4.52	113.26
1,1,2,2-Tetrafluoroethane	R134	$CHF_2–CHF_2$	1120	0	Low	Low	4.07	101.1
Fluorinated alkenes (HFOs) and alkynes								
Fluoroethene	R1141	$CHF=CH_2$	<1	0	High	Low	5.1	54.7
1,1,2-Trifluoroethene	R1123	$CF_2=CHF$	3	0	High	Low	–	–
2,3,3,3-Tetrafluoroprop-1-ene	R1234yf	$CH_2=CF–CF_3$	<1	0	High	Low	3.38	94.7
(E)-1,2-Difluoroethene	R1132(E)	CHF=CHF	1	0	–	–	–	–
3,3,3-Trifluoroprop-1-ene	R1243zf	$CH_2=CH–CF_3$	<1	0	High	Low	–	–
(E)-1,3,3,3-Tetrafluoroprop-1-ene	R1234ze(E)	$CHF=CH–CF_3$	<1	0			3.63	109.36
Fluorinated oxygenates								

(continued)

Table 2.2 (continued)

IUPAC name	ASHRAE designation	Structure	GWP_{100}	ODP	Flammability	Toxicity	Critical pressure (MPa)	Critical temperature (°C)
Trifluoro(methoxy)methane	RE143a	CF_3–O–CH_3	523	0	High	Low	3.64	104.77
Inorganic compounds								
Carbon dioxide	R744	CO_2	1	0	None	None	7.377	30.98
Ammonia	R717	NH_3	<1	0	High	Medium	11.33	132.25

Table 2.3 Existing refrigerants, future alternatives, and their applications (Shengchuan et al. 2020)

Applications	Existing refrigerants	Future alternatives
Household and building air-conditioning system	R22	HFC mixture refrigerants, R410A, R32
Large centrifugal chiller	R12, R500	R134a
	R22	HFC mixture refrigerants
Low-temperature freezing and cold storage unit	R12	R134a
Cold room	R502, R22	HFC/HCFC mixture refrigerants, CO_2
	NH_3	NH_3
Refrigerator, automotive air conditioning system	R12	R134a, CO_2, R1234yf

(f) Price. Expensive alternatives prevent themselves from wide refrigeration applications.

2.3 Natural Working Fluids

As the above has explained, the revival of natural working fluids emerges as refrigerants evolve into phase IV during which zero ODP and low GWP are the objectives. Typical natural fluids include ammonia (NH_3), air, water (H_2O), hydrocarbons (HCs), and carbon dioxide (CO_2). The following will focus on the review of the former four fluids serving as refrigerants. CO_2 will be discussed comprehensively in the next part.

2.3.1 Ammonia

Ammonia, with the refrigerant name as R717, has been employed in refrigeration technology for over 130 years, even before the invention of electricity. 80% of the ammonia produced every year is used in agriculture, approximately 18% in industry, and no more than 2% serves as a refrigerant.

R717 is environmental-benign and possesses excellent physical, transport and refrigeration performance properties, as shown in Table 2.4, of which the standard boiling temperature is low (−33.33 °C) and the pressures in the condenser and evaporator are moderate (The evaporating pressure at −15 °C is 0.24 MPa, and the condensing pressure at 30 °C is 1.17 MPa). Cold capacity per unit volume is comparatively high. Ammonia has high thermal conductivity, high latent heat (the evaporating heat is 8.12 times that of R12, and 6.04 times that of R22 at −15 °C), low throttling loss, and high COP. In other words, the compressor, condenser of refrigeration systems based on ammonia occupies less footprint. Due to the strongly irritating smell, it is easy to detect the leakage even without a professional detector. A moisture content lower than 0.2% is allowed in ammonia-based systems, which translates into that form where dry requirements are less restricted than CFCs-based systems. By using the ease of solubility in water of ammonia, we can treat leaked ammonia with water in cases of emergencies. Moreover, the ammonia–water mixture can be recycled to be a fertilizer. In all, ammonia is the most popular medium-temperature environmental-friendly refrigerant in China, entailing the properties like mature synthesis process, and relatively low price.

Pure ammonia imposes basically no corrosive effects on steel. But once mixed with even a little moisture, they can corrode zinc, copper, and bronze. As a result, pipes, meters, and valves in ammonia-based systems should avoid copper and copper-alloy, which is called "Copper Avoidance Requirement" (Wang 2007). The possibility of explosion when ammonia leaking into the air (16–25%) is the major hindrance to its wider applications. Experiments have demonstrated that ammonia would not

Table 2.4 Standard parameters of R717 (Abas et al. 2018; Riffat et al. 1997)

Type	Parameters	R717
Environment	ODP	0
	$GWP_{100\ year}$	0
Physical	T_{FP} (°C) (freezing point temperature)	−77.73
	T_{NBP} (°C) (normal boiling point temperature)	−33.33
	T_{CP} (°C) (critical point temperature)	132.25
	T_{IG} (°C) (auto ignition temperature)	850
	P_{Evap} (bar) (evaporating pressure)	2.3617
	P_{Conds} (bar) (condensing pressure)	11.672
	P_{Crit} (bar) (critical pressure)	113.33
	$h_{fg\ Evap}$ (kJ/kg) (heat of vaporization in evaporator)	1313.2
	V_{sp} (m^3/kg) (specific volume)	0.50868
Transport properties at 30 °C	μ_l (μPa s)	125.45
	μ_v (μPa s)	9.9953
	K_l (mW/m K)	471.35
	K_v (mW/m K)	26.846
	ρ_l (kg/m^3)	595.17
	ρ_v (kg/m^3)	9.0533
Performance properties	C_p (298 K) (J/mol K)	35.637
	M (kg/kmol)	17.03
	CR (compression ratio)	4.94
Chemical	ASHREA-34	B2L
	Miscibility	N
	Reactivity	Y

burn or explode in an outdoor environment if controlled properly. Say, among several earthquakes in Sendai, Japan, ammonia-based refrigeration units have been witnessed destroyed, whilst no fire or explosion has been reported. The risk of contamination from ammonia is higher than burning or explosion, which is believed to be the biggest problem of ammonia-based refrigeration systems. For instance, fat in food is easily destructed by ammonia oxidation. Even a little leakage of ammonia indoors will lead to air pollution, hence ammonia-based air conditioning is restricted in civil buildings. Ammonia does not mix with mineral-based lubricating oil, nor PAO lubricating oil, but can dissolve in PAG lubricating oil.

To sum up, the peril of ammonia has been exaggerated in the past, particularly in civil building air conditioning. Thanks to the ban on CFCs refrigerants, ammonia has been re-recognized and re-evaluated. It has widened its scope (central air-conditioning, pantry in the supermarket). Up to now, to better enlarge the application scope of ammonia refrigerant, some things must be done. They are: efficient ammonia compressor, ammonia pump, optimization design, leakage avoidance, unit's footprint reduction, and a higher COP.

2.3.2 Air

The Air refrigeration cycle is not new to people. Early in the 1900s, air (R729) has been providing freezing and refrigerating in marine transportation, food manufacturers, and retails. However, vapor compression refrigeration gradually prevailed over air refrigeration by high efficiency. With the rapid development in engineering manufacturing, the air refrigeration cycle has been revived since the 1990s (Miller et al. 1997).

Cold energy is acquired through the adiabatic expansion of compressed air in an air refrigeration cycle. Its main features can be concluded below (Wang 2007; Yanji et al. 2000):

(a) Low COP. Due to the absence of phase change, larger compression work, larger temperature difference, and poor heat transfer behavior, the air refrigeration cycle is less efficient, and it works in special circumstances, for example, aircraft, high-speed train air conditioning.
(b) Environmental-friendly performance. It's universally acknowledged that air is clean, free, easily accessible, and has zero ODP and GWP.
(c) Suitability for low-temperature operation. A demand temperature range of 0–140 °C can be satisfied through the air refrigeration cycle. Moreover, once evaporation temperature on the cold side falls below −72 °C, COP of the air refrigeration cycle excels that of the vapor compression cycle. Hence, better performance at a low temperature can be expected in an air refrigeration cycle.
(d) Availability for an open layout. R729 can be effectively adopted in an open cycle, a closed cycle, and a semi-open/closed cycle. In addition, simplicity and reliability can be attained through an open layout.

After years of development, air cycle refrigeration has set its foot in air conditioning (aircraft, residence, train), cryogenics (refrigerating and freezing of food, liquefaction of natural gas, pharmacy, petroleum industry) (Kinsell et al. 1977). Owing to the excellent cryogenic performance and stable off-design characteristics (Park et al. 2012), it is said that the food processing industry will be the biggest market for air cycle refrigeration. To alleviate the constraints related to low COP, advanced turbine machinery, inter-cooling, recuperation, and optimization of operating parameters can be employed in air cycle refrigeration.

2.3.3 Water

Precious natural resources like water are meat and potatoes for human beings. There goes without saying that water is non-toxic, non-flammable, and environmentally benign. As a refrigerant, low price, stable property, large heat latent, and high COP feature water (R718). The biggest issues related to water are that the saturation temperature under atmospheric pressure is high, up to 100 °C, and saturation pressure at normal temperature is low. In a water-based refrigerator, evaporation temperature must stay above 0 °C, which restricts water in refrigeration. Until now, water has been adopted in dehumidification refrigeration, absorption refrigeration, and vapor compression refrigeration. A high exhaust temperature can be achieved by a water compression refrigeration cycle, as a result, it is a preference for water to act as working fluid in a medium–high temperature heat pump. Though compared with CFCs, R718 behaves with a high COP (Kilicarslan and Müller 2005), ten times more volumetric flow for a given refrigeration capacity is still required, which translates into larger axial or centrifugal compressors (Calm 2008; Li et al. 2011). High critical temperature (373.95 °C) and pressure (221 bar), the high initial cost of larger axial or centrifugal compressors, large compression process, use of direct contact heat exchangers are all limiting factors, rendering R718 a less attractive option in comparison with the existing halocarbon-based vapor compression refrigeration systems (Abas et al. 2018; Lachner et al. 2007).

2.3.4 Hydrocarbons

Hydrocarbon refrigerants are known as ozone-friendly, negligible GWP, stable property, material compatibility, excellent miscibility with synthetic oil, and low refrigerant charge (Abas et al. 2018). Typical hydrocarbon refrigerants encompass methane (R50), ethane (R170), propane (R290), butane (R600), isobutene (R600a), ethylene (R1150), and propylene (R1270). Their properties are tabulated in Table 2.5. It has been experimentally demonstrated that hydrocarbons (R290, R600a, R1270) prevail over R-22 in terms of local condensing heat transfer coefficient and local evaporating heat transfer coefficient (Lee et al. 2006). R50 and R170 are cryogenic liquids, of which the low boiling point is −162 and −88.58 °C, and they can be used for extremely low-temperature refrigerants (−80 °C). Previous research studies have found that R290 has a higher cooling capacity than R12 and similar COP when tested in a propriety vapor compression refrigeration unit (Halimic et al. 2003). R1270 can serve as the alternative of R22, but with even higher capacity and COP (Chang et al. 2000; Saleh and Wendland 2006). Though HCs offer interesting refrigerant alternatives for energy-efficient and environmental-benign refrigerating systems and heat pumps, safety cautions must be addressed because HCs are usually flammable (Granryd 2001). Ignition temperatures of hydrocarbon refrigerants are in the range of 420–600 °C (Lemmon et al. 2002; Robinson and Smith 1984).

HC refrigerants are employed in domestic refrigerators, freezers, air conditioning, and commercial applications (beverage cabinets, ice-cream machines, truck-mounted refrigerators, heat pumps, and chillers) (Palm 2008). To be specific, R290 and R600a can be adopted in domestic refrigerators and air conditioners, R170 and R50 serve in low-temperature applications (Bolaji and Huan 2013; Protocol and Layer 2002). A

Table 2.5 Standard parameters of hydrocarbon refrigerants

Type	Parameters	R170	R290	R600	R600a	R1270
Environment	ODP	0	0	0	0	0
	$GWP_{100\ year}$	20	20	20	20	20
Physical	T_{FP} (°C)	−183	−188	−140	−159.6	−59
	T_{NBP} (°C)	−88.58	−42.11	−0.49	−11.75	−47.62
	T_{CP} (°C)	32.17	96.74	151.98	134.66	91.06
	T_{IG} (°C)	505	470	420	460	460
	P_{Evap} (bar)	16.29	2.916	0.563	0.890	3.6302
	P_{Conds} (bar)	46.55	10.79	2.834	4.0472	13.05
	P_{Crit} (bar)	48.72	42.512	37.96	36.29	45.55
	$h_{fg\ Evap}$ (kJ/kg)	350.6	394.65	398.23	367.97	399.09
	V_{sp} (m^3/kg)	0.0334	0.1538	0.6375	0.39904	0.12815
Transport properties at 30 °C	μ_l (μPa s)	292.42	92.188	151.07	143.43	91.479
	μ_v (μPa s)	8.15	8.4628	7.5182	7.6308	9.1707
	K_l (mW/m K)	153.12	91.409	102.66	87.489	110.14
	K_v (mW/m K)	14.561	19.724	17.114	17.371	19.722
	ρ_l (kg/m^3)	779.02	484.39	566.98	544.31	497.45
	ρ_v (kg/m^3)	0.9	23.451	7.1366	10.48	27.718
Performance properties	C_p (298 K) (J/mol K)	52.456	73.307	98.442	96.598	64.376
	M (kg/kmol)	30.07	44.096	58.122	58.122	42.08
	CR	2.85	3.7	5.033	4.54	3.59
Chemical	ASHREA-34	A3	A3	A3	A3	A3
	Miscibility	Y	Y	Y	Y	Y
	Reactivity	N	N	N	N	N

mixture of R290 and R600a (50, 50%) even performs better refrigeration in comparison with R134a in a home refrigerator, resulting in 4.4% consumed energy saving and 40% reduction of refrigerant mass (Jwo et al. 2009).

2.4 Carbon Dioxide (CO_2) Fluid

2.4.1 Thermodynamic and Transport Properties

Span et al. raised a new equation of state for CO_2 in the form of Helmholtz energy after analyzing and summarizing published CO_2 thermodynamic experimental data (Span and Wagner 1996). This new fundamental equation is valid for almost an entire fluid domain (216–1100 K, 0–800 MPa), through which one can get density, enthalpy, entropy, C_v, C_p, speed of sound, etc. at each state point. Earlier, the transport properties of CO_2, containing thermal conductivity and viscosity, have been compiled by representative equations based upon a body of experimental data (Vesovic et al. 1990). Commercial software, like REFPROP, has adopted the above equations to calculate the thermodynamic and transport properties of CO_2.

Natural fluids overshadow man-made refrigerants by their environmental friendliness, and CO_2 stands out due to its excellent refrigeration performance that resulted from thermodynamic and transport properties. This section will illustrate the property characteristics of CO_2, especially in the supercritical region and the subcritical region. We will discuss the dramatic variations of properties near the pseudocritical line where R744 may go through in a gas cooler in a transcritical refrigeration cycle. A comparative study concerning typical properties, say, saturated pressure, surface tension, density, viscosity, thermal conductivity, and C_p between subcritical R744 and existing refrigerants will be implemented to illustrate the strength of R744 in evaporator and pipes.

2.4.1.1 Supercritical CO_2

Once the temperature and pressure exceed the critical parameters (304.13 K, 7.377 MPa), CO_2 evolves into the supercritical state where the obvious interface between the liquid and vapor disappears, shown in Fig. 2.2. In contrast to two-phase CO_2 where temperature and pressure are not independent, supercritical CO_2 has two degrees of freedom. Generally speaking, supercritical CO_2 combines the features of liquid CO_2 and vapor CO_2. It is high-density, like liquid, while acts like gas in terms of viscosity and diffusivity, resulting in excellent fluidity and transmission performance.

Figure 2.3 depicts typical thermodynamic and transport properties of CO_2 varied with temperature and pressure, containing specific heat, density, dynamic viscosity, and thermal conductivity. An important characteristic of the supercritical fluid can be

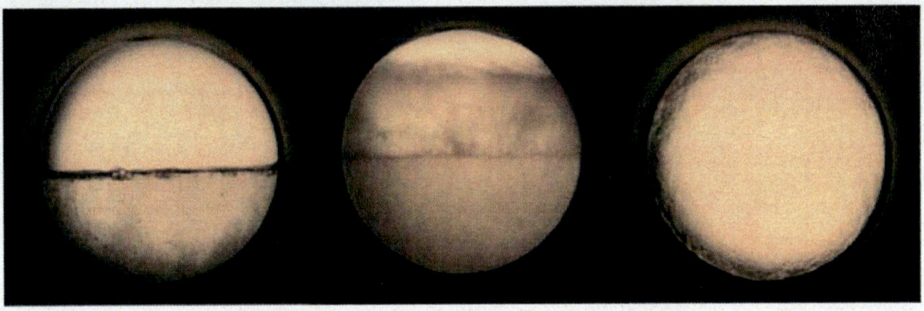

Fig. 2.2 Demonstration photographs of CO_2 from two-phase state to supercritical state (Oakes et al. 2001)

witnessed: thermophysical properties undergo dramatic variations in the near-critical zone. A peak occurs in specific heat and thermal conductivity, and a sharp drop of density, dynamic viscosity, thermal conductivity appears. In regions far away from the critical point, only smooth declination can be observed as temperature increases.

To be more specific, a term, named pseudocritical temperature t_{pc}, is associated with the corresponding point where C_p reaches its peak at supercritical pressure. The pseudocritical temperature grows as pressure increases, illustrated in Fig. 2.3a. The higher t_{pc} is, the lower peak C_p becomes. Even in the supercritical region without a noticeable phase interface, fluid possesses liquid-like characteristics at a temperature lower than t_{pc}, say, high density and viscosity, while behaves like gas at temperatures

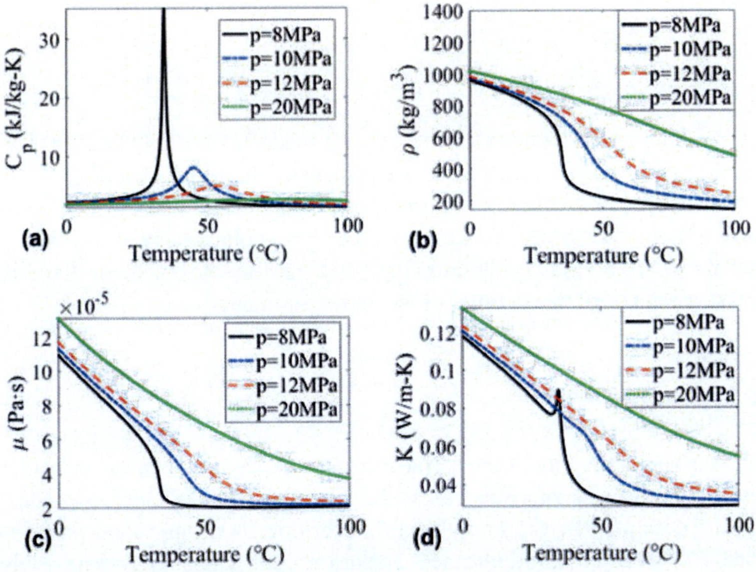

Fig. 2.3 Thermodynamic and transport properties of CO_2 varied with temperature and pressure. **a** Specific heat, **b** density, **c** dynamic viscosity, **d** thermal conductivity

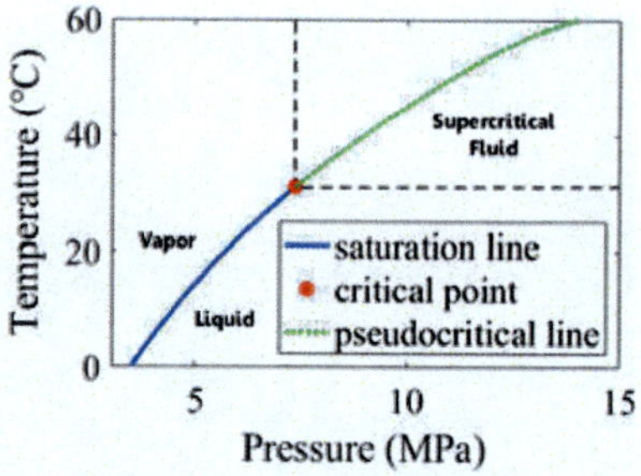

Fig. 2.4 Saturation line, critical point, and pseudocritical line in the phase diagram

higher than t_{pc}, as shown in Fig. 2.3. In other words, a crossing from pseudocritical temperature also means a transition of supercritical fluid from a liquid-like state to a gas-like state. Accordingly, minor deviation from pseudocritical points can lead to dramatic changes in thermodynamic and transport properties of supercritical fluids. The pseudocritical line can be obtained by connecting the pseudocritical point. Liao et al. fitted the pseudocritical temperature t_{pc} of CO_2 as a function of pressure, as shown below (Liao and Zhao 2002):

$$t_{pc} = -122.6 + 6.124p - 0.1657p^2 + 0.01773p^{2.5} - 0.0005608p^3 \quad 75 \le p \le 140 \tag{2.1}$$

where t_{pc} is in °C and p is in bar. According to the algebraic equation, the saturation line, critical point, and the pseudocritical line can be depicted in the phase diagram of CO_2, illustrated in Fig. 2.4. Ma ever defined the pseudocritical zone as (Yitai et al. 2017):

$$0.7t_{pc} \le t \le 1.3t_{pc} \tag{2.2}$$

From an engineering perspective, say, in the design of a CO_2 gas cooler, a large C_p will lead to improved heat transfer performance. Figure 2.5 describes C_p of CO_2 varied with enthalpy at supercritical pressure. One can see that large C_p at varied supercritical pressure suitable for the design of gas cooler nearly exists in the same enthalpy zone. As a result, this enthalpy zone can be chosen to be fixed design parameters, despite different cooler pressure requirements.

2.4.1.2 Subcritical CO_2

In an R744-based transcritical refrigeration system, CO_2 runs in a supercritical state in a gas cooler and acts as a subcritical fluid in other components, say, evaporator and pipes. Subsequently, the knowledge concerning thermophysical properties of subcritical CO_2 is of equal importance. To sum up, subcritical CO_2 behaves as high saturation pressure, low surface tension, high vapor density, and low density ratio of

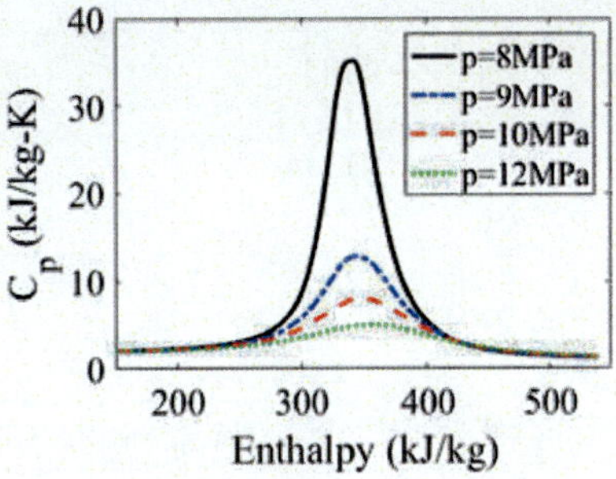

Fig. 2.5 C_p varied with enthalpy

saturated liquid to saturated vapor. As for saturated CO_2 liquid, low viscosity, high $C_{p,}$ and large thermal conductivity separate R744 from other traditional refrigerants.

A comparison between subcritical R744 and traditional refrigerants like R22 and R134a has been conducted, as shown in Fig. 2.6. The saturated pressure of CO_2 is higher than that of R22 and R134a, and expresses a larger variation rate in discussed temperature region. Thus, on condition of fixed pressure drop in an evaporator, the R744-based system has minor evaporation temperature change. Surface tension declines as temperature grows, tending to be zero at the critical temperature. R744 has much less surface tension when compared with R22 and R134a, demonstrated in Fig. 2.6b. It's widely acknowledged that tiny surface tension can reduce superheat in nucleation and bubble development, which will contribute to the production of bubbles and therefore enhance heat transfer performance. Figure 2.6c tells us that R744 has a lower ratio of liquid density to vapor density. A lower density ratio usually translates into a uniform distribution in the evaporator. As temperature elevates, the ratio of liquid viscosity to vapor viscosity declines in all 3 refrigerants. R744 still has the lowest ratio, which means the viscosity of liquid CO_2 is minor. On one hand, minor viscosity will curtail the flow resistance. Furthermore, the vitality of bubbles in a low-viscosity fluid will be enhanced, as well as the heat transfer coefficient. Another strength of liquid R744 is its higher thermal conductivity, demonstrated in Fig. 2.6e. A smaller resistance and a higher heat transfer coefficient in two-phase flow can be obtained in a high thermal conductivity working fluid. In all, compared with traditional refrigerants, subcritical R744 exhibits better boiling heat transfer performance via larger liquid thermal conductivity, smaller density, viscosity, and surface tension.

2.4.2 Applications of R744-based Refrigeration Systems

CFCs once superseded CO_2 by virtue of excellent performance in refrigeration cycles, but with the emergence of related environmental issues caused by CFCs, and its consequent effects to the human community, natural fluid CO_2 distinguished itself by dint of environmental friendliness and attractive thermodynamic and transport properties. Lorentzen highly praised CO_2 as "The most promising refrigerant in

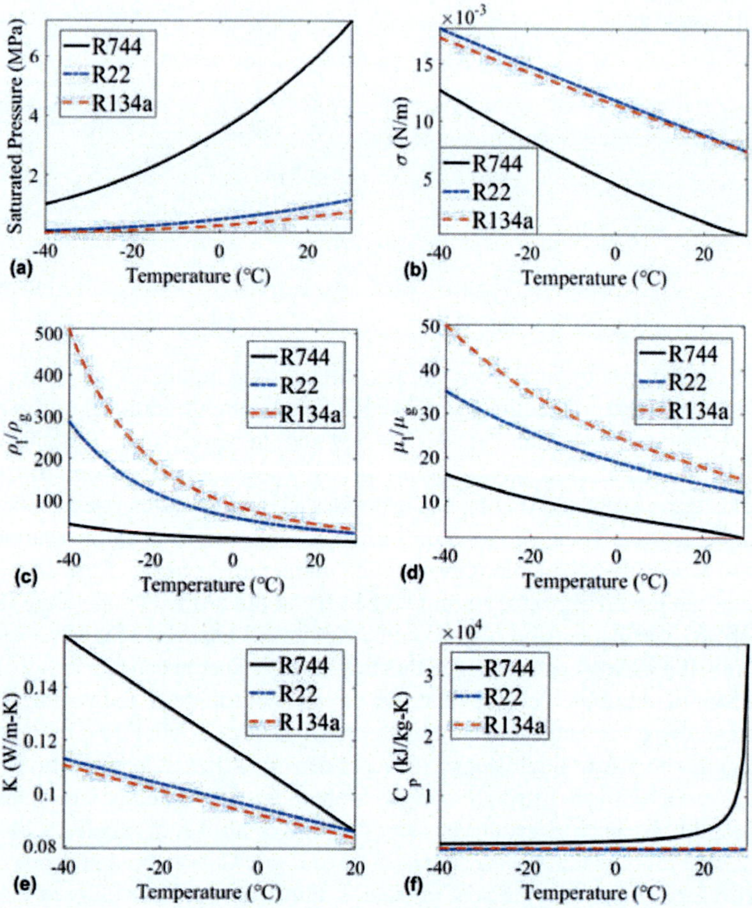

Fig. 2.6 Typical properties of subcritical CO_2 varied with temperature. **a** Saturated pressure, **b** surface tension, **c** ratio of liquid density to vapor density, **d** ratio of liquid viscosity to vapor viscosity, **e** thermal conductivity of liquid CO_2, **f** specific heat of liquid CO_2

21st" and proposed a transcritical CO_2 cycle because of its low critical temperature (30.98 °C) (Lorentzen 1994, 1995). He believed that transcritical CO_2 cycles could play an essential role in automotive air-conditioning and heat pump systems. Compared to other refrigerants, the refrigeration system driven by CO_2 characterizes the following patterns:

(a) Environmental friendliness. CO_2 exists in our planate as a natural medium, making up about 0.04% of the earth's atmosphere composition (Lüthi et al. 2008). Given that CO_2 is typically a byproduct of chemical-engineering processes, adopting CO_2 as a refrigerant can further reduce its global warming potential. It is a crucial part of life, as it is the main product of respiration and the main carbon source for plants during photosynthesis."0.04% of the earth's atmosphere.39".

(b) Low price. CO_2 is ubiquitous. Though it undergoes industrial purification and manufacturing process before it can be used in a refrigeration system. But in comparison with HCs and HCFCs, it's relatively cheaper.

(c) Inert property. CO_2 imposes little erosion on stainless steel or copper alloy, even mixing with water. Moreover, once water content lowers than 8 ppm, CO_2 can be transferred using carbon steel.

(d) Non-flammable and Non-toxic. Though CO_2 behaves non-toxic under most circumstances, if CO_2 content in the air exceeds 2%, it may hurt respiratory organs and even cause suffocation.

(e) Excellent thermophysical properties matching refrigeration cycle and equipment. The evaporation latent heat of CO_2 is high, resulting in a relatively large cooling capacity per unit volume, to be specific, cooling capacity per unit volume of R744 at 0 °C is 1.58 times that of R717, 5.12 times that of R22, and 8.25 times that of R12, accordingly, compressors and other components can decrease in size. The dynamic viscosity of saturation liquid CO_2 at 0 °C is 5.2% that of R717, and 23.8% that of R12, which yields less compressor work. R744 possesses a high thermal conductivity and a low ratio of liquid density to vapor density, consequently, the distribution of the refrigerant in loops after throttling can be uniform, which translates into an improved flow and heat transfer performance, further, the pipes and heat exchangers can be more compact.

Though CO_2 set itself apart via the above superiorities, it still has problems that need to overcome. For instance, high critical pressure and low critical temperature (seen in Figs. 2.4 and 2.7) can be impediments, neither subcritical cycles nor transcritical cycles will operate at pressures similar to conventional refrigeration systems, which means that new designs are required for system and components. Fortunately, some blocks have been removed and the R744-based refrigeration system has to gain colossal momentum in recent decades around the world.

R744-based refrigeration systems, transcritical CO_2 system, in particular, has been deployed in a variety of applications for many years and witnessed a colossal growing, as shown in Table 2.6 and Fig. 2.8. Specifically speaking, it covers convenience stores, commercial applications (supermarkets/retail), industrial applications (refrigerated warehousing, wineries, breweries, bakeries), food and drinks processing (food processing, meat processing, fruit processing, fish processing), and other niche applications (ice rinks, ski slopes, cruise ships, fast food, pharmaceutical processing and laboratories, product testing). The following will showcase recent transcritical CO_2 installations for varied purposes in different regions.

2.4.2.1 Applications in Convenience Stores

In October 2017, retail Carrefour has installed a full-CO_2 transcritical remote unit at a Carrefour City store in Vannes, Brittany, providing service for a commercial surface

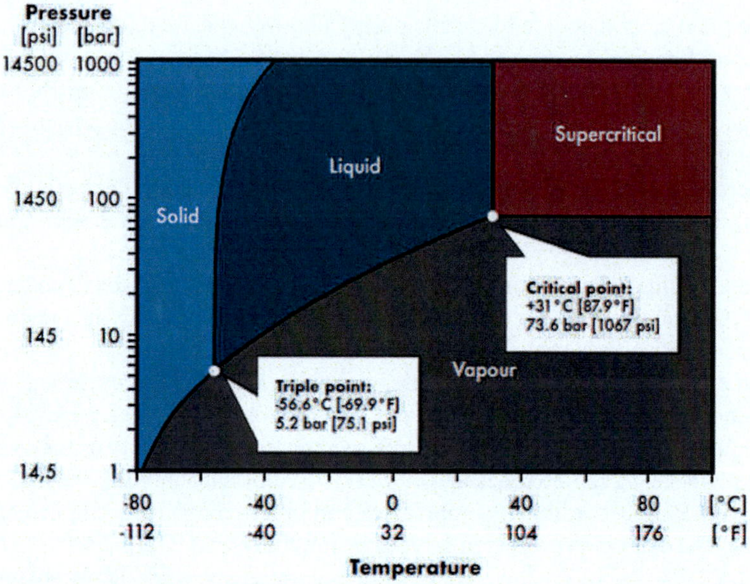

Fig. 2.7 Phase diagram of CO_2

Table 2.6 CO_2 transcritical installations in major regions in recent two decades (shecco 2020)

Region	Number in 2008	Number in 2018	Number in May 2020
Europe	140	>16,000	29,000
U.S.	0	>370	650
Canada	0	>245	340
Japan	0	>3530	5000
Australia	0	>20	95
New Zealand	0	>40	100
South Africa	0	>110	>220
China Mainland	0	2	3

of 293 m^2 (Battesti 2018). Delhaize convenience store also chose CO_2 condensing units to serve as medium-temperature cabinets and frozen food cabinets in a franchised Shop & Go with 250 m^2 footprint in Brussels recently (Williams 2018). Apart from Europe, Panasonic in Japan has installed a CO_2 refrigeration system in a Lawson convenience store in 2014. It was estimated that a 50% reduction of energy consumption can be achieved compared to the standard HFC system (Dusek 2014).

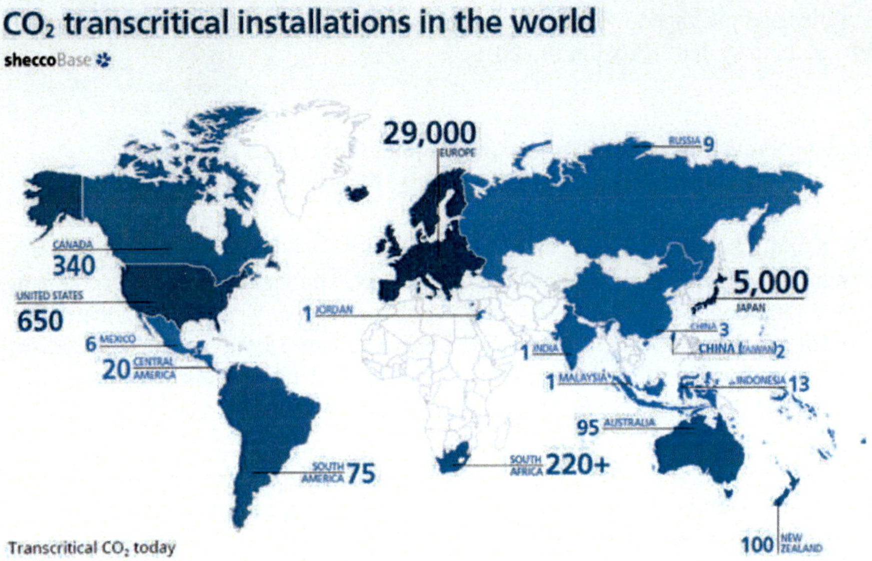

Fig. 2.8 CO_2 transcritical installations in the world (Status by May 2020). From ref. (Skačanová and Battesti 2019; shecco 2020)

2.4.2.2 Commercial Applications (Supermarkets/Retail)

When it comes to bigger stores, namely, supermarkets, transcritical CO_2 systems have been deployed in New Jersey and Florida, US (Garry 2019b, 2020b). In addition, remarkable energy savings have been observed when compared with HFC- and HFO-based systems. Thanks to the low-GWP natural fluid, a significant reduction in carbon footprint is projected for the new IGA Supa retail and liquor store, opened in Creswick, Australia, which installed a transcritical CO_2 system for heating and hot water (Koegelenberg 2019a). China's second transcritical CO_2 system has been installed in a remodeled CSF store in Beijing in 2019, as a part of a three-month store renovation project. The system adopted heat recovery, making the system save more energy when compared with the replaced R22 system (Yoshimoto 2019a).

2.4.2.3 Industrial Applications (Refrigerated Warehousing, Wineries, Breweries, Bakeries)

CO_2-based systems can also be expanded in industrial occasions. Hamamatsu Itaku Soko, a Japanese cold storage operator, reduced energy consumption by up to 35% at its Yonezu Cold Center facility after replacing a 22 years-old R22 system with a transcritical CO_2 system (Garry 2020a). Companies in Japan such as Asahi Breweries, which makes one of the most well-known lagers, along with margarine production facilities and packaged ice manufacturers, are currently installing transcritical CO_2 refrigeration systems (Yoshimoto 2019c).

2.4.2.4 Food and Drinks Processing Applications (Food Processing, Meat Processing, Fruit Processing, Fish Processing)

The world's largest transcritical CO_2 refrigeration system has been installed and commissioned at Yosemite Foods, a California-based pork and meat supply company with a total cooling capacity of 4 MW. And the company recently relocated and expanded to the city of Stockton, where it opened a new 18,580 m^2 meat processing facility (Yoshimoto 2019d). Besides, in 2019, one of the leading fruit processing companies in Peru was supplied with a transcritical CO_2 system (Aleu 2019). In terms of fish processing, DFDS Logistics Ltd., a logistics and freight shipping company headquartered in Copenhagen, Denmark, announced in October 2019 that it had purchased 50 CO_2 refrigerated shipping containers to use in its short-sea (coastal) shipping service (Yoshimoto 2019b).

2.4.2.5 Niche Applications

Aside from the applications in convenience stores, supermarkets, and industry, CO_2-based systems also play an increasingly important role in niche applications. The Beijing 2022 Organizing Committee has officially announced its plan to use CO_2 refrigeration systems for speed skating, figure skating, and short track venues in the Beijing 2022 Winter Olympics (Fig. 2.9). And it is the first time for this technology to be adopted at the Olympic Games (Zhang 2023). Two cruise ships in China are going to be equipped with a transcritical CO_2 refrigeration system. All food and beverage on the ships will be refrigerated with this system. They are the first two cruise ships ever to be built in China, according to the manufacturer of the CO_2 system. The first ship will be delivered in 2023 (Garry 2019). U.S. fast-food chain Burger King has chosen a transcritical CO_2 system as the preferred condensing unit for its restaurants in Spain (Koegelenberg 2019a). At the site of a multinational biotechnology group in Basel, Switzerland, a transcritical CO_2 double-stage is used for cold storage rooms for pharmaceuticals stored at -20 °C. The system primarily runs in subcritical mode, harnessing groundwater to cool the CO_2 and improve the efficiency (Williams 2019).

2.5 Historical Choice and Future of CO_2 Fluid

Between the late 1800s and 1930s, CO_2, ammonia, and SO_2 were widely adopted as refrigerants. CO_2 set itself apart among the above choices by no toxicity and non-inflammability. Correspondingly, it was favored in marine refrigeration and civil refrigeration.

Fig. 2.9 National speed skating oval in Beijing 2022 Winter Olympics

Early in 1850, an American, named Alexander Twining, patented a vapor compression cycle driven by CO_2 (Bodinus 1999). The first attempt to deploy CO_2 in a commercial unit was fulfilled by Thaddeus S.C. Lowe, who demonstrated the feasibility of CO_2 as a refrigerant (Thévenot 1979). He designed and made an ice machine, and a marine system to convey frozen meat in the Gulf of Mexico. Carl Linde developed a CO_2-based refrigerator for F. Krupp Inc. in Germany in 1882. At the same time, W. Raydt and J. Harrison patented a system refrigerated with CO_2 respectively.

After that, R744 has gained significant energy. J&E Hall company in England purchased a CO_2 compressor patent from a German Franz Windhausen and put it into production after improvements (Donaldson and Nagengast 1994). Hall's compressor rapidly superseded the original air compressor in ships. British ships generally installed CO_2 compressors in the 1940s. America began to adopt CO_2 in refrigeration systems in the 1890s. Kroeschell Bros. company manufactured CO_2 compressors, condensers, coolers valves, and high-pressure CO_2.

It is not until the 1920s when CO_2 was used in air conditioning. For example, CO_2-based air conditioning was deployed in stores in 1919, churches in 1920, offices in 1927, and residential buildings in 1930. However, CO_2 finally got replaced by R12 in 1931, a non-toxic, non-flammable, moderate pressure, and high COP synthetic refrigerant.

Revival of natural working fluids has emerged since the 1990s when the related environmental issues resulted from CFCs and HCFCs became a threat to humans.

Among them, CO_2 is widely viewed as the most prosing alternative, known as "refrigerant in 21st". Until now, the CO_2-based refrigeration system has gained increasing momentum all around the world, as has been illustrated in the above.

2.6 Conclusion

This chapter presents a fundamental and latest refrigerant review of natural fluid, CO_2 in particular. In the first part, four phases of refrigerants are divided. Ozone depletion and global warming resulted from CFCs, HCFCs, HFCs have drawn people's attention to natural fluids, among which CO_2 is universally acknowledged to be the most promising refrigerant alternative. A summary of future promising alternatives for sole refrigerants or blend components is offered. Requirements for working fluids in refrigeration systems are concluded based on the former phase division. To satisfy these requirements, examples of natural working fluids, containing ammonia (R717), air (R729), Water (R718) and Hydrocarbons, are discussed in the following. In the fourth part, the first level focuses on the thermophysical properties of supercritical and subcritical CO_2, from which one can see the dramatic variations of thermophysical properties near pseudocritical point and strengths of subcritical CO_2 over existing refrigerants. The second level addresses recent applications of R744-based refrigeration system around the world, including convenience stores, commercial applications (supermarkets/retail), industrial applications (refrigerated warehousing, wineries, breweries, bakeries), food and drinks processing (food processing, meat processing, fruit processing, fish processing) and other niche applications (ice rinks, ski slopes, cruise ships, fast food, pharmaceutical processing and laboratories, product testing). In the final part, the history of CO_2 serving as the refrigerant and outlook are summarized and analyzed.

References

Abas N, Kalair AR, Khan N, Haider A, Saleem Z, Saleem MS (2018) Natural and synthetic refrigerants, global warming: a review. Renew Sustain Energy Rev 90:557–569

Aleu P (2019) Hillphoenix supplies 4th transcritical CO_2 industrial system in Latin America. http://r744.com/articles/9064/hillphoenix_supplies_4th_transcritical_CO2_industrial_system_in_latin_

Battesti M (2018) Carrefour's first CO_2 transcritical convenience store. https://r744.com/articles/8184/carrefourandrsquo_s_first_CO2_transcritical_convenience_store

Bodinus WS (1999) The rise and fall of carbon dioxide systems: the first century of air conditioning. ASHRAE J 41(4):37

Böhringer C (2003) The Kyoto protocol: a review and perspectives. Oxf Rev Econ Policy 19(3):451–466

Bolaji B, Huan Z (2013) Ozone depletion and global warming: case for the use of natural refrigerant—a review. Renew Sustain Energy Rev 18:49–54

Calm JM (2008) The next generation of refrigerants—historical review, considerations, and outlook. Int J Refrig 31(7):1123–1133

Chang Y, Kim M, Ro S (2000) Performance and heat transfer characteristics of hydrocarbon refrigerants in a heat pump system. Int J Refrig 23(3):232–242

Dilshad S, Kalair AR, Khan N (2020) Review of carbon dioxide (CO_2) based heating and cooling technologies: past, present, and future outlook. Int J Energy Res 44(3):1408–1463

Donaldson B, Nagengast B (1994) Heat and cold: mastering the great indoors: a selective history of heating, ventilation, air-conditioning and refrigeration from the ancients to the 1930s. American Society of Heating

Dusek J (2014) Lawson's green flagship convenience store opens in Osaka, promises 50% energy reductions. http://www.r744.com/articles/5044/lawson_s_green_flagship_convenience_store_opens_in_osaka_promises_50_energy_reductions

Evans O (1805) The abortion of the young steam engineer's guide: containing an investigation of the principles, construction and powers of steam engines… illustrated with five engravings

Garry M (2019a) GEA to equip two Chinese cruise ships with transcritical CO_2. http://r744.com/articles/9128/gea_to_equip_two_chinese_cruise_ships_with_transcritical_CO2

Garry M (2019b) Weis Markets reports dramatic energy savings with transcritical CO_2. http://www.r744.com/articles/9096/weir_markets_reports_dramatic_energy_savings_with_transcritical_an

Garry M (2020a) Japanese cold storage operator cuts energy by 35% with CO_2. http://r744.com/articles/9392/japanese_cold_storage_operator_cuts_energy_by_35_with_CO2

Garry M (2020b) Transcritical CO_2 in Warm, Muggy Florida. http://r744.com/articles/9384/transcritical_CO2_in_warm_muggy_florida

Granryd E (2001) Hydrocarbons as refrigerants—an overview. Int J Refrig 24(1):15–24

Halimic E, Ross D, Agnew B, Anderson A, Potts I (2003) A comparison of the operating performance of alternative refrigerants. Appl Therm Eng 23(12):1441–1451

Jwo C-S, Ting C-C, Wang W-R (2009) Efficiency analysis of home refrigerators by replacing hydrocarbon refrigerants. Measurement 42(5):697–701

Kilicarslan A, Müller N (2005) A comparative study of water as a refrigerant with some current refrigerants. Int J Energy Res 29(11):947–959

Kinsell RC, Noe JC, Byrne JP (1977) Air cycle air conditioning system for vehicles. Google Patents

Koegelenberg I (2019a) Prioritizing sustainability, IGA store chooses CO_2. http://www.r744.com/articles/9162/prioritizing_sustainability_iga_store_chooses_CO2

Koegelenberg I (2019b) Burger king starts roll out of CO_2 condensing units in Spain. http://r744.com/articles/9902/ultra_low_temperature_air_cycle_machine_on_double_duty_as_heat_pump

Lachner BF Jr, Nellis GF, Reindl DT (2007) The commercial feasibility of the use of water vapor as a refrigerant. Int J Refrig 30(4):699–708

Lee H-S, Yoon J-I, Kim J-D, Bansal P (2006) Characteristics of condensing and evaporating heat transfer using hydrocarbon refrigerants. Appl Therm Eng 26(10):1054–1062

Lemmon EW, Huber ML, McLinden MO (2002) NIST reference fluid thermodynamic and transport properties. REFPROP

Li Q, Piechna J, Müller N. Thermodynamic potential of using a counter rotating novel axial impeller to compress water vapor as refrigerant. Int J Refrig 34(5):1286–1295

Liao SM, Zhao TS (2002) Measurements of heat transfer coefficients from supercritical carbon dioxide flowing in horizontal mini/micro channels. J Heat Transfer 124(3):413–420

Lorentzen G (1994) Revival of carbon dioxide as a refrigerant. Int J Refrig 17(5):292–301

Lorentzen G (1995) The use of natural refrigerants: a complete solution to the CFC/HCFC predicament. Int J Refrig 18(3):190–197

Lüthi D, Le Floch M, Bereiter B et al (2008) High-resolution carbon dioxide concentration record 650,000–800,000 years before present. Nature 453(7193):379–382

McLinden MO, Brown JS, Brignoli R, Kazakov AF, Domanski PA (2017) Limited options for low-global-warming-potential refrigerants. Nat Commun 8(1):1–9

Miller JP, Smith C, Allam RJ, Topham AK (1997) Refrigeration system. Appl Therm Eng 2(17):XII–XIII

Molina MJ, Rowland FS (1974) Stratospheric sink for chlorofluoromethanes: chlorine atom-catalysed destruction of ozone. Nature 249(5460):810–812

Oakes RS, Clifford AA, Rayner CM (2001) The use of supercritical fluids in synthetic organic chemistry. J Che Soc, Perkin Trans 1 (9):917–941

Palm B (2008) Hydrocarbons as refrigerants in small heat pump and refrigeration systems—a review. Int J Refrig 31(4):552–563

Paris Agreement (2015) Report of the conference of the parties to the United Nations framework convention on climate change, 21st Session, 2015, Paris

Park SK, Ahn JH, Kim TS (2012) Off-design operating characteristics of an open-cycle air refrigeration system. Int J Refrig 35(8):2311–2320

Perkins J (1834) Apparatus for producing ice and cooling fluids. British Patent 6662

Protocol M (1987) Montreal protocol on substances that deplete the ozone layer, vol 26. US Government Printing Office, Washington, pp 128–136

Protocol M Layer O (2002) Refrigeration, air conditioning and heat pumps technical options committee

Riffat S, Afonso C, Oliveira A, Reay D (1997) Natural refrigerants for refrigeration and air-conditioning systems. Appl Therm Eng 17(1):33–42

Robinson C, Smith D (1984) The auto-ignition temperature of methane. J Hazard Mater 8(3):199–203

Saleh B, Wendland M (2006) Screening of pure fluids as alternative refrigerants. Int J Refrig 29(2):260–269

shecco (2020) World guide to transcritical CO_2 refrigeration

Shengchuan L, Xue H, Baomin D, Zhili S, Jialiang N (2020) Natural working fluid CO_2 in refrigeration and heat pump technologies and applications. Tianjin University Press, Tianjin

Skačanová KZ, Battesti M (2019) Global market and policy trends for CO_2 in refrigeration. Int J Refrig 107:98–104

Span R, Wagner W (1996) A new equation of state for carbon dioxide covering the fluid region from the triple-point temperature to 1100 K at pressures up to 800 MPa. J Phys Chem Ref Data 25(6):1509–1596

Thévenot R (1979) History of refrigeration throughout the world

Vesovic V, Wakeham W, Olchowy G, Sengers J, Watson J, Millat J (1990) The transport properties of carbon dioxide. J Phys Chem Ref Data 19(3):763–808

Wang R (2007) Advances in refrigeration and HVAC. Science Press, Beijing, China

Williams A (2018) CO_2 at heart of new Delhaize convenience store. http://www.r744.com/articles/8422/CO2_at_heart_of_new_delhaize_convenience_store

Williams A (2019) The natural refrigerant treatment. https://issuu.com/shecco/docs/ae_1903_33d31f1e897942/22

Yanji Z, et al (2000) Economic benefits of the air cycle refrigeration and the vapor compression refrigeration in the climatic environmental simulation test. Cryogenics03:42–47+60

Yitai M, Minxia L, Hua T, Junlan Y, Shengchun L, Baomin D, et al (2017) Research and development on refrigeration and heat pump cycle with natural working fluid carbon dioxide. Science Press, Beijing

Yoshimoto D (2019a) Satisfying results observed in China's second trans critical CO_2 system. http://r744.com/articles/9245/satisfying_results_observed_in_china_s_second_transcritical_CO2andnbsp

Yoshimoto D (2019b) DFDS logistics switches to CO_2 containers for coastal shipping service. http://r744.com/articles/9204/dfds_logistics_switches_to_CO2_containers_for_coastal_shipping_service

Yoshimoto D (2019c) Japanese margarine, beer, ice makers adopting transcritical CO_2. http://r744.com/articles/9897/the_philippines_targets_10_15_annual_cold_chain_sector_growth_in_new_industry_roadmap

Yoshimoto D (2019d) World's largest transcritical CO_2 system commissioned in California. http://r744.com/articles/9042/world_s_largest_transcritical_CO2_system_commissioned_in_california

Yoshimoto D (2019e) Beijing 2022 Winter Olympics officially announces use of CO_2 for ice venues. http://r744.com/articles/9053/beijing_2022_winter_olympics_officially_announces_use_of_CO2_systems_f

Zhang Y (2023) Low Carbon Management of Beijing 2022 Olympic and Paralympic Winter Games. In *Annual Report on Actions to Address Climate Change (2019) Climate Risk Prevention.* Singapore: Springer Nature Singapore 219–231

Chapter 3
Transcritical CO_2 Refrigeration Cycle and Systems

Yi-Zhou Wang, Yi-Kun He, and Xin-Rong Zhang

In this chapter, the theories related to thermodynamic cycles in refrigeration are presented. This chapter first introduces the properties of CO_2 pure substances, which is a basis comprising CO_2 refrigeration cycles. After that, basic processes, phase change process, and compression and expansion process will be presented for the refrigeration thermodynamic cycles. This chapter will also give the reader a clear illustration of various CO_2 vapor-compression refrigeration thermodynamic cycles and their characteristics. Based on the above contents, the reader can have a strong base to further understand CO_2 refrigeration cycles and systems, also including their importance.

3.1 Properties of CO_2 Pure Substances

Carbon dioxide is a safe refrigerant with stable physical properties, non-toxic and non-flammable. As can be seen from Table 3.1, carbon dioxide has the following advantages compared with other refrigerants (Lorentzen 1994; Lorentzen and Pettersen 1992, 1993; Secretariat 2007):

(1) ODP = 0; GWP = 1, which is much smaller than CFC and HFC refrigerants and almost no damage to the environment;
(2) Capacity per unit volume is very large, can make the equipment more compact;

Y.-Z. Wang · Y.-K. He · X.-R. Zhang (✉)
Department of Energy and Resources Engineering, College of Engineering, Peking University, Beijing 100871, China
e-mail: xrzhang@pku.edu.cn

Beijing Engineering Research Center of City Heat, Beijing 100871, China

X.-R. Zhang and T. M. Eikevik (eds.), *CO_2 Refrigeration Cycle and Systems*, Lecture Notes in Energy 96, https://doi.org/10.1007/978-3-031-22512-3_3

Table 3.1 Comparison of the main properties of common refrigerants

Refrigerant	R12	R22	R134a	R410a	R717	R744
Molecular weight	120.93	86.48	102	72.58	17.03	44
The critical temperature (°C)	112	96	101.7	72.5	133	31.1
The critical pressure (MPa)	4.11	4.97	4.06	4.95	11.42	7.37
ODP	1	0.055	0	0	0	0
GWP	10,600	1700	1300	1730	0	1
Volume cooling capacity at 0 °C (kJ/m^3)	2740	4344	2860	6700	4360	22,600
Safety level	A1	A1	A1	A1	B2	A1

(3) The critical temperature is low, the general carbon dioxide system is operated under transcritical conditions, so the pressure and temperature of the refrigerant are not related to each other under high-pressure condition, they can be individually adjusted to the most optimal operating state;
(4) Carbon dioxide has high thermal conductivity, low viscosity, small surface tension, and it is easy to flow in turbulent flow, which can improve the characteristics of heat transfer and pressure drop.

Generally speaking, except the fact that the problems such as strength and sealing need to be taken into consideration because of the high working pressure of the system. Carbon dioxide has obvious advantages in many aspects such as environmental protection and cost, making it a very good alternative to replace the traditional working fluids.

3.2 Refrigeration Machine and Heat Pump

A refrigeration machine is a type of equipment that brings heat from the low-temperature source to the high-temperature heat source by refrigeration cycle. It can be used for building cooling, storage and transform heat from the environment to indoor for building heating. The former is named the refrigeration machine and the latter is named the heat pump. There is no difference between them in principle, just in purpose.

Assume that 1 kg of refrigerant in the refrigeration device absorbs heat q_2 from the refrigerating space at low temperatures, consumes mechanical work, and releases heat q_1 to the outside world. According to the law of conservation of energy (Fig. 3.1).

$$q_1 = q_2 + w \tag{3.1}$$

For the refrigeration machine, refrigeration coefficient equal to absorbed heat q_2 divided by compression work:

Fig. 3.1 Refrigeration machine

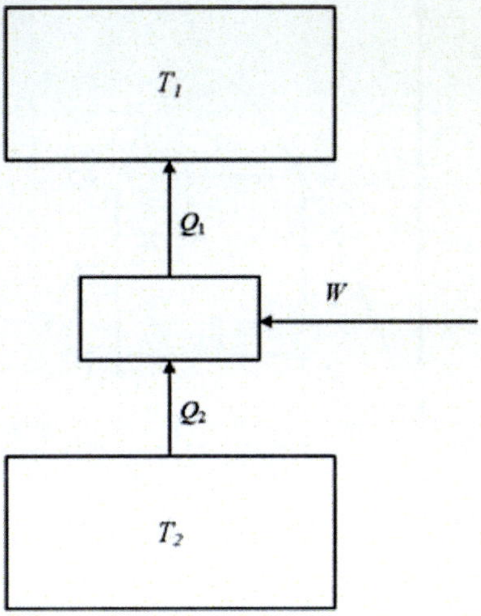

$$\varepsilon = \frac{q_2}{w} \tag{3.2}$$

For the heat pump, heating coefficient is equal to releases heat q_1 divided by compression work:

$$\varepsilon' = \frac{q_1}{w} \tag{3.3}$$

When $q_1 = q_2$:

$$\varepsilon' = \varepsilon + 1 \tag{3.4}$$

It means that the heating efficiency is greater than 1, therefore, using a heat pump is more efficient than electric heating. However, due to losses in actual operation, the actual operating efficiency may be less than 1.

3.3 Reversed Carnot Cycle

The Reversed Carnot cycle is an ideal refrigeration or heat pump cycle, where the cycle is plotted in the temperature versus entropy and pressure versus enthalpy diagrams in Fig. 3.2. The Reversed Carnot cycle consists of four thermodynamic processes:

1-2: Isentropic compression
2-3: Isothermal-isobaric condensation and heat release
3-4: Isentropic throttling
4-1: Isothermal-isobaric evaporation and heat absorption.

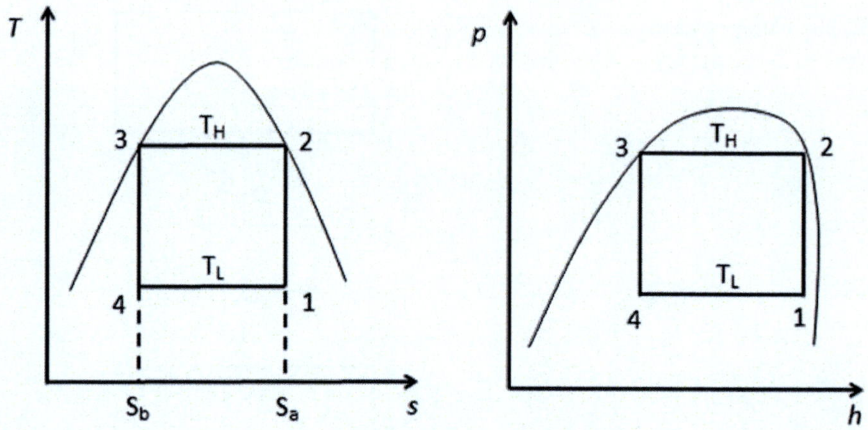

Fig. 3.2 Cycle in the temperature versus entropy diagram and the pressure versus enthalpy diagram

For a unit mass of working fluid, the amount of heat absorbed from the low-temperature heat source is the cooling capacity:

$$q_L = T_L(s_a - s_b) \tag{3.5}$$

Release heat to the high-temperature heat source:

$$q_H = T_H(s_a - s_b) \tag{3.6}$$

The required compression work (power consumption) is:

$$w = q_H - q_L = (T_H - T_L)(s_a - s_b) \tag{3.7}$$

Refrigeration coefficient (COP-Coefficient of Performance) is:

$$\text{COP} = \frac{T_L}{T_H - T_L} \tag{3.8}$$

It can be seen from the formula (3.4) that the refrigeration coefficient of the Reversed Carnot cycle has nothing to do with refrigerant properties, but only depends on high and low-temperature heat source temperature. When T_L increases or T_H decreases, the cycle performance can be improved.

3.4 Basic Characteristics of CO_2 Refrigeration Cycles

3.4.1 Classification of CO_2 Refrigeration Cycles

The critical temperature of CO_2 is close to the ambient temperature, and the following three cycles can be realized according to the external conditions of the cycle.

(1) Subcritical refrigeration cycle: The working process is the same as the general vapor compression refrigeration cycle, and the working process 1-2-3-4-1 is shown in Fig. 3.3. At this phase, the condensing temperature and the condensing pressure are both lower than the critical temperature and the critical pressure, and the heat exchange process mainly relies on releasing latent heat to complete. This cycle is used in the low-temperature stage of cascade refrigeration systems for low-temperature engineering.

(2) Transcritical refrigeration cycle: The working process is different from the general vapor compression refrigeration cycle, and the working process 1-5-6-4-1 is shown in Fig. 3.3. The endothermic process of this cycle is carried out under subcritical conditions, but the discharge pressure of the compressor, that is, the gas cooler pressure is higher than the critical pressure, the heat exchange process in the gas cooler is completed by releasing sensible heat. This cycle is the most common system of CO_2 refrigeration cycles.

(3) Supercritical refrigeration system: This cycle is completely different from the general vapor compression refrigeration cycle, and the working process 7-8-9-10-7 is shown in Fig. 3.3. System operation is above the critical point, there is no phase change in the working cycle, and is gaseous in circulation. The cycle is used for power generation.

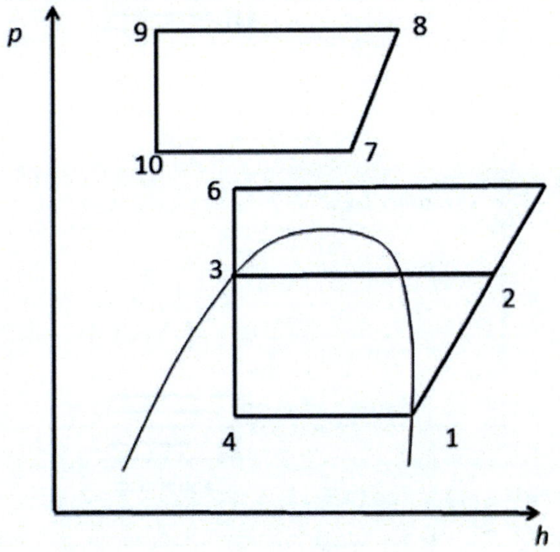

Fig. 3.3 Cycles in the pressure versus enthalpy diagram

3.4.2 *Transcritical CO_2 Refrigeration Cycle*

The schematic diagram and corresponding pressure enthalpy diagram of this cycle are illustrated in Fig. 3.4. This is a basic cycle that consists of an evaporator, a compressor, a gas cooler, a throttling valve. The ideal working processes of this cycle are described as follows:

1-2: Isentropic compression where the saturated vapor from the evaporator is compressed to the high temperature and high-pressure gas by the compressor.
2-3: Isobaric heat release where high temperature and high-pressure gas discharged from the compressor is cooled by air in the gas cooler.
3-4: Isentropic throttling where CO_2 refrigerant is throttled down to the evaporating pressure through the throttle valve, then the temperature is reduced and becomes a gas–liquid mixture.
4-1: Isobaric heat absorption where wet steam absorbs heat into saturated steam in the evaporator, then is sucked by the compressor to complete the cycle.

System unit cooling capacity is:

$$q = h_1 - h_4 \tag{3.9}$$

Unit mass compression work is:

$$W = h_2 - h_1 \tag{3.10}$$

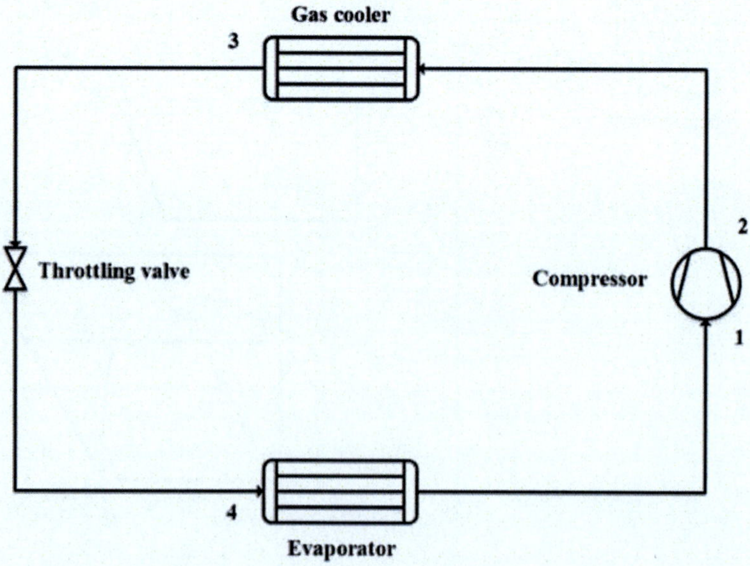

Fig. 3.4 The schematic diagram and corresponding pressure enthalpy diagram of the basic cycle

System performance coefficient is:

$$\mathrm{COP} = \frac{q}{w} = \frac{h_1 - h_4}{h_2 - h_1} \tag{3.11}$$

Compared with the traditional vapor compression cycle, the transcritical CO_2 refrigeration cycle has the following characteristics:

(1) The high-pressure side belongs to the supercritical area, and its temperature and pressure are independent of each other, which makes the system have a more controllable parameter, and the relevant parameters can be adjusted and optimized with the best performance of the system.
(2) There is no phase change in the gas cooler, so the temperature is constantly changing during an exothermic process, and has a large temperature slip, which enables a good temperature matching between the working fluid and the heat medium.
(3) The supercritical fluid of the system is directly throttled to the two-phase zone. On the one hand, the dryness of the two-phase working fluid is high, which is not conducive to the improvement of the cooling capacity; On the other hand, the isenthalpic line of the throttling process deviates greatly from the direction of entropy increase, resulting in the loss of the throttling process being much larger than the throttling loss of the conventional refrigerant.

3.5 Various CO_2 Refrigeration Thermodynamic Cycles

For the case where the throttling loss of transcritical CO_2 refrigeration cycle is large and the system efficiency is low, some modified cycles are proposed to improve system efficiency as follow.

3.5.1 The Single-Stage Cycle with an Internal Heat Exchanger (SCI)

Compared with the basic cycle, the SCI cycle adds an internal heat exchanger to the system, as shown in Fig. 3.5. The internal heat exchanger can not only overheat the saturated steam from the evaporator to make the compressor operation safe but also increase the degree of super-cooling for the refrigerant from the gas cooler to increase the cooling capacity and system efficiency. The saturated vapor is sucked by and compressed by the compressor, then both the temperature and pressure are raised. Afterward, it is cooled by water in the gas cooler. Before entering the throttling value, it exchanges heat with the refrigerant out of the evaporator. Finally, the fluid flows back to the compressor.

Ignore external heat dissipation for the internal heat exchanger, the energy equation is:

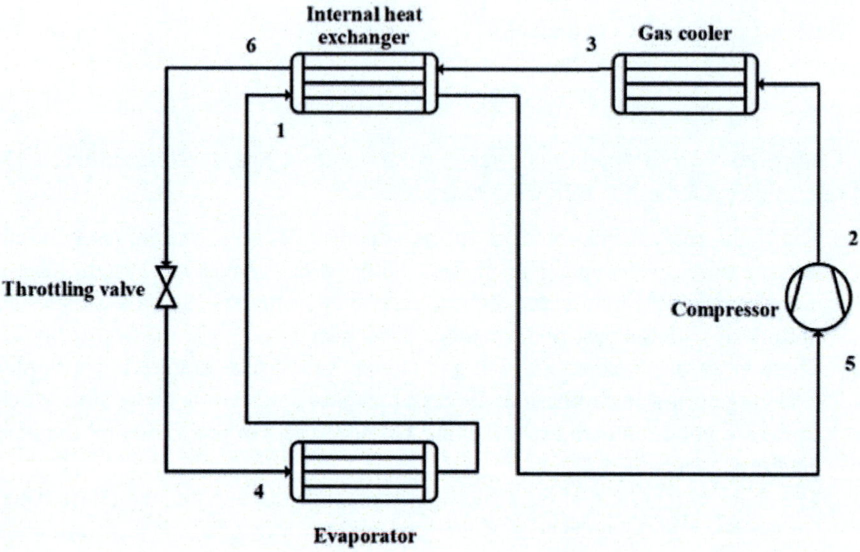

Fig. 3.5 The schematic diagram and corresponding pressure enthalpy diagram of the SCI cycle

$$h_5 - h_1 = h_3 - h_6 \tag{3.12}$$

The cooling capacity is:

$$q = h_1 - h_4 \tag{3.13}$$

The compression work is:

$$w = h_2 - h_1 \tag{3.14}$$

System performance coefficient is:

$$\mathrm{COP} = \frac{q}{w} = \frac{h_1 - h_4}{h_2 - h_5} \tag{3.15}$$

3.5.2 *The Single-Stage Cycle with an Expander (SCE)*

The throttling loss of the transcritical CO_2 cycle is not to be neglected. Under the same equivalent condensation temperature, the efficiency of the transcritical CO_2 cycle is 20–30% lower than the conventional working fluid, which offsets its environmental advantages. In 1994, Lorentzen (1995) proposed to use expanders instead of throttle valves to improve the performance of transcritical CO_2 cycle systems. This cycle

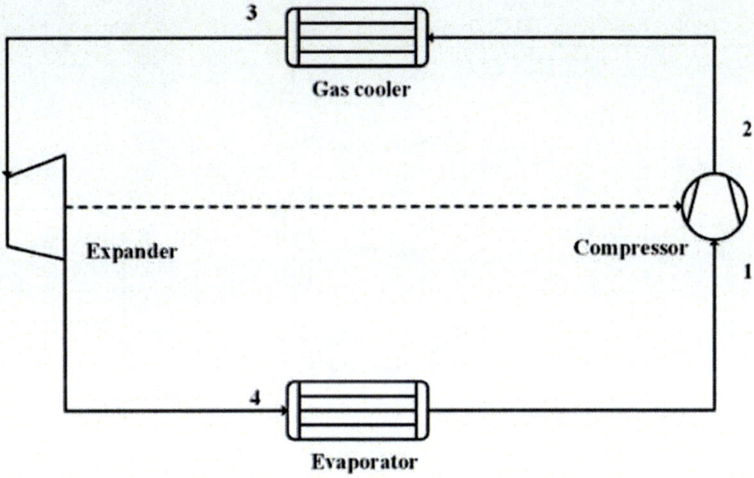

Fig. 3.6 The schematic diagram and corresponding pressure enthalpy diagram of the SCE cycle

is similar to the SC cycle except with the expander replacing the throttle valve and recovering work to drive the compressor, which can improve the system efficiency. The expansion process of the SCE cycle is expressed by the broken line in Fig. 3.6.

The cooling capacity is:

$$q = h_1 - h_{4s} \tag{3.16}$$

The work of recovery is:

$$w_{exp} = h_3 - h_{4s} \tag{3.17}$$

The compression work is:

$$W_{comp} = (h_2 - h_1) - (h_3 - h_{4s}) \tag{3.18}$$

System performance coefficient is:

$$\text{COP} = \frac{q}{w_{comp}} = \frac{h_1 - h_{4s}}{(h_2 - h_1) - (h_3 - h_{4s})} \tag{3.19}$$

3.5.3 The Double-Stage Cycle with a Gas Intercooler (DC)

This cycle is a method to improve the operating conditions of the system and improve the coefficient of performance. The schematic diagram and corresponding pressure enthalpy diagram of this cycle are illustrated in Fig. 3.7. This cycle consists of an

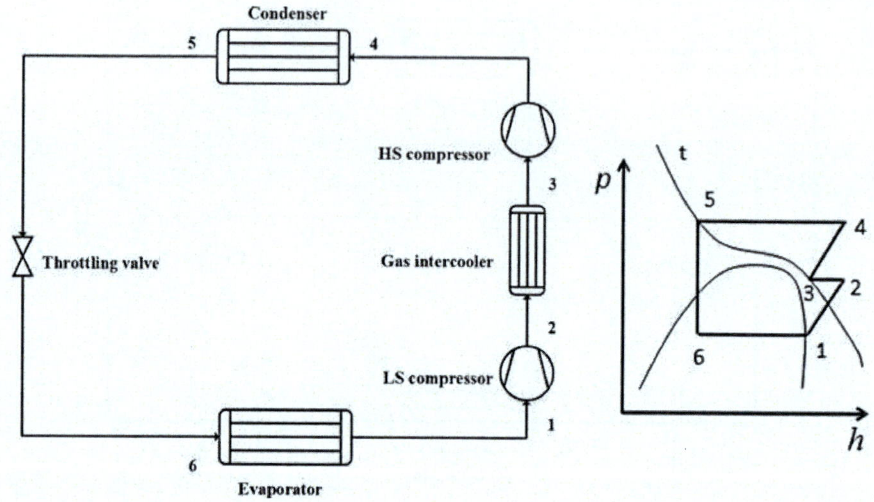

Fig. 3.7 The schematic diagram and corresponding pressure enthalpy diagram of the DC cycle

evaporator, two compressors (LS compressor and HS compressor), a gas intercooler, a gas cooler, a throttling valve. Compared with the basic cycle, the DC cycle can effectively reduce compressor discharge temperature.

The working process of the DC cycle can be expressed as follows:

1-2: Isentropic compression where the saturated vapor from the evaporator is compressed by the LS compressor.
2-3: Isobaric heat release where the gas refrigerant discharged from the LS compressor is cooled by air in the gas intercooler.
3-4: Isentropic compression where the superheated vapor from the gas intercooler is compressed by the HS compressor.
4-5: Isobaric heat release where high temperature and high-pressure gas discharged from the HS compressor is cooled by air in the gas cooler.
3-4: Isentropic throttling where the CO_2 refrigerant is throttled down to the evaporating pressure through the throttle valve, then the temperature is reduced and becomes a gas–liquid mixture.
4-1: Isobaric heat absorption where wet steam absorbs heat into saturated steam in the evaporator, then is sucked by the LS compressor to complete the cycle.

The cooling capacity is:

$$q = h_1 - h_6 \tag{3.20}$$

The compression work is:

$$w = h_2 - h_1 + (h_4 - h_3) \tag{3.21}$$

System performance coefficient is:

$$\text{COP} = \frac{q}{w} = \frac{h_1 - h_6}{h_2 - h_1 + (h_4 - h_3)} \tag{3.22}$$

3.5.4 *The Double-Stage Cycle with Two Evaporators (DW)*

The double-stage cycle with two evaporators is often used in the supermarket (Beshr et al. 2015; Ge and Tassou 2011a, b). It is characterized by having a high-temperature evaporator and a low-temperature evaporator, which can provide refrigeration requirements with different evaporating temperatures at the same time. The schematic diagrams and of this cycle are illustrated in Fig. 3.8.

The cooling capacity is:

$$w = (h_2 - h_1) + (h_4 - h_3) \tag{3.23}$$

The cooling capacity is:

$$q = q_{le} + q_{he} = (h_7 - h_6) + (h_1 - h_8) \tag{3.24}$$

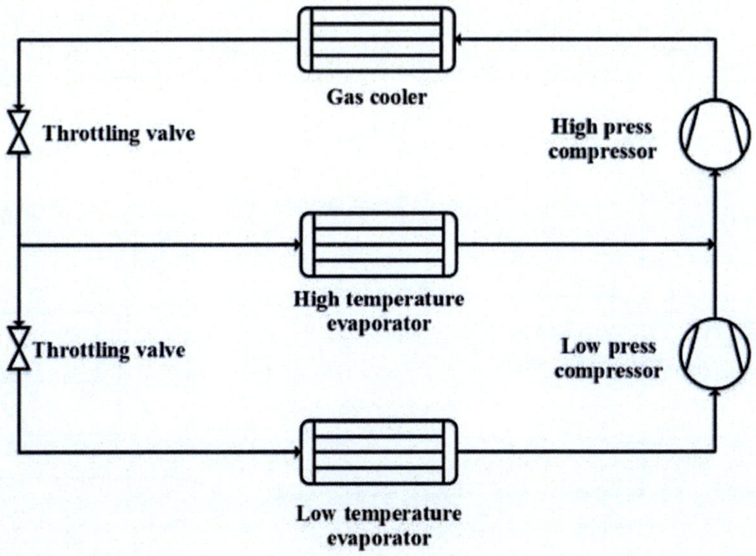

Fig. 3.8 The schematic diagram of the DW cycle

System performance coefficient is:

$$\mathrm{COP} = \frac{q}{w} = \frac{(h_7 - h_6) + (h_1 - h_8)}{(h_2 - h_1) + (h_4 - h_3)} \tag{3.25}$$

3.5.5 The Double-Stage Cycle with a Closed Flash Intercooler (DCFI)

The cycle is shown by the continuous line in Fig. 3.9. The high-pressure refrigerant after the gas cooler is divided into two streams: one of them is cooled by the liquid refrigerant of the closed flash intercooler and is throttled down to the evaporating pressure through the throttle valve B and then fed to the evaporator. The other one is throttled down to the intermediate pressure through the throttle valve A and flashes into vapor by absorbing heat in the closed flash intercooler, then mixes with the discharged high-temperature refrigerant from the LS compressor in the pipeline. So the superheated vapor is sucked by the HP compressor.

For the closed flash intercooler:

$$m_L(h_5 - h_7) = (m_H - m_L)(h_9 - h_6) \tag{3.26}$$

For the mixing process in the pipeline:

$$m_L h_2 + (m_H - m_L)h_9 = m_H h_3 \tag{3.27}$$

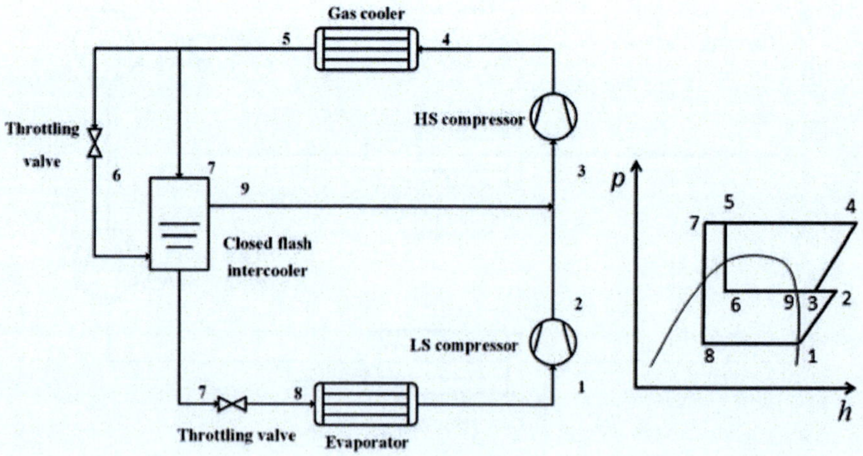

Fig. 3.9 The schematic diagram and corresponding pressure enthalpy diagram of the DCFI cycle

The cooling capacity is:

$$q = m_L(h_1 - h_8) \tag{3.28}$$

The compression work is:

$$w = m_L(h_2 - h_1) + m_H(h_4 - h_3) \tag{3.29}$$

System performance coefficient is:

$$\mathrm{COP} = \frac{q}{w} = \frac{m_L(h_1 - h_8)}{m_L(h_2 - h_1) + m_H(h_4 - h_3)} \tag{3.30}$$

3.5.6 The Double-Stage Cycle with an Open Flash Intercooler (DOFI)

The cycle is represented by the continuous line in Fig. 3.10. The intermediate pressure refrigerant in the open flash intercooler is divided into saturated vapor and saturated liquid: the saturated vapor mixes with the discharged high-temperature refrigerant from the LS compressor in the pipeline, then the resulting mixed superheat vapor is sucked by the HP compressor. The saturated liquid is expanded in the throttle valve B and then fed to the evaporator.

For the open flash intercooler:

$$m_H h_6 = (m_H - m_L)h_9 + m_L h_7 \tag{3.31}$$

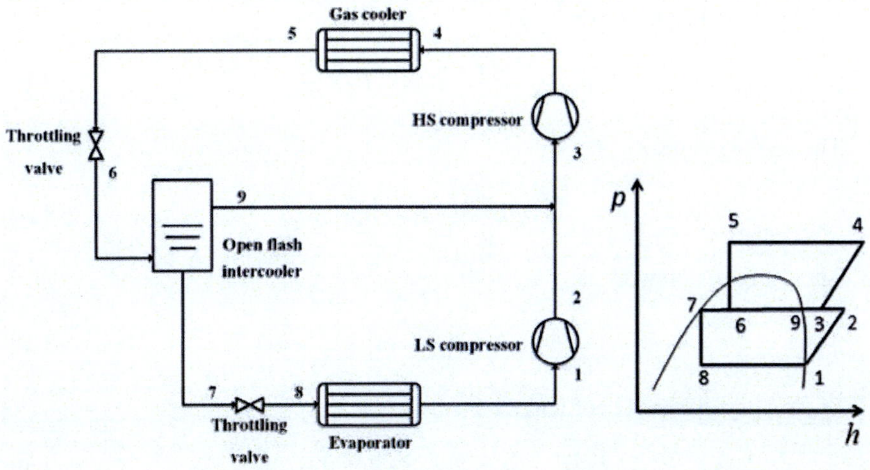

Fig. 3.10 The schematic diagram and corresponding pressure enthalpy diagram of the DOFI cycle

For the mixing process in the pipeline:

$$m_L h_2 + (m_H - m_L) h_9 = m_H h_3 \tag{3.32}$$

The cooling capacity is:

$$q = m_L (h_1 - h_8) \tag{3.33}$$

The compression work is:

$$w = m_L (h_2 - h_1) + m_H (h_4 - h_3) \tag{3.34}$$

System performance coefficient is:

$$\text{COP} = \frac{q}{w} = \frac{m_L (h_1 - h_8)}{m_L (h_2 - h_1) + m_H (h_4 - h_3)} \tag{3.35}$$

The advantages of the DCFI cycle and DOFI cycle are as follows:

(1) Small compression ratio and low energy loss;
(2) Reducing the discharge temperature;
(3) Increasing the cooling effects.

3.5.7 *The Cascade System*

Figure 3.11 shows the cascade system in which CO_2 is used as a refrigerant for the low-temperature stage, another refrigerant such as ammonia for the high-temperature stage. The high-temperature stage is connected with the low-temperature stage by an intermediate heat exchanger.

For intermediate heat exchanger:

$$m_L (h_2 - h_3) = m_H (h_5 - h_8) \tag{3.36}$$

The cooling capacity is:

$$q = m_L (h_1 - h_4) \tag{3.37}$$

The compression work is:

$$w = m_L (h_2 - h_1) + m_H (h_6 - h_5) \tag{3.38}$$

System performance coefficient is:

$$\text{COP} = \frac{q}{w} = \frac{m_L (h_1 - h_4)}{m_L (h_2 - h_1) + m_H (h_6 - h_5)} \tag{3.39}$$

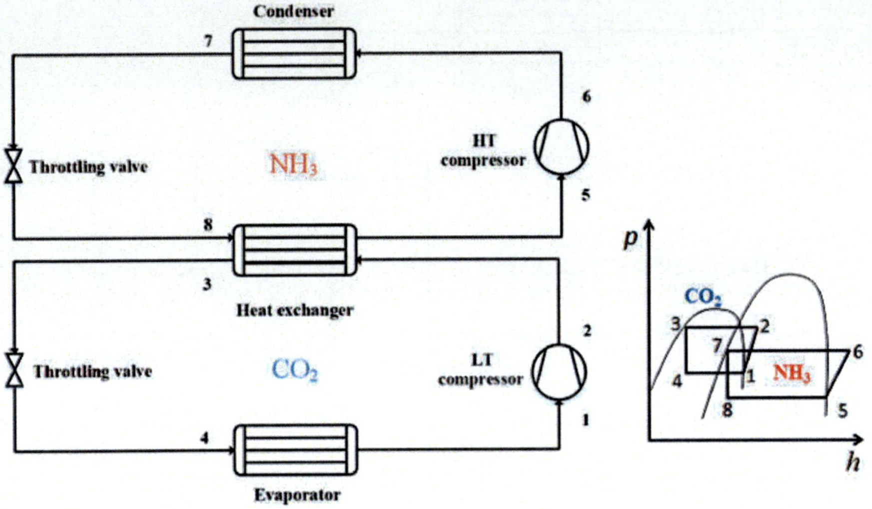

Fig. 3.11 The schematic diagram and corresponding pressure enthalpy diagram of the cascade cycle

3.5.8 *Transcritical CO_2 Reversible System*

Reversible systems are commonly used in residential buildings to supply annual heating and cooling load (Dai et al. 2019, 2020). The conversion of heating (heat pump)/cooling mode (refrigeration machine) is realized by switching the two four-way reversing valves (Fig. 3.12).

In recent studies, economics is the most problem of transcritical CO_2 reversible systems. Especially in the price of four-way reversing valves. Compare with the traditional system, the high press of the transcritical CO_2 system is above 7.38 MPa, values must reconsider how to under that press. This will greatly increase the cost. Designing valves with both higher-pressure resistance and good economics is the current research direction.

3.5.9 *Subcooling*

CO_2 subcooling has resulted in a method to upgrade the performance of CO_2 refrigeration plants in recent years (Llopis et al. 2018; Song and Cao 2018). Revision of the state of the art shows that considering as baseline system the CO_2 cycle without improvements, the possibility to enhance the overall performance reaches 12% (Torrella et al. 2011) using internal heat exchangers, 22% (Cavallini et al. 2005) using economizers, 25.6% (Sarkar 2013) using thermoelectric systems, 21.3% (Gullo and Cortella 2016) using integrated mechanical subcooling systems and 30.3% (Llopis et al. 2016) using dedicated mechanical subcooling systems. Most of the

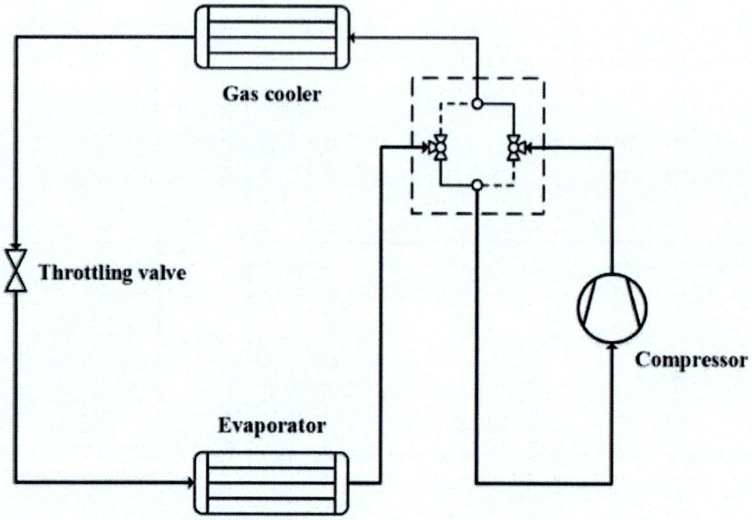

Fig. 3.12 Schematic layout of transcritical CO_2 reversible system

review research is at an initial stage and there is room for improvement in some of the methods (Wang et al. 2019) (Fig. 3.13).

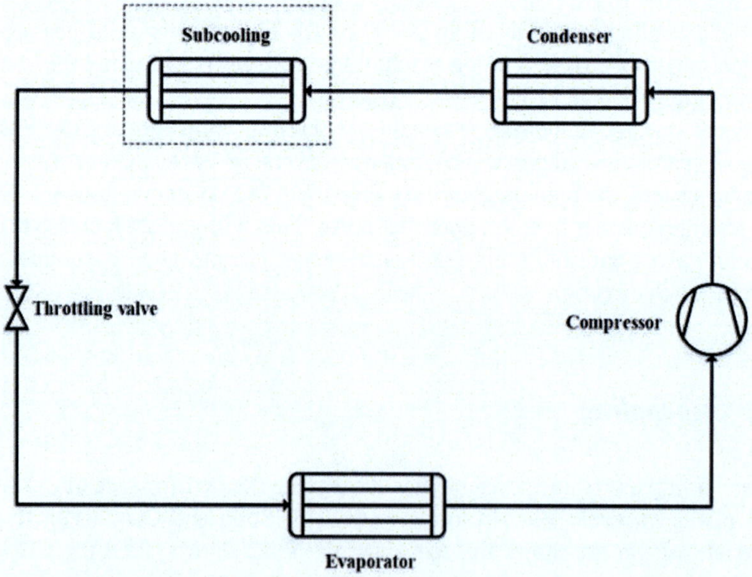

Fig. 3.13 Schematic layout of a CO_2 refrigeration system with subcooling system

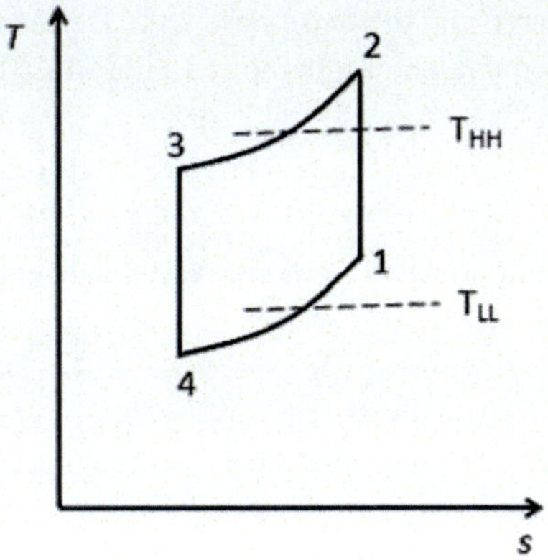

Fig. 3.14 Cycle in the temperature versus entropy diagram

In recent years, the use of subcooling methods has been re-searched and different developments have shown that subcooling of CO_2 at the exit of the gas-cooler/condenser presents numerous advantages to artificial refrigerant cycles, which makes it an improvement to be considered to enhance the performance of such cycles.

3.6 Equivalent Temperature Method

3.6.1 *Lorentzen Cycle*

The Lorentz cycle is the cycle with the largest refrigeration coefficient under the temperature change of the heat source. The cycle is an inverse reversible cycle consisting of two variable processes of heat transfer without temperature difference with heat source and two isentropic processes, as shown in Fig. 3.14.

The refrigeration coefficient (COP) of the Lorentzen cycle is:

$$\text{COP} = \frac{T_{LL}}{T_{HH} - T_{LL}} \tag{3.40}$$

where T_{HH} represents the average exothermic temperature, T_{LL} represents the average endothermic temperature. The refrigeration coefficient of the Lorentzen cycle is equivalent to Reversed Carnot cycle Between constant temperature heat source T_{HH} and T_{LL}.

3.6.2 *Equivalent Temperature*

How to change the variable temperature heat transfer process to a constant temperature heat transfer process? We can take the average temperature of the overheating area and the two-phase area. Define the temperature corresponding to the process as equivalent temperature (Yitai Ma 2000):

$$\mathrm{ET} = \frac{\int_1^2 T\,ds}{\int_1^2 ds} = \frac{\int_1^2 T\,ds}{\Delta s} \tag{3.41}$$

For constant pressure–variable specific heat:

$$\mathrm{ET} = \frac{\int_1^2 c_p dT}{\Delta s} \tag{3.42}$$

For constant pressure–constant specific heat:

$$\mathrm{ET} = \frac{\int_1^2 c_p dT}{\int_1^2 \frac{c_p}{T} dT} = \frac{c_p \int_1^2 dT}{c_p \int_1^2 \frac{1}{T} dT} = \frac{T_2 - T_1}{\ln\left(\frac{T_2}{T_1}\right)} \tag{3.43}$$

When calculating the performance coefficient, the same equivalent condensation temperature and equivalent evaporation temperature are used as the basis for comparison.

Re-analyzing the vapor compression cycle based on the equivalent temperature will help eliminate the defects of the analysis method.

References

Beshr M, Aute V, Sharma V, Abdelaziz O, Fricke B, Radermacher R (2015) A comparative study on the environmental impact of supermarket refrigeration systems using low GWP refrigerants. Int J Refrig 56:154–164

Cavallini A, Cecchinato L, Corradi M, Fornasieri E, Zilio C (2005) Two-stage transcritical carbon dioxide cycle optimisation: a theoretical and experimental analysis. Int J Refrig 28(8):1274–1283

Dai B et al (2019) Evaluation of transcritical CO_2 heat pump system integrated with mechanical subcooling by utilizing energy, exergy and economic methodologies for residential heating. Energy Convers Manag 192:202–220

Dai B, et al (2020) Energetic, exergetic and exergoeconomic assessment of transcritical CO_2 reversible system combined with dedicated mechanical subcooling (DMS) for residential heating and cooling. Energy Convers Manag 209

Ge Y, Tassou S (2011a) Performance evaluation and optimal design of supermarket refrigeration systems with supermarket model "SuperSim". Part I: model description and validation. Int J Refrig 34(2):527–539

Ge Y, Tassou S (2011b) Performance evaluation and optimal design of supermarket refrigeration systems with supermarket model "SuperSim". Part II: model applications. Int J Refrig 34(2):540–549

Gullo P, Cortella G (2016) Comparative exergoeconomic analysis of various transcritical R744 commercial refrigeration systems, pp 19–23

Llopis R, Nebot-Andrés L, Cabello R, Sánchez D, Catalán-Gil J (2016) Experimental evaluation of a CO_2 transcritical refrigeration plant with dedicated mechanical subcooling. Int J Refrig 69:361–368

Llopis R, Nebot-Andrés L, Sánchez D, Catalán-Gil J, Cabello R (2018) Subcooling methods for CO_2 refrigeration cycles: a review. Int J Refrig 93:85–107

Lorentzen G (1994) Revival of carbon dioxide as a refrigerant. Int J Refrig 17(5):292–301
Lorentzen G (1995) The use of natural refrigerants: a complete solution to the CFC/HCFC predicament. Int J Refrig 18(3):190–197
Lorentzen G, Pettersen J (1992) New possibilities for non-CFC refrigeration
Lorentzen G, Pettersen J (1993) A new, efficient and environmentally benign system for car air-conditioning. Int J Refrig 16(1):4–12
Sarkar J (2013) Performance optimization of transcritical CO_2 refrigeration cycle with thermoelectric subcooler. Int J Energy Res 37(2):121–128
Secretariat UO (2007) Report of decisions adopted by the nineteenth meeting of the parties to the Montreal Protocol on substances that deplete the ozone layer, Nairobi
Song Y, Cao F (2018) The evaluation of the optimal medium temperature in a space heating used transcritical air-source CO_2 heat pump with an R134a subcooling device. Energy Convers Manag 166:409–423
Torrella E, Sánchez D, Llopis R, Cabello R (2011) Energetic evaluation of an internal heat exchanger in a CO_2 transcritical refrigeration plant using experimental data. Int J Refrig 34(1):40–49
Wang G-B, Zhang X-R, Management (2019) Thermoeconomic optimization and comparison of the simple single-stage transcritical carbon dioxide vapor compression cycle with different subcooling methods for district heating and cooling. Energy Convers Manag 185:740–757
Yitai Ma KW (2000) Thermodynamic analysis of CO_2 transcritical reverse cycle with expander. J Eng Thermophys

Chapter 4
Theoretical Analysis of Expansion Process and Components in CO_2 (Transcritical) Refrigeration System

Min-Qiang Zeng, Xin-Rong Zhang, Xue-Lai Zhang, and Yi-Wei Yan

Nomenclature

d, D	Diameter (mm)
F	Sectional area (m^2)
g	Gravitational acceleration ($m\ s^{-2}$)
h	Specific enthalpy ($kJ\ kg^{-1}$)
k	Heat capacity ratio
m	Section shrinkage ratio, Mass flow rate
P	Pressure (kPa)
ΔP	Pressure drop of the secondary nozzle
Q	Refrigeration capacity (kW)
r	Unit weight
W	Average flow velocity at fluid cross section ($m\ s^{-1}$)
C	Discharge coefficient
s	Specific entropy ($kJ\ kg^{-1}\ K^{-1}$)
z	EEV opening
x	Quality

M.-Q. Zeng
Nanchang Innovation Institute, Peking University, Nanchang 330096, China

X.-R. Zhang (✉)
Department of Energy and Resources Engineering, College of Engineering, Peking University, Beijing 100871, China
e-mail: xrzhang@pku.edu.cn

Beijing Engineering Research Center of City Heat, Beijing 100871, China

X.-L. Zhang · Y.-W. Yan
Merchant Marine College, Shanghai Maritime University, Shanghai 201306, China

X.-R. Zhang and T. M. Eikevik (eds.), *CO_2 Refrigeration Cycle and Systems*, Lecture Notes in Energy 96, https://doi.org/10.1007/978-3-031-22512-3_4

Greek Letters

α	The flow coefficient
ρ	Density (kg m^{-3})
u	Velocity (m s^{-1})
v	Specific volume (m^3 kg^{-1})
ω	Ejector entrainment ratio
η	Isentropic efficiency
μ	Shrinkage coefficient
ε	Expansion coefficient
λ	Coefficient(s) of friction

Subscripts

b	Receiving chamber
c	Critical
d	Outlet of diffuser
v	Volume
e, eva	Evaporator
ej	Ejector
gc	Gas cooler
gc	Gas cooler
is	Isentropic process
imp	Improvement
m	Motive stream at the nozzle inlet
mix	The outlet of the mixing section
in	Inlet
out	Outlet
s	Suction stream at nozzle inlet, Saturation
sc	Subcooling
TP	Two-phase
1, 2…	Cycle locations

Acronyms

COP	Coefficient of performance
Exp	Experiment
EEV	Electronic expansion valve
EVCS	Ejector-expansion vapor compression system
IHX	Internal heat exchanger

ORC	Organic Rankine cycle
PLR	Pressure lift ratio
SEVCS	Standard ejector-expansion vapor compression systems
Sim	Simulation
CVCS	Conventional vapor compression systems

Chemical Compounds

CO_2	Carbon dioxide

CO_2 has become a research hotspot in the context of refrigerant replacement. Due to the low critical temperature point of CO_2, CO_2 needs to operate in the supercritical region under most conditions. The expansion process is an essential part of the CO_2 transcritical refrigeration cycle, and the energy loss here is relatively large. Therefore, we need to have a more comprehensive understanding of the expansion process. This chapter introduces the expansion process in the CO_2 transcritical cycle and the commonly used expansion devices. The first part introduces the basic principles of throttling expansion. The relevant content of the application of CO_2 in expansion valves, capillaries, and ejectors, as well as the current theoretical and experimental research progress, are summarized in the second to fourth parts. Finally, the expansion device in the CO_2 transcritical refrigeration cycle is presented and outlooked.

4.1 Introduction

The natural working fluid CO_2 has a long history in the refrigeration cycle. Especially in the face of the high consumption of fluorinated refrigerants and the replacement of refrigerants in the refrigeration industry, CO_2 is expected to become one of the answers to this problem. CO_2 is a non-flammable, non-toxic fluid with a GWP of 1. Its production is considered simple and cheap, besides being recoverable (Holling et al. 2013). The main drawback of CO_2 is associated with its high discharge pressure, which exceeds the critical point in warm climates. Nonetheless, the main technical constraints related to the pressure are already overcome (Yu et al. 2019). Because the CO_2 critical temperature is low, the refrigeration cycle usually runs under transcritical conditions. The CO_2 transcritical refrigeration cycle usually includes a compressor, a gas cooler, a throttling device, and an evaporator (Bellos and Tzivanidis 2019). In the refrigeration cycle, in order to achieve the target temperature, throttling expansion is indispensable, and there is a large energy loss in the throttling part (Ma et al. 2013; Chen et al. 2015). How to reduce the throttling loss has also been continuously studied (Chunnanond and Aphornratana 2004). To describe the throttling expansion more clearly and comprehensively, the following will discuss the basic principles of throttling expansion, the types of throttling devices (expansion valve, capillary tube, ejector), principles, and current research progress. Finally, the development history and prospects of the throttling expansion device in the carbon dioxide transcritical cycle are summarized and analyzed.

4.2 CO_2 Expansion Fundamental

Throttle expansion has a very important position in the refrigeration system. It is well known that the expansion valve in a traditional vapor compression refrigerating cycle only represents a mass flow rate control that conforms to the operative constraints set by the other main components of the circuit. Its function is to throttle and reduce pressure, adjust flow and pressure, and maintain normal system superheat. The throttling process is generally approximated as an adiabatic process. Its basic principle is that when the fluid flows through the pipe, the cross-section of the pipe suddenly becomes smaller. At this time, the kinetic energy of the fluid increases and the pressure decreases as shown in Fig. 4.1. When the refrigerant does not pass through the orifice, the cross-sectional area is F1. When the fluid enters the orifice, the cross-sectional area becomes F0. At this time, the fluid velocity rises from W1 to W0. The pressure drops to P0, and then the fluid continues to shrink to the minimum cross-section F2, and then the fluid gradually fills the tube. The figure also shows the pressure change trend during the throttling process.

In a transcritical cycle, the function of the expansion valve differs; as a matter of fact, such a cycle has one freedom degree more, being the pressure of the gas cooler not related to the heat transfer conditions: the upper cycle pressure is a parameter

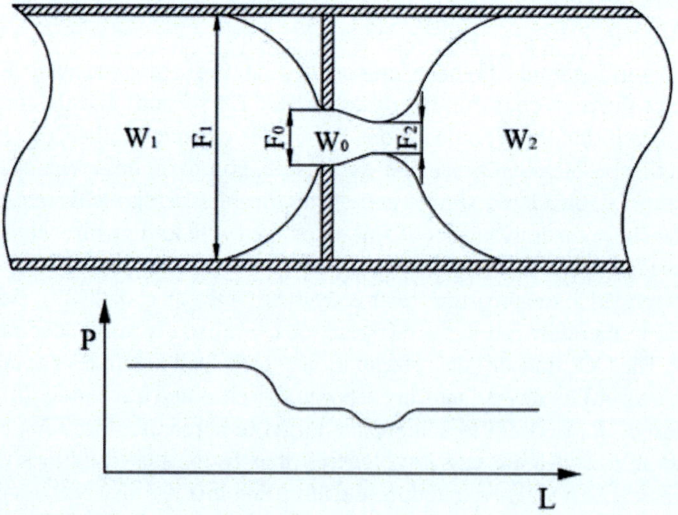

Fig. 4.1 Schematic diagram of throttling process

subjected to optimization, and its value is set by a suitable controller, acting on the flow resistance of the valve. Therefore, the throttling device changes according to the changes in the cooling load demand of the system, and the system can work more effectively and normally. Because the inlet of the throttling device in the CO_2 transcritical cycle is in a supercritical state, supercritical CO_2 has certain characteristics, especially near the critical area, the physical properties of CO_2 change greatly. So the CO_2 expansion process has its uniqueness. Because of the transcritical operation of the CO_2 cycle, the CO_2 flowing in the expansion process is a supercritical fluid, and the flow patterns of the supercritical CO_2 flowing through an expansion valve are not the same as the conventional refrigerants. Therefore, there is no surface tension and degree of subcooling for the supercritical CO_2. The correlations for subcritical refrigerants cannot be used to predict the mass flow rate of supercritical CO_2. Therefore, it is necessary to deeply study the characteristics of supercritical CO_2 flowing through an expansion process.

Below we will discuss the CO_2 expansion devices one by one to understand the principles and characteristics of their respective expansion processes.

4.3 Expansion Valves

4.3.1 Introduction

The CO_2 electronic expansion valve (EEV) is the most commonly used throttling device with a simple structure. The purpose of the expansion valve is to control the flow of refrigerant from the high-pressure condensing side of the system into the low-pressure evaporator. In most cases, the pressure reduction is achieved through a variable flow orifice, either modulating or two-position. Due to the excellent performance and the accurate control, EEV becomes popular in transcritical CO_2 refrigeration systems (Shanwei et al. 2005; Zhifang et al. 2008).

An EEV can be regarded as an expansion device with flexible local flow resistance. The typical EEV flow area is shown in Fig. 4.2. When the supercritical CO_2 flows through an EEV, the velocity of the flow increases with the decrease of the flow passage area (Hou et al. 2014a). The shrink surface can be adjusted by changing the EEV opening or by the pressure difference between the EEV inlet and EEV outlet. The static pressure of the supercritical CO_2 will reach the minimum value in the shrink surface when the velocity of the flow reaches maximum. At this moment, if the minimum pressure is smaller than the corresponding saturation pressure, the supercritical CO_2 flow becomes a two-phase flow. According to different inlet temperatures, inlet pressures, and outlet pressures, the general flow pattern of the supercritical CO_2 flow through an EEV can be classified as flashing flow, cavitation flow, and choked flow.

At the inlet of the CO_2 throttling device, CO_2 is maintained at the supercritical state. Thus it is quite different to determine the mass flow rate characteristics

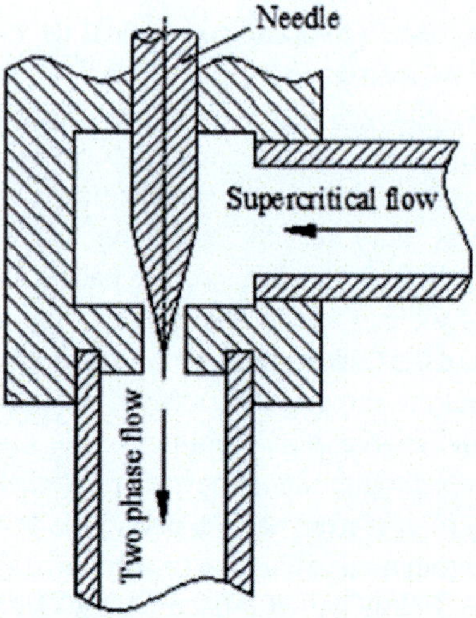

Fig. 4.2 The typical flow passage of EEV (Hou et al. 2014a)

as other refrigerants. Therefore, the mass flow characteristics of CO_2 in electronic expansion valves have become a research topic. The following will describe the flow characteristics and development of the electronic expansion valve.

4.3.2 Flow Characteristics of Refrigerant in Expansion Valves

4.3.2.1 Incompressible Flow

For incompressible fluids, in the absence of heat exchange, the following formula can be listed according to the conservation of energy:

$$\frac{P_1}{r_1} + \frac{{W_1}^2}{2g} = \frac{P_2}{r_1} + \frac{{W_2}^2}{2g} + \lambda\frac{{W_2}^2}{2g} \quad (4.1)$$

In the above formula:

$\frac{P_1}{r_1}$ and $\frac{P_2}{r_1}$ are the pressure energy per unit mass of fluid at the fluid cross section;

$\frac{{W_1}^2}{2g}$ and $\frac{{W_2}^2}{2g}$ are the kinetic energy of the unit mass fluid at the fluid cross section;

$\lambda\frac{{W_2}^2}{2g}$ is the energy loss of unit mass fluid in the flow process;

W_1 and W_2 are the average flow velocity at the fluid cross section;

m is used to represent the ratio of the cross-sectional area of the orifice F_0 to the cross-sectional area of the fluid before the throttling F_1, which is called the ratio of the shrinkage section:

$$m = \frac{F_0}{F_1} = \frac{d^2}{D^2} \quad (4.2)$$

μ is used to express the ratio of the minimum cross-sectional area F_2 to the cross-sectional area F_0 of the orifice after the fluid flows through the orifice, which is called the shrinkage coefficient:

$$\mu = \frac{F_2}{F_0} \tag{4.3}$$

From Eq. 4.2 to Eq. 4.3:

$$F_0 = mF_1;\ F_2 = \mu F_0 \tag{4.4}$$

$$W_1 = mW_0;\ W_2 = \frac{W_0}{\mu} \tag{4.5}$$

$$W_2 = \frac{W_1}{\mu m} \tag{4.6}$$

Putting Eq. 4.6 into Eq. 4.1 to sort out:

$$\frac{P_1 - P_2}{r_1} = \frac{{W_1}^2}{2g}\left(\frac{1 - \mu^2 m^2 + \lambda}{\mu^2 m^2}\right) \tag{4.7}$$

$$W_1 = \frac{\mu}{\sqrt{1 - \mu^2 m^2 + \lambda}} \times m\sqrt{2g\frac{P_1 - P_2}{r_1}} \tag{4.8}$$

Call Eq. 4.9 the flow coefficient α:

$$\alpha = \frac{\mu}{\sqrt{1 - \mu^2 m^2 + \lambda}} \tag{4.9}$$

The volume flow rate is expressed as:

$$Q_V = \alpha F_0\sqrt{2g\frac{P_1 - P_2}{r_1}} \tag{4.10}$$

The mass flow rate is expressed as:

$$Q_m = \alpha F_0\sqrt{2\rho(P_1 - P_2)} \tag{4.11}$$

4.3.2.2 Compressible Fluid

Equations 4.10 and 4.11 mentioned above are incompressible fluid flow formulas. For compressible fluids (such as refrigerant CO_2), the changes in pressure P_1 and P_2 before and after the fluid when the fluid flows through the orifice must be considered. The following is the calculation formula for a compressible fluid.

When there is no heat conduction, the pressure energy is converted into kinetic energy for an instantaneous adiabatic process, which is obtained by thermodynamics:

$$P_0 {v_0}^K = P_1 {v_1}^K = constant \tag{4.12}$$

In the above formula:
v_1 and v_0 are the specific volume at the fluid cross section;
Substituting the specific volume $v = \frac{1}{r}$ on both sides of Eq. 4.12 at the same time:

$$\frac{1}{r} = P_1^{\frac{1}{k}} v_1 P^{-\frac{1}{k}} \tag{4.13}$$

From the fluid energy equation $\frac{dP}{r} + \frac{W}{g} dW = 0$ and $v = \frac{1}{r}$, the following conservation differential equations can be obtained:

$$-P_1^{\frac{1}{k}} v_1 g \times P^{-\frac{1}{k}} dP = W dW \tag{4.14}$$

Integrate both sides of Eq. 4.14 from fluid position 1 to position 2 at the same time:

$$-P_1^{\frac{1}{k}} v_1 g \frac{k}{k-1}\left(P_2^{\frac{k-1}{k}} - P_1^{\frac{k-1}{k}}\right) = \frac{1}{2}\left(W_2^2 - W_1^2\right) \tag{4.15}$$

In compressible fluid:

$$\frac{W_1}{W_2} = \mu m \frac{r_2}{r_1} \tag{4.16}$$

Equations 4.15 and 4.16 combined with the thermodynamic formula $\frac{r_2}{r_1} = \left(\frac{P_2}{P_1}\right)^{\frac{1}{k}}$ to calculate:

$$W_2 = \frac{1}{\sqrt{1-\mu^2 m^2 \left(\frac{P_2}{P_1}\right)^{\frac{2}{k}}}} \times \sqrt{P_1 v_1 2g \frac{k}{k-1}\left[1 - \left(\frac{P_2}{P_1}\right)^{\frac{k-1}{k}}\right]} \tag{4.17}$$

Equation 4.16 can express the flow rate as the following formula:

$$Q_2 = F_2 W_2 \tag{4.18}$$

From Eq. 4.3 to Eq. 4.18 and $\frac{r_2}{r_1} = \left(\frac{P_2}{P_1}\right)^{\frac{1}{k}}$, we can get:

$$Q_1 = \frac{\mu}{\sqrt{1-\mu^2 m^2 \left(\frac{P_2}{P_1}\right)^{\frac{2}{k}}}} \sqrt{\frac{k}{k-1}\left(\frac{P_2}{P_1}\right)^{\frac{2}{k}}\left[1 - \left(\frac{P_2}{P_1}\right)^{\frac{k-1}{k}}\right]} \times F_0 \sqrt{2g \frac{P_1}{r_1}} \tag{4.19}$$

Substitute Eq. 4.20 into Eq. 4.19:

$$P_1 = \frac{P_1 - P_2}{1 - \frac{P_2}{P_1}} \tag{4.20}$$

The fluid equation of the compressible fluid is:

$$Q_1 = \frac{\mu}{\sqrt{1-\mu^2 m^2}} \times \sqrt{\frac{1-\mu^2 m^2}{1-\mu^2 m^2\left(\frac{P_2}{P_1}\right)^{\frac{2}{k}}} \times \frac{1}{1-\frac{P_2}{P_1}} \times \frac{k}{k-1}\left(\frac{P_2}{P_1}\right)^{\frac{2}{k}}\left[1-\left(\frac{P_2}{P_1}\right)^{\frac{k-1}{k}}\right]} \times \times F_0\sqrt{2g\frac{P_1-P_2}{r_1}} \tag{4.21}$$

The two differences between Eqs. 4.21 and 4.8 are, on the one hand, the flow coefficient α, which does not consider the influence of friction when the density changes.

$$\alpha = \frac{\mu}{\sqrt{1-\mu^2 m^2}} \tag{4.22}$$

On the other hand, the coefficient added by taking into account the change in fluid velocity caused by the influence of throttling expansion and the change of fluid density during throttling is called the expansion coefficient.

$$\varepsilon = \sqrt{\frac{1-\mu^2 m^2}{1-\mu^2 m^2\left(\frac{P_2}{P_1}\right)^{\frac{2}{k}}} \times \frac{1}{1-\frac{P_2}{P_1}} \times \frac{k}{k-1}\left(\frac{P_2}{P_1}\right)^{\frac{2}{k}}\left[1-\left(\frac{P_2}{P_1}\right)^{\frac{k-1}{k}}\right]} \tag{4.23}$$

From Eq. 4.22 to Eq. 4.23, the flow equation of compressible fluid is:

Volume flow:

$$Q_V = \alpha\varepsilon F_0\sqrt{2g\frac{P_1-P_2}{r_1}} \tag{4.24}$$

Mass flow:

$$Q_m = \alpha\varepsilon F_0\sqrt{2\rho(P_1-P_2)} \tag{4.25}$$

The above theoretical analysis shows that the incompressible fluid flow formula is a special case of the compressible fluid flow formula. At least the friction between the fluids and the expansion coefficient $\varepsilon = 1$ are not considered.

4.3.2.3 Empirical Formula

In actual calculations, it is difficult to measure α and ε, because μ and m are difficult to measure. Therefore, for compressible fluids, α and ε are unified into the discharge coefficient C, and the flow formula becomes:

$$Q_m = CF_0\sqrt{2\rho_1(P_1-P_2)} \tag{4.26}$$

The empirical formula for the discharge coefficient C of different refrigerants is different. The following will introduce the research progress of CO_2 empirical correlation.

4.3.3 Theoretical and Experimental Studies

Due to EEV's wider working range, higher control accuracy, and faster response, EEV becomes more popular in transcritical refrigeration systems. The transcritical CO_2 refrigeration system showed lower performance than conventional air-conditioners due to large expansion losses and high irreversibility during the gas cooling process (Hwang and Radermacher 1999; Hwang et al. 2001). Generally, the performance of the CO_2 system was more sensitive to the gas cooler pressure and outdoor temperature than that of conventional refrigerant systems (Cho et al. 2005). Therefore, it is required to investigate the effects of control parameters such as compressor frequency and EEV opening on the transcritical CO_2 system in the cooling mode. There have been many scholars who continue to study the mass flow characteristics of EEV and modify the empirical correlation formula through theoretical and experimental methods. Table 4.1 listed some studies on the characteristics of electronic expansion valves in the transcritical CO_2 cycle.

Cho et al. (2007) measured and analyzed the cooling performance of a variable speed CO_2 cycle with an EEV, the optimum EEV opening determined at the maximum coefficient of performance (COP) at a given compressor frequency increased with compressor frequency. Baek et al. (2013) measured the cooling performance of the CO_2 heat pump by varying the refrigerant charge amount, EEV opening, compressor frequency, and outdoor fan speed at various outdoor temperatures. The effects of the EEV opening and the outdoor fan speed on the gas-cooler pressure and the COP were analyzed by using the experimental data. In the standard cooling condition at the compressor frequency of 45 Hz, the optimum gas-cooler pressure and the maximum COP were 9.2 MPa and 3.04, respectively, at the optimum EEV opening of 41% and the outdoor fan speed of 500 rpm. As the compressor frequency increased from 45 to 55 Hz at the standard cooling condition, the optimum outdoor fan speed increased from 500 to 700 rpm.

To investigate the influence of EEV opening on the performance of the transcritical CO_2 refrigeration system, Hou et al. (2014b) established an experimental test rig of the transcritical CO_2 system. The system operating parameters such as temperature and pressure were measured with different EEV openings, when the inlet temperatures of the gas-cooler water and the evaporator water were set to 30 and 15 °C, respectively. The schematic diagram of the transcritical CO_2 refrigeration system is shown in Fig. 4.3. The experiment result shows that the EEV has a large influence on the compressor discharge pressure, discharge temperature, and the gas-cooler outlet pressure, and small influence on the evaporating pressure, and the gas-cooler outlet temperature, which are strongly dependent on the heat transfer condition of the two heat exchangers. The compressor power decreases with the increase of EEV

Table 4.1 Research on characteristics of EEV in transcritical CO_2 cycle

Authors	Year	Method	Type	Main conclusion (empirical correlation)	Relative deviation
Cho et al. (2007)	2007	Exp	EEV parameters	For each compressor frequency, the CO_2 system showed the maximum COP at the optimum normalized charge of 0.282; The optimal opening increases with the increase of compressor frequency	–
Baek et al. (2013)	2013	Exp	EEV parameters	The optimum normalized charge was determined as 0.712 to achieve the maximum COP at the standard cooling condition with the compressor frequency of 45 Hz and the superheat of 5 °C	–
Hou et al. (2014a)	2014	Sim	Correlation formula	$C_d = 1.1075(z)^{0.4436}\left(\frac{\rho_1}{\rho_S}\right)^{-1.4971}\left(\frac{P_1 - P_s}{XP_1}\right)^{-1.4971}$	±10%
Hou et al. (2014b)	2014	Exp	EEV parameters	The compressor input power decreases with the increase of EEV opening	–
Liu et al. (2016a)	2016	Exp	EEV parameters	CO_2 density is very important for the empirical correlation of mass flow	–
Zhang et al. (2017)	2017	Exp	EEV parameters	An optimal coefficient of performance (COP) was found that corresponded to a specific refrigerant charge and a specific EEV opening	–
Liu et al. (2018a)	2018	Sim	Correlation formula	$m = 1.1632 \times 10^{33}(P_{in} - P_s)^{0.5}(P_{in} - P_{out})^{-0.276} \times T_{in}^{16.4875} \times C_{P_{in}}^{0.8789} \times h_{in}^{-6.1854} \times x^{0.1393} \times \mu_{in}^{4.3565} \times D^{-4.0573} \times \alpha^{1.9775} \times P_c^{3.40425} \times T_c^{-15.6086} \times \rho_{in}^{-6.98475} \times D_e^{1.7008}$	−5.6–6.9%

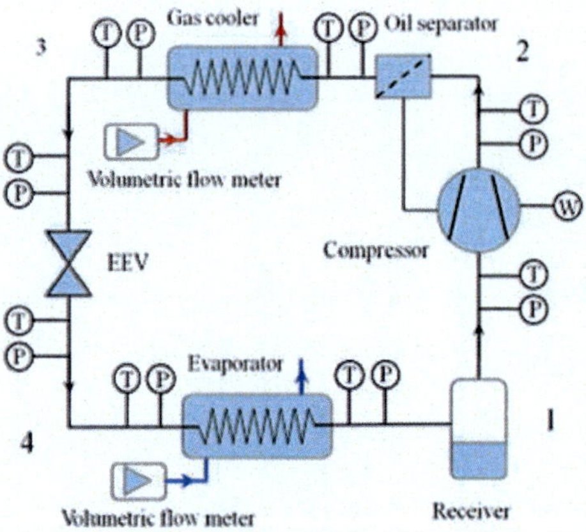

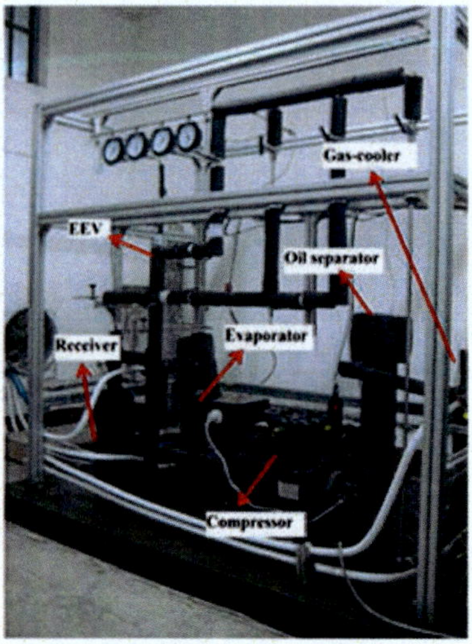

Fig. 4.3 Schematic diagram of the transcritical CO_2 refrigeration system

opening, and it can be attributed to the tradeoff between the increase of the mass flow rate and the decrease of the discharge pressure. Furthermore, the maximum system cooling capacity and the maximum system COP were achieved under different EEV openings of 60 and 40%, respectively.

Subsequently, Hou et al. (2014a) proposed the mass flow rate correlation of supercritical CO_2 flowing through an EEV. The mass flow rate predicted by the correlation in this paper shows a good agreement with experimental data. The maximum relative errors are within 10%. The average and standard deviations are 0.76 and 5.9%, respectively.

4.3.4 Summary of characteristics of CO_2 EEV

It is worth noting from these studies that because the refrigerant CO_2 is a supercritical fluid before it enters the expansion valve in the CO_2 transcritical cycle, its physical properties vary greatly in the near-critical region. Therefore, its mass flow characteristics are different from other refrigerants. Figure 4.4 briefly shows the difference between CO_2, HCFCs, and HFCs when using EEV. First, the fluid state at the inlet and outlet is different, but both become two-phase after passing through the EEV. Secondly, the biggest difference between the two is the need to modify the coefficient C in the flow equation based on different refrigerants based on experiments. The third is that the opening of the EEV changes dynamically according to different refrigerants and different working conditions. Therefore, the expansion valve should be designed reasonably according to the actual situation when applying the expansion valve.

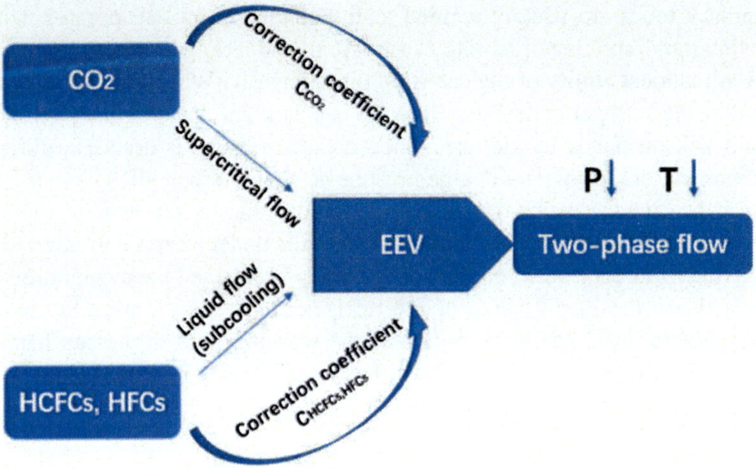

Fig. 4.4 The difference of CO_2, HCFCs, HFCs in EEV

4.4 Capillary Tube and Analysis

4.4.1 Introduction

As a throttling device, the capillary tube has been widely used in various refrigeration and heat pump systems. It was born in the 1930s and was patented in 1942 (Liu et al. 2018b, Liu et al. 2020; Meng 2009; Kurosawa et al. 2021; Ding and Liu 2015; Zeng et al. 2019). The capillary tube is a hollow copper tube, usually with an inner diameter of 0.5–3 mm, and the length varies according to different operating systems, generally between 400 and 5000 mm. According to different operating conditions, the capillary is divided into two types: non-adiabatic capillary and adiabatic capillary. The non-adiabatic capillary tube welds the reservoir and the capillary tube together to form a counterflow heat exchanger. The counterflow heat exchanger integrates heat transfer and throttling. Its advantages are that it can increase the refrigeration capacity of the system, and can ensure that the compressor suction is overheated to prevent liquid from entering the compressor cavity. Adiabatic capillary tubes are generally exposed to the air and are often used in small and medium-sized refrigeration systems (Wei 1993).

4.4.2 Capillary Characteristics

The main characteristics of the capillary tubes are as follows:

① It is usually made of red copper tube, with a simple structure and low price.

② Capillary tubes are usually welded to the condenser and evaporator, without moving parts, so it is not easy to cause failure and leakage.
③ The adjustment ability of the capillary tubes is weak. When the actual working point of the heat pump deviates from the design point, the efficiency of the heat pump will decrease. In addition, when a capillary tube is used as a throttling component, the refrigerant charge must be accurate (Xiong 2014).
④ It has the characteristics of self-compensation, that is, the flow of the liquid working fluid through the capillary tube is stable under a certain pressure difference (the difference between the condensing pressure and the evaporation pressure). When the load change of the refrigeration system causes the pressure difference to increase, the flow rate of the working fluid in the capillary tube also increases, so that the pressure difference returns to a stable value, but this compensation ability is small (Wang et al. 2011).

4.4.3 Flow Characteristics of Refrigerant in Capillary Tubes

4.4.3.1 Flow Characteristics of CFCs and HCFCs Refrigerants

The pressure of the refrigerant in the capillary tube changes with the tube length as shown in Fig. 4.5.

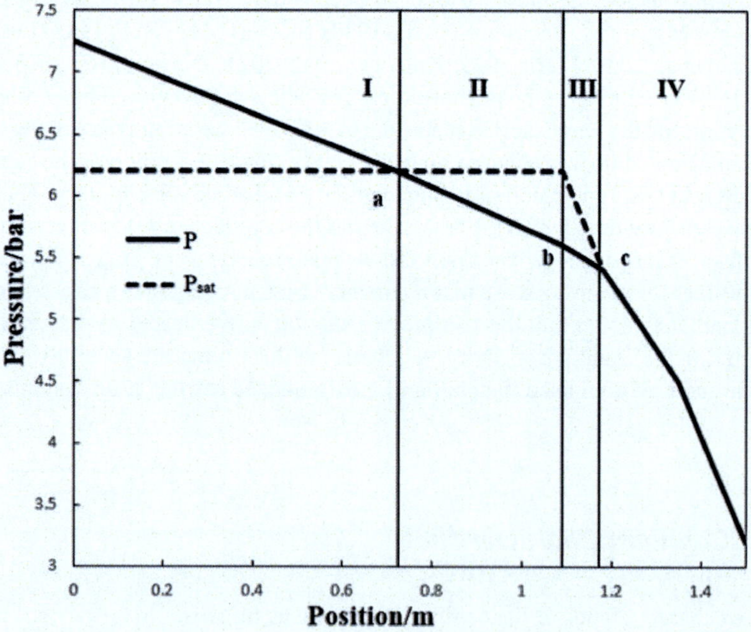

Fig. 4.5 Distribution diagram of pressure (P) and saturation pressure (Psat) along the capillary

The flow of refrigerant in the capillary tube can be divided into four stages as shown in Fig. 4.5 (Cao 2005):

Supercooled section: single-phase liquid, the pressure drops linearly, and the temperature remains at a constant value;

Sub-cold section: single-phase liquid, the temperature remains the same value, and the pressure drops below the corresponding saturation pressure, since the generation and growth of bubbles are based on satisfying the thermodynamic equilibrium and the mechanical equilibrium, the refrigerant liquid does not flash immediately, so an "overheated" liquid zone II is formed in the capillary tube. This state is very unstable, once it is affected by a small disturbance, the balance will be destroyed, so it is called the metastable liquid zone (Kim et al. 2001);

Metastable section: gas–liquid two-phase, when the pressure between the gas and the liquid phase meets the mechanical conditions for bubble generation, the liquid in the tube begins to flash. As the vapor phase increases, the flow speed increases rapidly and the frictional resistance of the flow increases, the pressure of the refrigerant decreases rapidly, and the formation of bubbles tends to be violent and disorderly. The requirements for mechanical equilibrium are weakened, and the thermodynamic equilibrium potential is dominant, so it quickly transitions to the vapor–liquid two-phase flow under thermodynamic equilibrium (Dong and Pan 2012);

Thermal balance section: in the gas–liquid two-phase, the pressure drops non-linearly, and the temperature always corresponds to the saturation temperature of the pressure at each point (Park et al. 2007).

4.4.3.2 Flow Characteristics of Carbon Dioxide

The carbon dioxide transcritical refrigeration cycle is shown in Fig. 4.6, 1–2–3–4 points represent the isenthalpic expansion process of carbon dioxide in the adiabatic capillary tube (Wang et al. 2011).

The capillary tube can be divided into three distinct flow regions, namely, supercritical flow region 1–2, transcritical flow region 2–3, and the subcritical flow region 3–4 as shown in Fig. 4.5. Point '2' lies on the critical temperature line (Fig. 4.6). Therefore, in the region 2–3, the fluid is considered as a subcooled liquid, the specific volume of the supercooled liquid does not change much, and it is regarded as an incompressible fluid. According to the energy conservation equation, the enthalpy value does not change during this process, but the pressure decreases due to resistance (Cao et al. 2010). The inception of vapourization is near the critical point due to the unique thermodynamic properties of carbon dioxide where constant dryness fraction lines are very close unlike in subcritical cycles where the inception of vapourization is away from the critical point (Agrawal and Bhattacharyya 2007). On inception of vapourization, the temperature and pressure drop rapidly in a nonlinear trend, and the influence of the viscosity was found to be insignificant unlike the cases of R22 and R134a. This can be attributed to a relatively more homogeneous two-phase flow of carbon dioxide because the viscosity of carbon dioxide liquid is relatively less

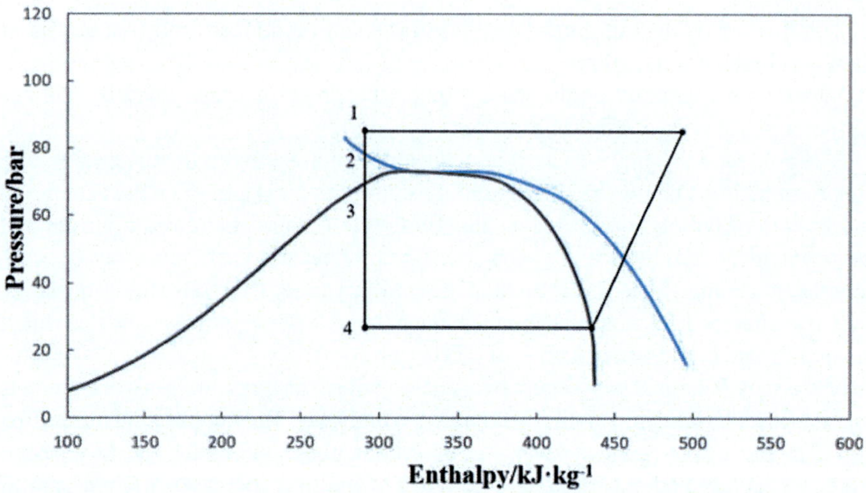

Fig. 4.6 P–h diagram of CO_2 throttling process in adiabatic capillary

than the traditional refrigerants and a low ratio of liquid to vapor density (Cao et al. 2010).

The flow of carbon dioxide in the capillary tube is mainly in the single-phase zone (gas and liquid zone). The dryness of the refrigerant in the two-phase zone increases rapidly, making the length of the two-phase zone very short; as the inlet temperature increases, the length of the gas phase zone gradually increases, while the length of the liquid phase zone decreases. Although the length of the two-phase zone has a change process that first increases and then shortens, the magnitude of the change is very small and almost constant. When other parameters are fixed, the total length of the capillary increases proportionally with the inlet pressure, but inversely proportional to the inlet temperature (Madsen et al. 2005). The flow rate of the inlet and outlet of the capillary increases with the increase of the outdoor dry bulb temperature, and decreases with the increase of the capillary length; the throttling pressure difference increases with the increase of the outdoor dry bulb temperature and decreases with the increase of the capillary length and inner diameter (Xiong 2014).

4.4.4 Calculation of Capillary Length

The capillary tubes in the heat pump are usually adiabatic capillary tubes, and the flow includes two stages: the supercooling section and the two-phase section (Li et al. 1990; Deodhar et al. 2015; Xie 2011).

The formula for calculating the length of the supercooling section is (Guo et al. 2003):

$$L_{SC} = \frac{2\Delta p_{sc} D_1}{f_{sc} v_{sc} G_R^2} \tag{4.27}$$

In Eq. 4.27, L_{SC} is the length of the capillary subcooling section, m; Δp_{sc} is the pressure drop of the supercooling section, Pa; D_1 is the inner diameter of the capillary, m; f_{sc} is the average friction resistance coefficient of the supercooled section, dimensionless; v_{sc} is the specific volume of the supercooled liquid, m³/kg; G_R is the mass flow rate of the working fluid, kg/(m²·s).

The formula for calculating the length of the two-phase section is (Li et al. 2009):

$$L_{TP} = -\frac{2D_1}{f_{tp}}\left\{\ln(\frac{v_{tp0}}{v_{tp1}}) - \frac{p_{TP1}}{v_{tp1} G_R^2 (1-k_1)}\left[\frac{p_{TP0}}{p_{TP1}} - 1 - \frac{k_1}{1-k_1}\ln\left(\frac{p_{TP0} v_{tp0}}{p_{TP1} v_{tp1}}\right)\right]\right\} \tag{4.28}$$

$$k_1 = \left(\frac{p_{TP0} v_{tp0}}{p_{TP1} v_{tp1}}\right)^{0.928533}\left(\frac{1-\frac{v_{tp1}}{v_{tp0}}}{1-\frac{p_{TP0}}{p_{TP1}}}\right) \tag{4.29}$$

In Eq. 4.28, L_{TP} is the length of the capillary two-phase section, m; D_1 is the inner diameter of the capillary, m; f_{tp} is the average frictional resistance coefficient of the two-phase section, dimensionless; v_{tp0} is the specific volume of the working fluid at the outlet of the two-phase section, m³/kg; v_{tp1} is the two-phase The working fluid pressure at the inlet of the section, Pa; p_{tp0} is the working fluid pressure at the outlet of the two-phase section, Pa; G_R is the working fluid mass flow rate, kg/(m²·s); k_1 is the parameter of the equation, dimensionless.

The total length of the capillary is:

$$L = L_{SC} + L_{TP} \tag{4.30}$$

4.4.5 Analysis

Capillary tubes can only adjust the flow rate slightly, so it is more suitable for systems with a stable load. When the load changes greatly, the refrigerant flow cannot be changed effectively and timely, and the method of using capillary to throttle to control the high pressure of the cycle has been questioned by many researchers. But Neeraj Agrawal (2007), after studying the use of capillary tubes in the carbon dioxide transcritical heat pump system to control the high circulating pressure of the system, found that optimizing the length of capillary tubes can adjust the circulating high pressure of the transcritical cycle. As for how to optimize the length, diameter, roughness of capillary tubes and the matching of the system to adjust the circulating high pressure of the system, no more in-depth and detailed research has been done.

Through experimental (Sarkar and Bhattacharyya 2004), J. Sarkar also found that in a transcritical heat pump system with a small displacement, optimizing the outer diameter, length, and roughness of capillary tubes to adjust the mass flow of the refrigerant can change the superheat of the refrigerant outlet. So that the high-pressure cycle of the transcritical cycle system works near the optimal pressure point. This is mainly because at a constant speed if the length of capillary tubes is increased, the mass flow of the refrigerant decreases, the suction pressure decreases, the superheat of the refrigerant outlet increases, and the compressor discharge temperature increases, so the circulating high pressure also increases.

Therefore, it is feasible to use the capillary to control the high pressure of the system in the transcritical carbon dioxide cycle to work near the optimal pressure, but the adjustability of capillary tubes is too small, and it is only suitable for systems with very small displacement changes. As for how to match the basic parameters of the capillary tube with the transcritical system to control the circulating high pressure better, further research is needed.

4.5 CO_2 Ejectors

4.5.1 Introduction

Ever since the renaissance of transcritical CO_2 air-conditioning systems in the late 1980s, ejectors have been considered to improve energy efficiencies (Elbel 2011). The ejector requires limited maintenance, is low in cost, and does not impose any restrictions on the working fluid. Because of these advantages, ejector technology is very attractive in many applications (Besagni 2019; Tashtoush et al. 2019). Research studies on transcritical carbon dioxide refrigeration systems have drastically increased in recent years because carbon dioxide is being advocated as one of the natural refrigerants to replace CFCs and HCFCs in vapor compression systems. To recover the expansion losses of the basic transcritical CO_2 refrigeration cycle and increase its cycle efficiency, ejector expansion devices have been used to replace expansion valves (Besagni et al. 2016).

There are four main sections in this section, and each section has subsections. The first part introduces the principle of ejector technology. In the second part, the CO_2 performance and its influence on ejector performance are described in detail. The third part summarizes the theoretical and experimental research of CO_2 transcritical cycles. Finally, it summarizes the past, present, and future development trends of all ejector technologies.

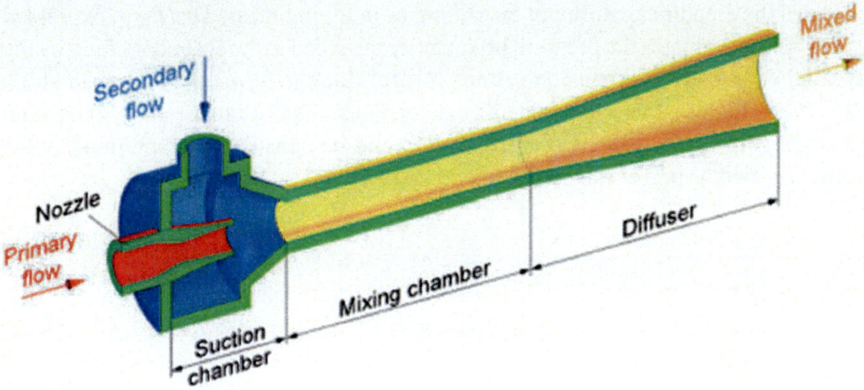

Fig. 4.7 Schematic diagram of the ejector (Tashtoush et al. 2019)

4.5.2 *Ejectors Technology*

4.5.2.1 Technology

In general, an ejector has six parts: a converging–diverging nozzle, suction chamber, mixing chamber, converging part, and a diffuser as shown in Fig. 4.7 (Tashtoush et al. 2019).

An ejector provides a threefold effect (namely, pressure lift, mixing, and entrainment). An ejector is a device that uses the expansion of a high-pressure (motive) fluid to entrain and compress a low-pressure (suction) fluid utilizing momentum transfer between the two streams of fluid. The motive fluid is expanded through a usually converging–diverging (though sometimes converging-only) nozzle to high velocity and low pressure. This high velocity and low pressure are used to entrain the suction fluid through the suction nozzle; the motive and suction fluids are then mixed in the mixing section. The high-speed mixed flow is then decelerated in the diffuser and static pressure is recovered, resulting in a pressure increase provided to the suction stream across the ejector (Cao and Brake 2020; Liu and Groll 2013; Palacz et al. 2018). The ejector is a flow device with two intake ports and one discharge that allows the primary high-pressure stream to entrain the secondary low-pressure stream, where both streams are being mixed inside the ejector and discharged at some intermediate pressure termed as backpressure. Thus, the ejector does pump effect, where the vacuum needed to create the suction, is being generated through accelerating the primary flow by the converging–diverging nozzle.

4.5.2.2 Performance Parameters

Mass entrainment ratio (ω), pressure lift ratio (P_{PLR}), and ejector efficiency are usually applied to depict the ejector performance. The ω, expressed in Eq. (4.31), evaluates the ejector capability of entraining or pumping mass. The P_{PLR}, expressed in Eq. (4.32), evaluates the pressure lift quantity provided to the secondary fluid by the ejector. For an ejector-expansion system, it is fulfilling to have both high ω and high P_{PLR}. Nevertheless, there is a tradeoff between ω and P_{PLR} in an ejector-expansion system, signifying that the two arguments should be considered synchronously for evaluating ejector performance (Zhang et al. 2020).

$$\omega = \frac{m_s}{m_p} \tag{4.31}$$

$$P_{\mathrm{PLR}} = \frac{P_d}{P_s} \tag{4.32}$$

The ejector efficiency is usually described as the ratio between the actual recovery energy and the maximal acquirable power in the primary stream. Several ejector efficiencies have been defined to evaluate the performance of ejectors in the literature, such as (Nakagawa et al. 1998; Arbel et al. 2003; Elbel and Hrnjak 2008). Lawrence and Elbel (Lawrence and Elbel 2015) found that the ejector efficiency usually ranges from 20 to 30% for transcritical systems using CO_2 as refrigerant, whereas this figure is usually less than 20% for subcritical systems using low-pressure refrigerants (R410A, R134a). Besides, it was also found that the best system performance may not occur when the efficiency is up to optimum for a given ejector geometry.

Concerning the ejector itself, there are many ways to define the ejector efficiency, ηejector. The efficiency used by ASHRAE is defined as the ratio between the actual recovered compression energy and the available theoretical energy in the motive stream (Little and Garimella 2011).

$$\eta_{\mathrm{ej}} = \omega \frac{h\left(P_{\mathrm{diff}}, s_{s,\mathrm{in}}\right) - h_{s,\mathrm{in}}}{h_{p,\mathrm{in}} - h\left(P_{\mathrm{diff}}, s_{p,\mathrm{in}}\right)} \tag{4.33}$$

The method of determining ejector component efficiencies is based on the ejector model and measured parameters (pressures, temperatures, and mass flow rates at the inlet to each of the motive and suction nozzles, and ejector discharge pressure). The two-phase flow ejector model was utilized to determine the efficiencies of the motive nozzle, suction nozzle, and mixing section using the measured data (Yadav et al. 2020). The COP of the ejector cycle is very sensitive to the ejector efficiency. However, there is very limited research on ejector efficiency (Domanski 1995). In most of the literature studies, values of 0.7–0.95 were assumed for the individual ejector component efficiencies as listed in Table 4.2.

Table 4.2 Summary of ejector efficiency using CO_2 transcritical cycle

Authors	Fluid	η_m	η_s	η_{mix}	η_d
Elbel and Hrnjak (2004)	CO_2	0.9	0.9	1	0.9
Li and Groll (2005)	CO_2	0.9	0.9	1	0.8
Ksayer and Clodic (2006)	CO_2	0.85	0.85	1	0.75
Deng et al. (2007)	CO_2	0.7	0.7	1	0.8
Sarkar (2008)	CO_2	0.8	0.8	1	0.75
Elbel and Hrnjak (2008)	CO_2	0.8	0.8	1	0.8
Sun and Ma (2011)	CO_2	0.9	0.9	1	0.8
Eskandari Manjili and Yavari (2012)	CO_2	0.7	0.7	0.95	0.8
Ahammed et al. (2014)	CO_2	0.85	0.85	–	0.85
Zheng et al. (2016)	CO_2	0.8	0.8	0.9	0.9
Taleghani and Sorin (2018)	CO_2	0.9	0.9	1	0.8
Eskandari Manjili and Cheraghi (2019)	CO_2	0.9	0.9	–	0.8
Zare and Rostamnejad Takleh (2020)	CO_2	0.9	0.9	0.88	0.85
Peris Pérez et al. (2021)	CO_2	0.9	0.9	0.95	0.8

4.5.3 Transcritical CO_2 Ejector-Expansion Refrigeration System

4.5.3.1 Exploratory Researches of Adjustable Measures

Since the fixed geometry ejector is hypersensitive to load variation and to operating conditions, the ejector has to operate in limited working conditions to ensure efficient operation owing to its innate working characteristics. However, practical refrigeration and heat pump systems require to adapt for variations of load and working conditions. Also, a fixed geometry ejector cannot actively control high-side pressure on its own for transcritical CO_2 ejector-expansion vapor compression systems (EVCS). As previously mentioned, Menegay (1991) and Guangming et al. (2010) adjusted the primary mass flow rate by changing the amounts of hot gas bypass flow between the compressor outlet and the primary nozzle inlet. But the system performance decreases markedly by this scheme. Thus it is significant to exploit new adjustable measures that can be adapted to the variation of the operation conditions. Figure 4.8 summarized four adjustable measures for ejectors in the literature. Figure 4.8a shows a schematic of the needle adjustable measure (Elbel and Hrnjak 2008; Liu et al. 2012a), in which a needle is inserted into the primary nozzle throat and used to control the effective area of the primary nozzle by changing the needle position. Figure 4.8b shows parallel multi-ejectors adjustable measure (Armin et al. 2012), in which usually various-sized, fixed ejectors are assembled and turned on or off independently for obtaining the appropriate effective nozzle size under certain conditions. Figure 4.8c shows a series–parallel valve adjustable measure (Lawrence

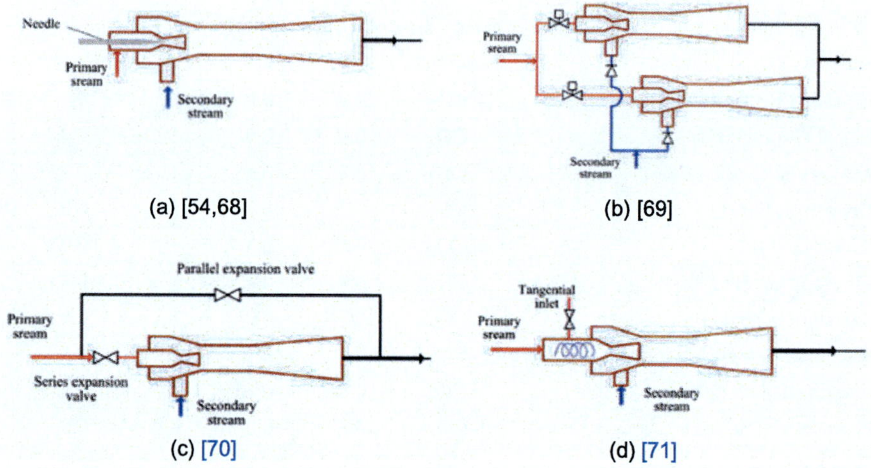

Fig. 4.8 Schematic of the adjustable measures for ejector

and Elbel 2019), which uses a throttle valve upstream of the primary nozzle (to adjust the mass flow rate of the primary stream or raise P_{gc} for transcritical CO_2 systems) or in parallel with the ejector (to reduce P_{gc} for transcritcial CO_2 systems). Figure 4.8d shows a primary inlet vortex adjustable measure (Zhu and Elbel 2020), where the ratio of mass flow rate through the two inlets is adjusted by valves installed at the primary stream axial and tangential inlets, thereby changing the vortex strength, thus primary stream can be adjusted.

4.5.3.2 Theoretical Analysis

Owing to the high throttle losses and large COP enhancement potential, many scholars investigated ejector utilization in the transcritical CO_2 refrigeration system. The pieces of literature about theoretical analysis of transcritical SEVCSs discussed in this section are summarized in Table 4.3, indicating the working fluids, operating conditions, COP_{imp}, and system features.

The ejector-expansion refrigeration cycle was first proposed by Kornhauser (1990). As shown in Fig. 4.9a. Li and Groll (2005) proposed an ejector expansion transcritical CO_2 refrigeration cycle to improve the COP of the basic transcritical CO_2 cycle by reducing the expansion process losses. The effect of the entrainment ratio and the pressure drop in the receiving section of the ejector on the relative performance of the ejector expansion transcritical CO_2 cycle was investigated for typical air conditioning operation conditions. It was found that the COP of the ejector expansion transcritical CO_2 cycle can be improved by more than 16% over the basic transcritical CO_2 cycle for typical air conditioning operation conditions.

Bai et al. (2015a) proposed a vapor injection transcritical R744 ejector-expansion vapor compression system with a subcooler, as shown in Fig. 4.9b. The simulated

Table 4.3 Theoretical analysis of CO_2 (transcritical) refrigeration system

Authors	Year	t_{eva}/°C	$t_{gc,out}$/°C (P_{gc}/MPa)	Baseline	COP_{imp}(%)	System features
Li and Groll (2005)	2005	0–10	36–48 (8–12)	CVCS	7–18	Vapor feedback
Deng et al. (2007)	2007	0–10	36–48 (8–12)	CVCS	22	–
Sarkar (2008)	2008	−45–5	30–60	CVCS	9	–
Yari and Sirousazar (2008)	2008	10	40	CVCS	55.5	Two-stage SEVCS including IHX, ejector, and intercooler
Yari (2009)	2009	−30–5	33–55 (8–12)	Two-stage system	12.5–21	Two-stage SEVCS including IHX, ejector, and intercooler
Fangtian and Yitai (2011)	2011	−5–17	40–45 (9)	CVCS	30	
Liu et al. (2012b)	2012	22.3	27.8–37.8 (10)	CVCS	30.7	Component efficiencies of ejector were estimated using empirical correlations
Manjili and Yavari (2012)	2012	−15–5	36–55 (8–12.5)	One-stage SEVCS with IHX One-stage SEVCS without IHX	19.6 15.3	Two-stage SEVCS which includes two intercoolers
Zhang et al. (2013)	2013	0–10	40–50 (8.5–11)	CVCS	45.1	–
Bai et al. (2015a)	2015	−25 to −5	30–50	Vapor injection system	7.7	Ejector enhanced vapor injection heat pump system with subcooler
Bai et al. (2016)	2016	−5–5	35 (8–10)		43.44	–
Minetto et al. (2016)	2016	−6.2–9.2	35–42	CVCS	20–34	–

(continued)

Table 4.3 (continued)

Authors	Year	t_{eva}/°C	$t_{gc,out}$/°C (P_{gc}/MPa)	Baseline	$COP_{imp}(\%)$	System features
Megdouli et al. (2017)	2017	−25–20	32–50 (8–14)	SEVCS	12	Transcritical SEVCS with ORC
Nemati et al. (2018)	2018	−30–0	35–55	Two-stage transcritical SEVCS	10.75 8.37	Two-stage SEVCS with ORC
Taslimi et al. (2019)	2019	–	–	–	17	–
Liu et al. (2019)	2019	5	40	CVCS	39.34	Transcritical SEVCS with thermoelectric subcooler

results revealed that the proposed system increased the COP and volumetric Q_h by 7.7 and 9.5%, respectively compared with the traditional vapor injection system. The exergy efficiencies of the gas cooler and ejector were approximately 57.9 and 69.7% respectively, which were critical parts for improving system energy efficiency. In 2016, Bai et al. (2016) investigated a transcritical R744 refrigeration standard ejector-expansion vapor compression system (SEVCS) with exergy analysis and found that the compressor with the highest exergy loss should be given improving priority, the ejector second, the evaporator third. It was found that 43.44% of the system exergy loss could be prevented with component improvements.

Yari and Sirousazar (2008) proposed a transcritical R744 double-stage compression refrigeration SEVCS including an internal heat exchanger (IHX), ejector, and intercooler. The schematic and P–h diagram of the proposed SEVCS is depicted in Fig. 4.9c. The simulated results showed that in comparison with the corresponding conventional vapor compression systems (CVCS) and SEVCS, the COP and the energetic efficiency of the proposed system were increased by around 55.5 and 26% respectively when t_{eva} and $t_{gc,out}$ were 10 and 40 °C respectively. Subsequently, Yari (2009), Yari and Mahmoudi (2011) performed an optimizing analysis of the system by the first and the second laws of thermodynamics. The correlated formulas for predicting the designing specifications of the system were proposed. The maximal COP and exergetic efficiency of the system were increased by around 12.5–21% compared with the double-stage system without an ejector.

Megdouli et al. (2017) tried to recover the exhaust heat from the gas cooler of the transcritical R744 SEVCS through a transcritical R744 organic Rankine cycle. The power generated by ORC was applied to drive the compressor and the feed pump, thus lowering the input power consumption and enhancing the performance of the whole system. Figure 4.9d shows the schematic and P–h diagram of the system. The simulated results revealed that the application of the organic Rankine cycle resulted

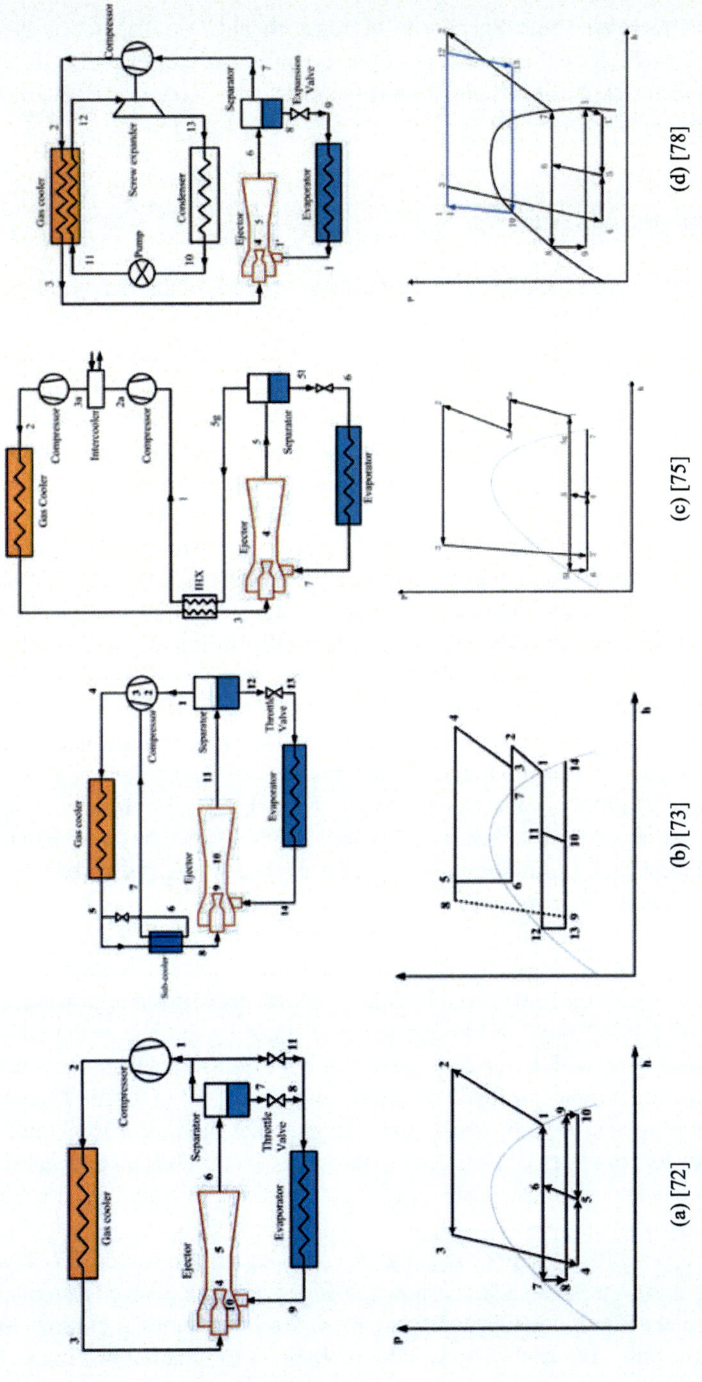

Fig. 4.9 Schematic diagram of transcritical carbon dioxide refrigeration cycle with ejector

in a COP improvement of 12% compared with the refrigeration SEVCS under the same working conditions.

In addition, there are some other forms of transcritical CO_2 refrigeration systems with ejectors, such as dual ejectors, multiple evaporators, etc. Figure 4.10 shows other novel ejector-expansion vapor compression systems (Zheng et al. 2016; Xing et al. 2014; Bai et al. 2015b, 2017).

4.5.3.3 Experimental Analysis

Many works of literature have developed ejectors with fixed geometry for transcritical R744 refrigeration systems and investigated the system performance as well as gas cooler pressure control measures. The literature about the experimental analysis of transcritical SEVCSs discussed in this section is summarized in Table 4.4. indicating the working fluids, operating conditions, Q_c, ω, COP_{imp}, and system features.

Lee et al. (2011) developed a fixed geometry ejector and tested it in a transcritical R744 water-to-water refrigeration SEVCS with an IHX. The gas cooler, evaporator, and IHX were all counter-flow copper co-axial double pipe heat exchangers with high-pressure fluid flowing in the interior tubes and low-pressure fluid flowing in the annulus gap. A semi-hermetic reciprocal compressor was used. During the test, the temperature of the water entering the gas cooler and the evaporator was 30 and 27 °C respectively. It was experimentally found that the COP was improved by about more than 15% in comparison with the CVCS at matched capacity owing to the ejector. A maximum mass entrainment ratio was found to be about 0.9. Lee et al. (2014) continued to investigate the variations of the maximum mass entrainment ratio under various compressor speeds and outdoor temperatures and found that the maximum mass entrainment ratio was generally between 0.7 and 0.9. The improvement of both COP and Q_C were expected to be about 6–9 and 2–5% respectively compared with the corresponding CVCS. Additionally, the control strategy was discussed to ensure that the SEVCS was superior to the corresponding CVCS.

Zhu et al. (2018) developed a transcritical R744 air source SEVCS water heater, as shown in Fig. 4.11. A convergent primary nozzle was used for the ejector prototype. The gas cooler was a concentric double pipe type with cooling water flowing in the interior pipe and R744 flowing through the ring-shaped gap. The evaporator was a micro-channel type with a variable-speed fan controlling the evaporator exiting superheat. The outlet temperature of tap water ranged from 50 to 90 °C. The system COP was found to achieve 4.6, which was 10.3% higher than the CVCS under the condition that the tap water exiting temperature was 70 °C. They also revealed that the use of an ejector was more profitable for the condition of high-temperature hot water production.

Boccardi et al. (2017) experimentally investigated an air source R744 water heater with a multi-ejectors pack, as shown in Fig. 4.12. The control system could enable each ejector individually and form 15 different schemes according to the working conditions. The results showed the probability to obtain a maximum COP by changing the ejector area when other dimensions remain unchanged, owing to

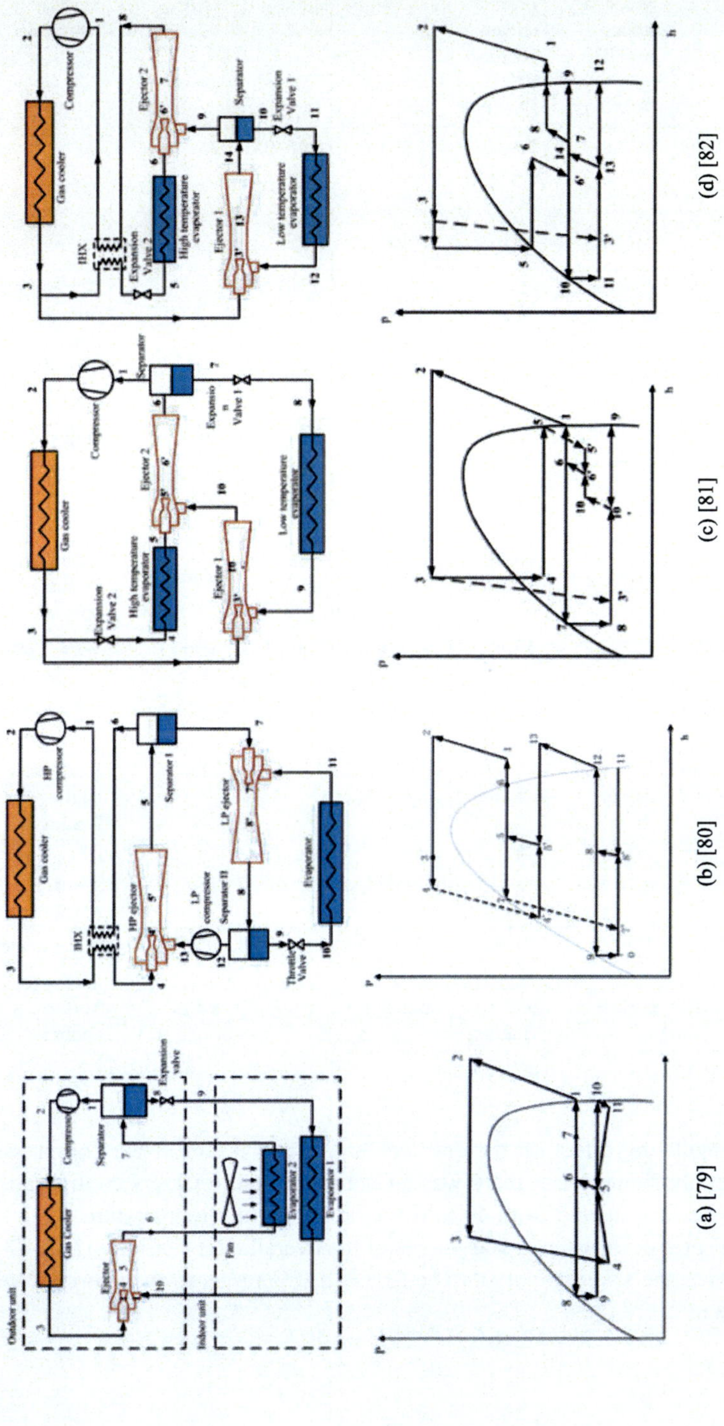

Fig. 4.10 Other novel ejector-expansion vapor compression systems

Table 4.4 Experimental analysis of CO_2 (transcritical) refrigeration system

Authors	Year	Secondary stream conditions t/°C (P/MPa)	Primary stream conditions t/°C (P/MPa)	Q/kW	ω	COP_{imp}(%)	System features
Elbel and Hrnjak (2008), Elbel (2011)	2008, 2011	(3–3.8)	26–40 (8–12.4)	4.3–5.1	0.45–0.6	7	
Guangming et al. (2010)	2010	10(3.3–4.5)	35 (7.4–9.8)	12	0.1–1		Converging primary nozzle
Lee et al. (2011)	2011	(4.3)	(8–10)	3.28–6.38	0.9	15	
Liu et al. (2012c)	2012	15.5–26.5 Indoor	27.5–37.5 Outdoor	10.8–16	0.3–0.55	36	Adjustable ejector
Banasiak et al. (2012)	2012	(3.55)	30.5 (8–11.5)		0.55–0.69	6–8	
Lucas and Koehler (2012)	2012	−10, −1 (2.6, 3.4)	30–40 (7–10.5)		0.38–0.65	17	Converging primary nozzle
Minetto et al. (2013)	2013	12.1–24.2	40–60	4–5.5	0.8–1.6	40.6	Water heater
Lee et al. (2014)	2014	30–40	27	3–6	0.7–0.9	6–9	
Liu et al. (2016b)	2016	2.8–26.7 Indoor	35–41.1 Outdoor	24.8–31.5	0.3–0.6	71.4	Simultaneous cooling and heating
Zhu et al. (2017)	2017	21 (3–3.7)	35 (8–10)	5	0.4–0.8	−11–18.9	Converging primary nozzle
Zhu et al. (2018)	2018	22	50–90	5	0.55–0.95	10.3	Water heater
Zhu and Elbel (2020)	2020	6.5–10.6	35 Outdoor	3.2–5.5		8.1	Vortex control

the ejector regulating effect on the entrance and exit pressure of the compressor. It experimentally showed that there was an optimum multi-ejectors configuration. When the water was heated from 40 to 60 °C at an outdoor temperature of 12 °C, and the multi-ejector throat area was 46.5% of the overall intersecting surface, COP and Q_h improvements were reported to be 13.8 and 20% respectively compared with the worst case.

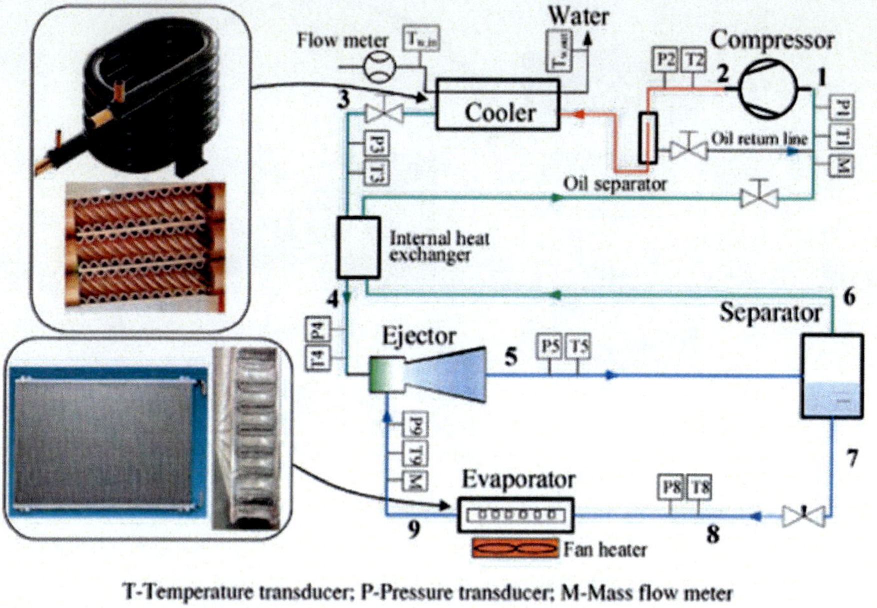

Fig. 4.11 Schematic of the transcritical R744 air source SEVCS water heater test setup (Zhu et al. 2018)

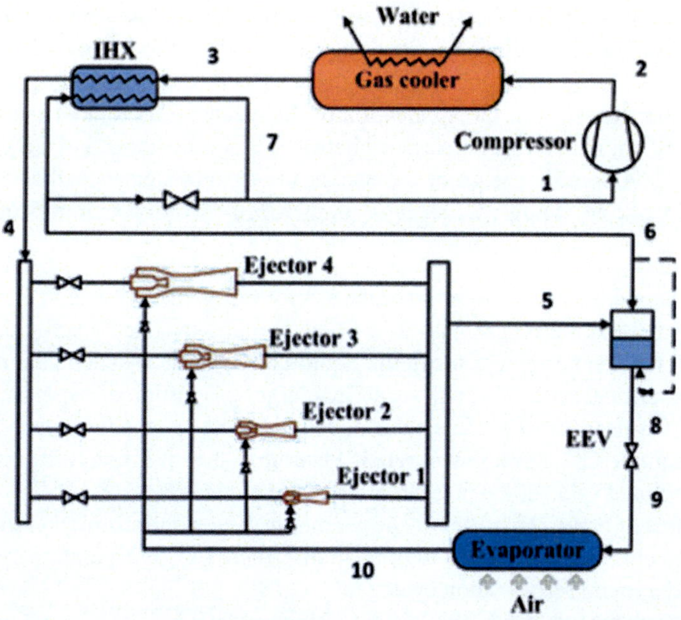

Fig. 4.12 R744 water heater parallel multi-ejector expansion vapor compression system (Boccardi et al. 2017)

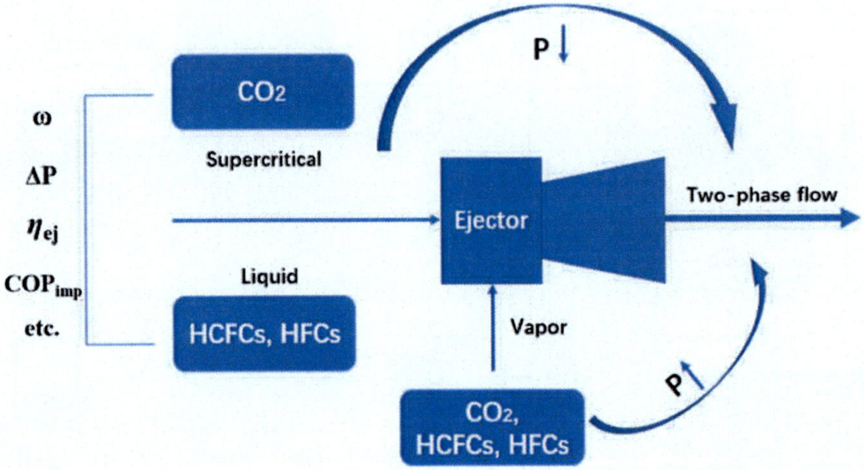

Fig. 4.13 The difference of CO_2, HCFCs, HFCs in ejector

4.5.4 Summary of Characteristics of CO_2 Ejector

It is worth noting from these studies that the ejector can reduce the compression work by reducing the throttling loss and excessive liquid feeding and increasing the compressor inlet pressure, thereby increasing the cycle COP. As shown in Fig. 4.13, because CO_2 enters the main inlet of the ejector as a supercritical fluid in the transcritical CO_2 refrigeration cycle, this results in that CO_2 and other Freon refrigerants have different characteristics in the application of the ejector. Throughout the literature review, there are not many comparisons between Freon injectors and CO_2 ejectors. In addition, the optimal design of the ejector needs to be determined according to the actual working conditions, so it is summarized as follows from a qualitative perspective:

(a) To ensure the normal operation of the ejector refrigeration cycle, it is necessary to ensure that the steam mass flow at the compressor inlet is sufficient. Freon refrigerants enter the ejector in the liquid phase, and CO_2 enters the ejector in a supercritical state. This will result in a larger proportion of vapor mass in the CO_2 two-phase flow at the ejector outlet under the same entrainment rate, and the required CO_2 entrainment rate is lower than that of Freon refrigerants.
(b) The optimal value of the pressure drop of the secondary nozzle of CO_2 is higher than that of Freon refrigerant in the geometric design optimization of the ejector.
(c) The ejector improves the COP of the transcritical CO_2 refrigeration cycle more than the Freon refrigeration cycle.
(d) CO_2 seems to achieve higher ejector efficiency, etc.

Attempts concerning the fabrication and application of CO_2 ejectors are still very limited, the majority of theoretical studies of CO_2 ejector refrigeration systems are limited to steady-state simulation. The challenges of modeling turbulent and non-equilibrium two-phase flows should be further investigated. Moreover, further efforts are needed to study the utilization potential of small-scale CO_2 ejectors.

4.6 Conclusions

The lower efficiency of the conventional transcritical CO_2 refrigeration cycle, compared to vapor compression systems using HFC refrigerants, is a major hindrance for the technology to make progress toward practical applications, and the loss during throttling is the largest. Since the inlet of the throttling device in the CO_2 transcritical cycle is in a supercritical state, it is necessary to re-optimize the design of the expansion device for the CO_2 fluid. On the one hand, the electronic expansion valve and capillary tube have the characteristics of simple structure and low cost, and the industry has become more mature, which makes them currently the most commonly used throttling devices in the CO_2 transcritical refrigeration cycle. The actual working conditions are always changing, and future research on them should be more intelligent and dynamic adjustment faster. On the other hand, the technology of ejector expansion isn't mature yet. Most of the ejector expansion vapor compression systems techniques have been limited to theoretical analysis and laboratory tests. There are still struggles in developing ejector-expansion vapor compression systems for commercial applications and markets.

First, the enhancement of COP proposed in the above studies is encouraging, but it usually occurs under certain conditions. Future research is required to study how to attain COP improvements not only under designing conditions but also under off-designing conditions. Secondly, a few unsteady characteristics of the system, such as the start-up procedure, the dynamic response to the variations of the working parameters, and the equilibration time, have been published. Finally, in future research on solutions to reduce throttling losses, CO_2 ejectors should attract more attention. Future research should not only focus on improving the energy efficiency of ejector-expansion vapor compression systems but also focus on effectively overcoming these problems in practical applications.

References

Agrawal N, Bhattacharyya S (2007) Adiabatic capillary tube flow of carbon dioxide in a transcritical heat pump cycle. Int J Energy Res 31(11):1016–1030

Ahammed ME, Bhattacharyya S, Ramgopal M (2014) Thermodynamic design and simulation of a CO_2 based transcritical vapour compression refrigeration system with an ejector. Int J Refrig 45:177–188

Arbel A, Shklyar A, Hershgal D, Barak M, Sokolov M (2003) Ejector irreversibility characteristics. J Fluids Eng 125:121–129

Armin H, Poppi S, Petter N, Silvia M, Eikevik TM (2012) Development of commercial refrigeration systems with heat recovery for supermarket building. In: 10th IIR Gustav Lorentzen conference on natural working fluids, Delft, The Netherlands, p 206–1

Baek C, Heo J, Jung J, Cho H, Kim Y (2013) Optimal control of the gas-cooler pressure of a CO_2 heat pump using EEV opening and outdoor fan speed in the cooling mode. Int J Refrig 36:1276–1284

Bai T, Yan G, Yu J (2015a) Thermodynamic analyses on an ejector enhanced CO_2 transcritical heat pump cycle with vapor-injection. Int J Refrig 58:22–34

Bai T, Yan G, Yu J (2015b) Thermodynamics analysis of a modified dual-evaporator CO_2 transcritical refrigeration cycle with two-stage ejector. Energy 84:325–335

Bai T, Yan G, Yu J (2017) Performance evolution on a dual-temperature CO_2 transcritical refrigeration cycle with two cascade ejectors. Appl Therm Eng 120:26–35

Bai T, Yu J, Yan G (2016) Advanced exergy analyses of an ejector expansion transcritical CO_2 refrigeration system. Energy Convers Manag 126:850–861

Banasiak K, Hafner A, Andresen T (2012) Experimental and numerical investigation of the influence of the two-phase ejector geometry on the performance of the R744 heat pump. Int J Refrig 35:1617–1625

Bellos E, Tzivanidis C (2019) A comparative study of CO_2 refrigeration systems. Energy Convers Manag X 1

Besagni G (2019) Ejectors on the cutting edge: the past, the present and the perspective. Energy 170:998–1003

Besagni G, Mereu R, Inzoli F (2016) Ejector refrigeration: A comprehensive review. Renew Sustain Energy Rev 53:373–407

Boccardi G, Botticella F, Lillo G, Mastrullo R, Mauro AW, Trinchieri R (2017) Experimental investigation on the performance of a transcritical CO 2 heat pump with multi-ejector expansion system. Int J Refrig 82:389–400

Cao HS, ter Brake HJM (2020) Progress and challenges in utilization of ejectors for cryogenic cooling. Appl Therm Eng 167

Cao X, Chen H, Feng M et al (2010) Simulation of the flow of carbon dioxide in the capillary tube in a transcritical refrigeration cycle. J Cent South Univ (Nat Sci Edn) 41(4):1604–1608

Cao Y (2005) Characteristic and experimental research of two-phase flow in adiabatic capillary. Zhejiang University

Chen J, Jarall S, Havtun H, Palm B (2015) A review on versatile ejector applications in refrigeration systems. Renew Sustain Energy Rev 49:67–90

Cho H, Ryu C, Kim Y, Kim HY (2005) Effects of refrigerant charge amount on the performance of a transcritical CO_2 heat pump. Int J Refrig 28:1266–1273

Cho H, Ryu C, Kim Y (2007) Cooling performance of a variable speed CO_2 cycle with an electronic expansion valve and internal heat exchanger. Int J Refrig 30:664–671

Chunnanond K, Aphornratana S (2004) Ejectors: applications in refrigeration technology. Renew Sustain Energy Rev 8:129–155

Deng J-q, Jiang P-x, Lu T, Lu W (2007) Particular characteristics of transcritical CO_2 refrigeration cycle with an ejector. Appl Therm Eng 27:381–388

Deodhar SD, Kothadia HB, Iyer KN, Prabhu SV (2015) Experimental and numerical studies of choked flow through adiabatic and diabatic capillary tubes. Appl Therm Eng 90:879–894

Ding Y, Liu C (2015) Research on application of combined system of shallow ground source heat pump and capillary air conditioning. Fluid Mach 43(01):82–86

Domanski PA (1995) Minimizing throttling losses in the refrigeration cycle. In: Proceedings of 19th international congress of refrigeration, vol 1Vb, pp 766–773

Dong Z, Pan Z (2012) Quartz capillary method to determine the solubility of ethanol in supercritical carbon dioxide. J Zhejiang Univ Technol 40(4):418–421

Elbel S (2011) Historical and present developments of ejector refrigeration systems with emphasis on transcritical carbon dioxide air-conditioning applications. Int J Refrig 34:1545–1561

Elbel S, Hrnjak P (2008) Experimental validation of a prototype ejector designed to reduce throttling losses encountered in transcritical R744 system operation. Int J Refrig 31:411–422
Elbel SW, Hrnjak PS (2004) Effects of internal heat exchanger on performance of transcritical CO_2 systems with ejector. In: Proceedings of 10th international refrigeration and air conditioning conference at Purdue, Purdue University, USA, Paper No. R166
Eskandari Manjili F, Cheraghi M (2019) Performance of a new two-stage transcritical CO_2 refrigeration cycle with two ejectors. Appl Therm Eng 156:402–409
Guangming C, Xiaoxiao X, Shuang L, Lixia L, Liming T (2010) An experimental and theoretical study of a CO_2 ejector. Int J Refrig 33:915–921
Guo C, He S, Xun Z (2003) Calculation and analysis of capillary length of different working fluids. Domest Technol 2:75–78
Holling B, Kandziora C, Ritter R (2013) CO_2 recovery from industrial hydrogen facilities and steel production to comply with future european emission regulations. Energy Procedia 37:7221–7230
Hou Y, Liu C, Ma J, Cao J, Chen S (2014a) Mass flowrate characteristics of supercritical CO_2 flowing through an electronic expansion valve. Int J Refrig 47:134–140
Hou Y, Ma J, Liu C, Cao J, Liu X (2014b) Experimental investigation on the influence of EEV opening on the performance of transcritical CO_2 refrigeration system. Appl Therm Eng 65:51–56
Hwang Y, Huff H, Preissner R, Radermacher R (2001) CO_2 tran-scritical cycles for high temperature application. In: Proceedings of 2001 ASME international mechanical engineering congress in New York (IMECE2001/AES-23630)
Hwang Y, Radermacher R (1999) Experimental investigation of the CO_2 refrigeration cycle. ASHRAE Trans 105(1):1219–1227
Kim SG, Kim MS, Ro ST (2001) Flow and pressure drop characteristics of R22 in adiabatic capillary tubes. KSME Int J 15(9):1328–1338
Kornhauser AA (1990) The use of an ejector as a refrigerant expander. In: Proceedings of the 1990 USNC/IIR—Purdue refrigeration conference, Purdue University, West Lafayette, IN, USA, pp 10–19
Ksayer EB, Clodic D (2006) Enhancement of CO_2 refrigeration cycle using an ejector: 1D analysis. In: Proceeding of international refrigeration and air conditioning conference at Purdue, Purdue University, USA, Paper No. R058
Kurosawa SGK, Almeida CAR et al (2021) An experimental study of refrigerant expansion inside coiled adiabatic capillary tubes and development of a general correlation. J Braz Soc Mech Sci Eng 43(2)
Lawrence N, Elbel S (2015) Analysis of two-phase ejector performance metrics and comparison of R134a and CO_2 ejector performance. Sci Technol Built Environ 21:515–525
Lawrence N, Elbel S (2019) Experimental investigation on control methods and strategies for off-design operation of the transcritical R744 two-phase ejector cycle. Int J Refrig 106:570–582
Lee JS, Kim MS, Kim MS (2011) Experimental study on the improvement of CO_2 air conditioning system performance using an ejector. Int J Refrig 34:1614–1625
Lee JS, Kim MS, Kim MS (2014) Studies on the performance of a CO_2 air conditioning system using an ejector as an expansion device. Int J Refrig 38:140–152
Li D, Groll EA (2005) Transcritical CO_2 refrigeration cycle with ejector-expansion device. Int J Refrig 28:766–773
Li M, Wang D, Shi Q, Zhang H (2009) Design and experimental research of capillary tube in carbon dioxide refrigeration system. In: Proceedings of the 2009 annual conference of the Chinese society of refrigeration, vol 4. Chinese Society of Refrigeration
Li R-Y, Lin S, Chen ZY, Chen ZH (1990) Metastable flow of R12 through capillary tubes. Int J Refrig 13(5):181–186
Little AB, Garimella S (2011) A review of ejector technology for refrigeration applications. Int J Air-Cond Refrig 19:1–15
Liu C, Hou Y, Ma J, Liu X, Chen L (2016a) Experimental study on the CO_2 flow characteristics through electronic expansion valves in heat pump. Int J Refrig 69:106–113

Liu C, Zhang J, Gui Y, Li W, Shi J, Chen J et al (2018a) Mass flow characteristics and empirical modeling of R744 flow through electronic expansion device. Int J Refrig 86:82–88

Liu F, Groll EA (2013) Study of ejector efficiencies in refrigeration cycles. Appl Therm Eng 52:360–370

Liu F, Groll EA, Li D (2012a) Investigation on performance of variable geometry ejectors for CO_2 refrigeration cycles. Energy 45:829–839

Liu F, Groll EA, Li D (2012b) Modeling study of an ejector expansion residential CO_2 air conditioning system. Energy Build 53:127–136

Liu F, Li Y, Groll EA (2012c) Performance enhancement of CO_2 air conditioner with a controllable ejector. Int J Refrig 35:1604–1616

Liu F, Groll EA, Ren J (2016b) Comprehensive experimental performance analyses of an ejector expansion transcritical CO_2 system. Appl Therm Eng 98:1061–1069

Liu X, Fu R, Wang Z, Lin L, Sun Z, Li X (2019) Thermodynamic analysis of transcritical CO_2 refrigeration cycle integrated with thermoelectric subcooler and ejector. Energy Convers Manag 188:354–365

Liu Y, Chai G, Li Y (2020) Capillary assembly and refrigeration system. Anhui Province: CN211823288U, 30 October 2020

Liu Y, Meng D, Chen S (2018b) Defrosting performance of parallel capillary small carbon dioxide heat pump system. Heat Vent Air Cond 48(12):62–65

Lucas C, Koehler J (2012) Experimental investigation of the COP improvement of a refrigeration cycle by use of an ejector. Int J Refrig 35:1595–1603

Ma Y, Liu Z, Tian H (2013) A review of transcritical carbon dioxide heat pump and refrigeration cycles. Energy 55:156–172

Madsen KB, Poulsen CS, Wiesenfarth M (2005) Study of capillary tubes in a transcritical CO_2 refrigeration system. Int J Refrig 28:1212–1218

Manjili FE, Yavari MA (2012) Performance of a new two-stage multi-intercooling transcritical CO_2 ejector refrigeration cycle. Appl Therm Eng 40:202–209

Megdouli K, Tashtoush BM, Ezzaalouni Y, Nahdi E, Mhimid A, Kairouani L (2017) Performance analysis of a new ejector expansion refrigeration cycle (NEERC) for power and cold: exergy and energy points of view. Appl Therm Eng 122:39–48

Menegay P (1991) Experimental investigation of an ejector as a refrigerant expansion engine. Virginia Polytechnic Institute and State University. Virginia Polytechnic Institute and. State University, Blacksburg VA USA

Minetto S, Brignoli R, Banasiak K, Hafner A, Zilio C (2013) Performance assessment of an off-the-shelf R744 heat pump equipped with an ejector. Appl Therm Eng 59:568–575

Minetto S, Cecchinato L, Brignoli R, Marinetti S, Rossetti A (2016) Water-side reversible CO_2 heat pump for residential application. Int J Refrig 63:237–250

Nakagawa M, Takeuchi H, Nakajima K (1998) Performance of two-phase ejector in refrigeration cycle. In: Proceedings of the 3rd international conference on multiphase flow, Lyon, France, pp 1–8

Neeraj Agrawal SB (2007) Optimized transcritical CO_2 heat pumps: performance comparison of capillary tubes against expansion valves. Int J Refrig 1–8

Nemati A, Mohseni R, Yari M (2018) A comprehensive comparison between CO_2 and Ethane as a refrigerant in a two-stage ejector-expansion transcritical refrigeration cycle integrated with an organic Rankine cycle (ORC). J Supercrit Fluids 133:494–502

Palacz M, Haida M, Smolka J, Plis M, Nowak AJ, Banasiak K (2018) A gas ejector for CO_2 supercritical cycles. Energy 163:1207–1216

Park C, Lee S, Kang H, Kim Y (2007) Experimentation and modeling of refrigerant flowthrough coiled capillary tubes. Int J Refrig 30:1168–1175

Peris Pérez B, Expósito Carrillo JA, Sánchez de La Flor FJ, Salmerón Lissén JM, Morillo Navarro A (2021) Thermoeconomic analysis of CO_2 Ejector-Expansion Refrigeration Cycle (EERC) for low-temperature refrigeration in warm climates. Appl Therm Eng 188

Sarkar J (2008) Optimization of ejector-expansion transcritical CO_2 heat pump cycle. Energy 33:1399–1406
Sarkar J, Bhattacharyya J (2004) Optimization of transcrtical CO_2 heat pump cycle for simultaneous cooling and heating applications. Int J Refrig 27:830–838
Shanwei M, Chuan Z, Jiangping C, Zhiujiu C (2005) Experimental research on refrigerant mass flow coefficient of electronic expansion valve. Appl Therm Eng 25:2351–2366
Sun F, Ma Y (2011) Thermodynamic analysis of transcritical CO_2 refrigeration cycle with an ejector. Appl Therm Eng 31:1184–1189
Tashtoush BM, Al-Nimr MdA, Khasawneh MA (2019) A comprehensive review of ejector design, performance, and applications. Appl Energy 240:138–172
Taslimi Taleghani S, Sorin M, Poncet S (2018) Modeling of two-phase transcritical CO_2 ejectors for on-design and off-design conditions. Int J Refrig 87:91–105
Taslimi Taleghani S, Sorin M, Poncet S, Nesreddine H (2019) Performance investigation of a two-phase transcritical CO_2 ejector heat pump system. Energy Convers Manag 185:442–454
Wang D, Li M, Qi L, Qian F (2011) Design and experimental research of capillary tube in carbon dioxide refrigeration system. J Chem Ind 62(10):2753–2758
Wang J, Zhao Y, Li L, Cao F, Wang Z (2011) Research progress of capillary in transcritical CO_2 refrigeration (heat pump) system. Low Temp Eng (03):29–33+43
Wei F (1993) Characteristics and selection of capillary tube in refrigeration system. Refrig Technol 04:33–37
Wu M (2009) Carbon dioxide transcritical cycle characteristics and system control research. Central South University
Xie S (2011) Brief introduction of the method of determining the size of capillary tube. Sci Technol Innov Her 27:106
Xing M, Yu J, Liu X (2014) Thermodynamic analysis on a two-stage transcritical CO_2 heat pump cycle with double ejectors. Energy Convers Manag 88:677–683
Xiong T (2014) Study on the throttling characteristics of the dual capillary tube combination of carbon dioxide heat pumps and the thermal performance of the system. Kunming University of Science and Technology
Yadav SK, Murari Pandey K, Gupta R (2020) Recent advances on principles of working of ejectors: a review. Mater Today Proc
Yari M (2009) Performance analysis and optimization of a new two-stage ejector-expansion transcritical CO_2 refrigeration cycle. Int J Therm Sci 48:1997–2005
Yari M, Mahmoudi SMS (2011) Thermodynamic analysis and optimization of novel ejector-expansion TRCC (transcritical CO_2) cascade refrigeration cycles (Novel transcritical CO_2 cycle). Energy
Yari M, Sirousazar M (2008) Cycle improvements to ejector-expansion transcritical CO_2 two-stage refrigeration cycle. Int J Energy Res 32:677–687
Yu B, Yang J, Wang D, Shi J, Chen J (2019) An updated review of recent advances on modified technologies in transcritical CO_2 refrigeration cycle. Energy 189
Zare V, Rostamnejad Takleh H (2020) Novel geothermal driven CCHP systems integrating ejector transcritical CO_2 and rankine cycles: thermodynamic modeling and parametric study. Energy Convers Manag 205
Zeng F, Tian Z, Chen W, He Z (2019) Single system refrigeration system and refrigeration equipment. Shandong Province: CN109269160A, 25 January 2019
Zhang Z, Dong X, Ren Z, Lai T, Hou Y (2017) Influence of refrigerant charge amount and EEV opening on the performance of a transcritical CO_2 heat pump water heater. Energies 10
Zhang Z, Feng X, Tian D, Yang J, Chang L (2020) Progress in ejector-expansion vapor compression refrigeration and heat pump systems. Energy Convers Manag 207
Zhang Z-y, Ma Y-t, Wang H-l, Li M-x (2013) Theoretical evaluation on effect of internal heat exchanger in ejector expansion transcritical CO_2 refrigeration cycle. Appl Therm Eng 50:932–938
Zheng L, Deng J, Zhang Z (2016) Dynamic simulation of an improved transcritical CO_2 ejector expansion refrigeration cycle. Energy Convers Manag 114:278–289

Zhifang X, Lin S, Hongfei O (2008) Refrigerant flow characteristics of electronic expansion valve based on thermodynamic analysis and experiment. Appl Therm Eng 28:238–243
Zhu J, Elbel S (2020) Experimental investigation into the influence of vortex control on transcritical R744 ejector and cycle performance. Appl Therm Eng 164
Zhu Y, Huang Y, Li C, Zhang F, Jiang P-X (2018) Experimental investigation on the performance of transcritical CO_2 ejector–expansion heat pump water heater system. Energy Convers Manag 167:147–155
Zhu Y, Li C, Zhang F, Jiang P-X (2017) Comprehensive experimental study on a transcritical CO 2 ejector-expansion refrigeration system. Energy Convers Manag 151:98–106

Chapter 5
CO_2 Gas Cooler and Cooling Process

Yunting Ge

Nomenclature

a	Area (m^2)
C_p	Specific heat at constant pressure (J kg^{-1} K^{-1})
C	Capacity rate (W K^{-1})
d	Diameter(m)
D	Depth (m)
f	Friction factor
G	Mass flux (kg m^{-2} s^{-1})
h	Enthalpy(J kg^{-1})
H	Height (m)
i, j, k	Coordinates
$\dot{m}$	Mass flow rate (kg s^{-1})
P	Pressure (Pa)
$\dot{q}$	Heat transfer per square meter (W m^{-2})
$\dot{Q}$	Heat transfer (W)
R	Resistance (K W^{-1})
s	Perimeter of inner pipe (m)
T	Temperature (K)
u	Velocity (m s^{-1})
U	Overall heat transfer coefficient (W m^{-2} K^{-1})
Va, $\dot{V}$	Air velocity (m s^{-1})
W	Width (m)
z	Length (m)

Y. Ge (✉)
Centre for Civil and Building Services Engineering, London, UK
e-mail: yunting.ge@lsbu.ac.uk

School of the Built Environment and Architecture, London South Bank University, London, UK

X.-R. Zhang and T. M. Eikevik (eds.), *CO_2 Refrigeration Cycle and Systems*,
Lecture Notes in Energy 96, https://doi.org/10.1007/978-3-031-22512-3_5

Subscripts

a	Air
avg	Average
e	Evaporator
ex	Exit
f	Friction
gc	Gas cooler
h	Hot side
i	Inner, *i*th grid
In	Inlet
Is	Isentropic
j	*j*Th grid
k	*k*Th grid
min	Minimum
max	Maximum
o	Outer
opt	Optimal
pc	Pseudocritical
r	Refrigerant
sh	Superheating
wi	Inner pipe wall

Greek symbol

α	Heat transfer coefficient (W m^{-2} K^{-1})
η	Efficiency
Δ	Difference
ρ	Density (kg m^{-3})
τ	Shear stress (N m^{-2})
ε	Effectiveness

As a natural working fluid, CO_2 refrigerant has been widely applied in refrigeration and heat pump systems. Being the main component, a CO_2 gas cooler plays an important role in the system's highly efficient and safe operations. The CO_2 gas coolers, therefore, need to be well investigated in terms of optimal controls, heat transfer, and hydraulic analyses, as well as structural designs for different applications.

5.1 Optimal Heat Rejection Pressure

Compared to conventional HFC refrigerants such as R134a in vapour compression systems, the most extraordinary thermophysical properties of the CO_2 working fluid are its low critical temperature of 31.1 °C and extremely high critical pressure of 7.38 MPa. In such circumstances, for a CO_2 refrigeration or heat pump system, if the temperature of a heat rejection medium such as ambient air or cooling water is relatively high, there will be no condensation during the high-pressure heat rejection process. Considering the essential low-pressure evaporation process in the system, the CO_2 heat rejection process at the high-pressure side will subsequently operate at a supercritical cooling process starting from the compressor outlet. In that case, the heat exchanger for the CO_2 heat rejection will be called CO_2 gas cooler instead of CO_2 condenser. The operating supercritical pressure of CO_2 will thus be independent of CO_2 temperature.

A typical CO_2 transcritical refrigeration cycle can be demonstrated in Fig. 5.1. As depicted, there are four processes in the cycle, including isentropic compression 1–2 s by a compressor, isobaric gas cooling 2 s–3 by a gas cooler, isenthalpic expansion 3–4 by a thermostatic expansion valve, and isobaric evaporation 4–1 by an evaporator. The evaporator refrigeration effect q_0, compressor specific work w, cooling COP in the cycle, an isothermal line T_{ex} showing the CO_2 temperature at the gas cooler exit and an isentropic line for the compression process are also indicated and demonstrated in the diagram.

At a constant evaporating temperature T_0, constant CO_2 gas cooler exit temperature T_{ex} and a fixed refrigerant state at the compressor inlet, the refrigeration effect q_0 and compressor specific work w both increase with higher CO_2 pressures at the compressor outlet. However, due to the 'S' shape of the isothermal line T_{ex} and nearly

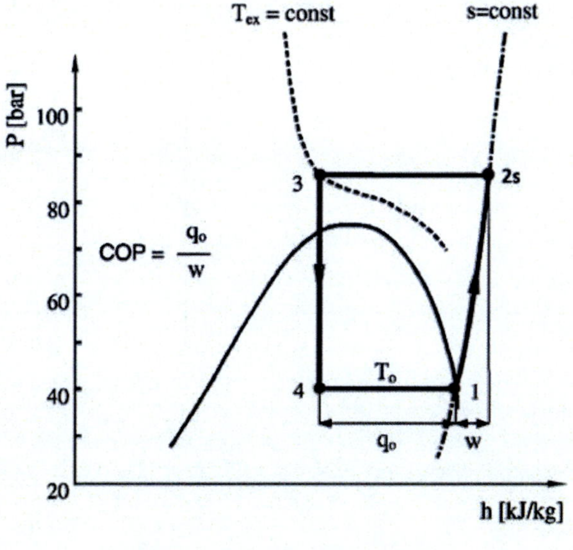

Fig. 5.1 A simplified CO_2 transcritical refrigeration cycle

linear shape of the isentropic line, the cooling COP will increase and decrease with the growth of heat rejection pressure. This means that there is an optimal heat rejection pressure in the gas cooler at which the cooling COP is maximized (Inokuty 1928; Pettersen and Skaugen 1994). This is different from that of a conventional vapour compression refrigeration cycle with both evaporation and condensation processes in which the cooling COP always decreases with higher condensing pressures. Subsequently, to enhance the performance of a CO_2 refrigeration or heat pump cycle, an important task is to determine and control the optimal high-side pressure at a specific operating condition.

5.2 Prediction of Optimal Heat Rejection Pressure

In an actual operation of a CO_2 refrigeration or heat pump system, the evaporator evaporating temperature will vary, while the compressor compression process will not be isentropic, and there will be some degrees of superheating at the evaporator outlet or the compressor inlet. In addition, there are some pressure drops during CO_2 gas cooling and evaporation processes. To reflect this, a practical CO_2 transcritical refrigeration cycle is shown in Fig. 5.2. To simplify the analysis, no pressure drop is assumed for either CO_2 gas cooling (2–3) or the evaporation process (4–1). In Fig. 5.2, point e is at the saturated vapour state along the evaporation line while the dot line 1–2 s indicates the isentropic compression process.

Accordingly, the cooling COP can be calculated as

$$COP = \frac{h_1 - h_4}{h_2 - h_1} = \frac{h_1 - h_3}{h_2 - h_1} = \frac{h_1 - h_3}{h_{2s} - h_1} \times \eta_{\text{is}} \tag{5.1}$$

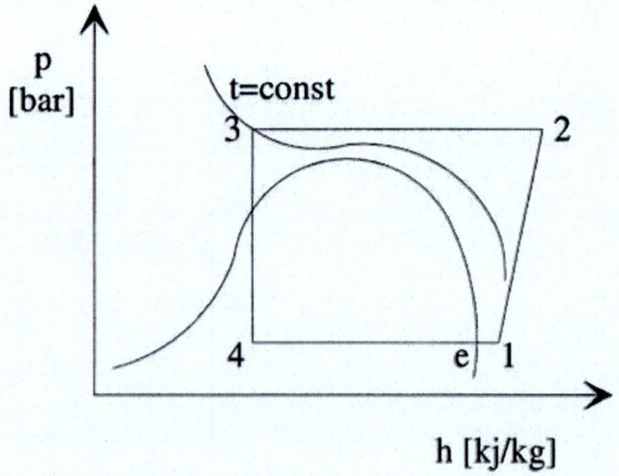

Fig. 5.2 A practical CO_2 transcritical cycle

The cooling COP is, therefore, a function of a number of parameters including heat rejection pressure p_{gc}, CO_2 gas cooler outlet temperature t_{ex}, evaporating temperature t_e, superheating at the compressor inlet Δt_{sh}, and isentropic efficiency of the compressor η_{is}:

$$COP = f\left(p_{gc}, t_{ex}, t_e, \Delta t_{sh}, \eta_{is}\right) \tag{5.2}$$

The optimal heat rejection pressure can be calculated when the COP is maximized and the following equation is satisfied:

$$\left[\frac{\partial COP}{\partial P_{gc}}\right]_{p_{gc}=p_{gc,opt}} = 0 \tag{5.3}$$

Based on a parametric simulation, superheating Δt_{sh} affects the optimal heat rejection pressure less compared to other parameters (Liao et al. 2000). If the isentropic efficiency could be considered as a constant, the optimal pressure will predominantly be the function of the gas cooler outlet temperature and evaporating temperature. Furthermore, of these two parameters, the effect of gas cooler outlet temperature is more significant (Kim et al. 2004a). As an example, the effects of the gas cooler outlet temperature on the optimal heat rejection pressure and cooling COP are calculated and shown in Fig. 5.3. Hence, a lower gas cooler outlet temperature is expected since it leads to higher cooling COP and lower optimal heat rejection pressure. This requires the gas cooler approach temperature to be as small as possible. The approach temperature is defined as the temperature difference between the gas cooler CO_2 outlet temperature and the inlet temperature of the heat rejection medium. There are a number of issues that can affect the approach temperatures including the heat exchanger types, designs, and heat transfer and hydraulic behaviours of CO_2 flow and heat rejection medium.

5.3 Heat Transfer and Hydraulic Analyses

5.3.1 Heat Transfer Coefficient

The measurements and correlations of CO_2 heat transfer coefficient and pressure drop for CO_2 gas coolers are mostly on internal supercritical cooling flow with both larger-diameter and microchannel tubes. The term 'microchannel' is used for flow channels with a hydraulic diameter of less than 1 mm.

There are two well-known correlations by Gnielinski (1976) and Pitla et al. (2002) for the calculations of in-tube supercritical CO_2 gas flow heat transfer coefficients. The Gnielinski correlation is based on the Nusselt number that is calculated using the thermophysical properties at the bulk temperature and is calculated as

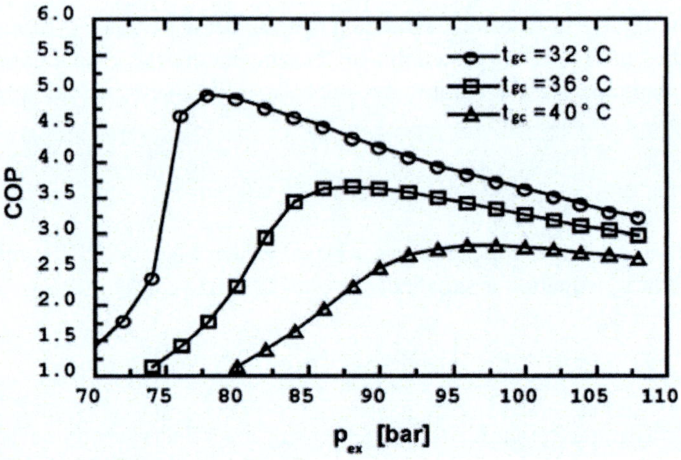

Fig. 5.3 The effect of CO_2 gas cooler outlet temperature on the optimal heat rejection pressure and cooling COP

$$\mathrm{Nu} = \frac{\xi/8(\mathrm{Re}-1000)\mathrm{Pr}}{12.7\sqrt{\frac{\xi}{8}}\left(\mathrm{Pr}^{\frac{2}{3}}-1\right)+1.07} \tag{5.4}$$

The friction factor ξ is a function of the Reynolds number only and is calculated as the correlation proposed by Krasnochekov et al. (1970)

$$\xi = (0.79\ln(\mathrm{Re}) - 1.64)^{-2} \tag{5.5}$$

The Reynolds number, Prandtl number, and Nusselt number are calculated in Eqs. (5.6)–(5.8), respectively.

$$\mathrm{Re} = \frac{\mathrm{Gd}}{\mu} \tag{5.6}$$

$$\mathrm{Pr} = \frac{\mu C_p}{\lambda} \tag{5.7}$$

$$\mathrm{Nu} = \frac{\alpha d}{\lambda} \tag{5.8}$$

The heat transfer coefficient α can, therefore, be calculated.

The Pitla correlation is based on mean Nusselt numbers that are calculated using the thermophysical properties at the wall and bulk temperatures. The mean Nusselt number is calculated as

$$\mathrm{Nu} = \left(\frac{\mathrm{Nu_{wall}}+\mathrm{Nu_{bulk}}}{2}\right)\frac{\lambda_{\mathrm{wall}}}{\lambda_{\mathrm{bulk}}}, \tag{5.9}$$

where $\mathrm{Nu_{wall}}$ and $\mathrm{Nu_{bulk}}$ are Nusselt numbers that are evaluated based on the thermophysical properties at the wall and bulk temperatures, respectively. In each case, the

Gnielinski correlation listed in Eq. (5.4) is applied to calculate the respective Nusselt number.

For the calculation of Re_{wall} in Eq. (5.4), it was found that the best fit was obtained by using the inlet velocity of CO_2 to compute the Reynolds number at the wall. As to the Re_{bulk}, it is calculated based on the local mean velocity. The mean velocity is calculated by some local parameters

$$\dot{V}_{avg} = \frac{\dot{m}}{A\rho_{bulk}} = \frac{G}{\rho_{bulk}} \tag{5.10}$$

Based on a detailed numerical model developed by Pitla et al. (2001a, b), the effects of CO_2 mass flow rate and supercritical pressure on the heat transfer coefficient were predicted and shown in Figs. 5.4 and 5.5, respectively. As shown in Fig. 5.4, at a constant CO_2 gas cooler inlet temperature (395 K) and pressure (10 MPa) and heat exchanger wall temperature (303 K), the heat transfer coefficient significantly increases with higher CO_2 mass flow rates. In addition, for a fixed CO_2 mass flow rate, the heat transfer coefficient increases during the gas cooling process until a maximum is reached. The maximum region in the heat transfer coefficient is called the pseudocritical region around the pseudocritical point and coincides with the region where the specific heat has a maximum, as shown in Fig. 5.6 (Pitla et al. 1998). The pseudocritical point is defined as the temperature at which the specific heat becomes a maximum for a given pressure. The heat transfer coefficient then drops suddenly as the fluid enters the liquid regime. The pseudocritical temperature and maximum isobaric specific heat of CO_2 can be shown in Fig. 5.4.

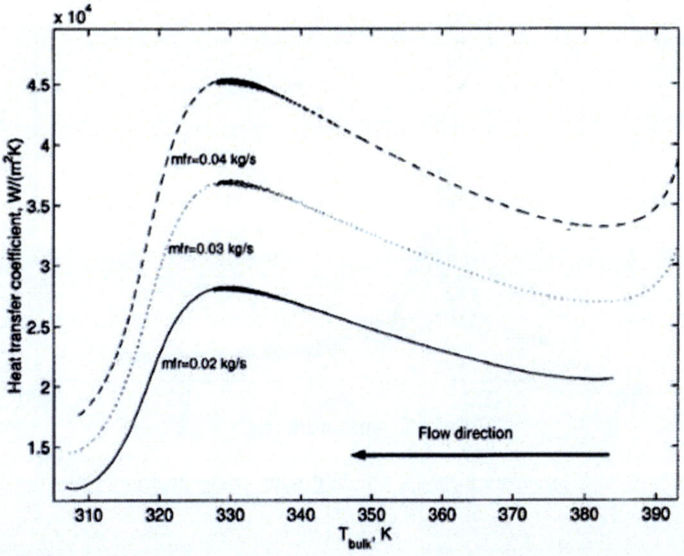

Fig. 5.4 The effect of mass flow rate on the heat transfer coefficient ($T_{in} = 395$ K, $P_{in} = 10$ MPa, $T_{wall} = 303$ K) (Pitla et al. 2002)

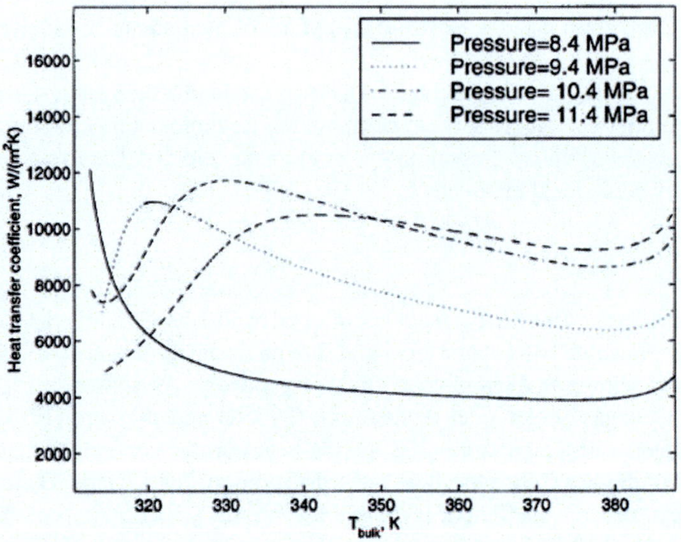

Fig. 5.5 The effect of pressure on the heat transfer coefficient (T_{in} = 390 K, mass flow rate = 0.04 kg/s, T_{wall} = 310 K). (Pitla et al. 2002)

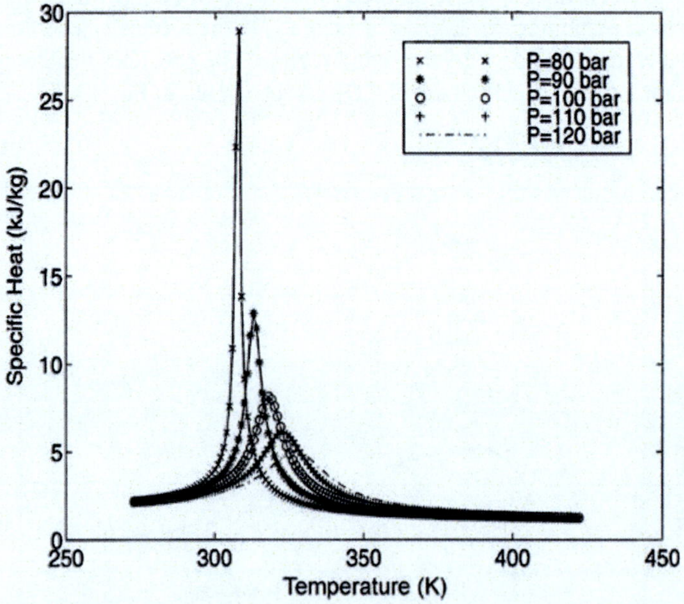

Fig. 5.6 Variation of CO_2 supercritical specific heat with temperature and pressure (Pitla et al. 1998)

Correspondingly, the pseudocritical temperature of CO_2 can be calculated with the following equation (Liao and Zhao 2002):

$$T_{pc} = -122.6 + 6.124P - 0.1657P^2 + 0.1773P^{2.5} - 0.0005608P^3,\quad 75 \le P \le 140 \tag{5.11}$$

where the temperature (T_{pc}) and pressure (P) are in °C and bar, respectively.

As depicted in Fig. 5.5, the numerical model was also used to predict the effect of the CO_2 inlet pressure on the CO_2 heat transfer coefficient along its flow direction. To clarify, the CO_2 inlet temperature, mass flow rate, and heat exchanger wall temperature were maintained at 390 K, 0.04 kg/s, and 310 K, respectively. The peak heat transfer coefficient value can be observed to shift to a higher temperature with increasing pressures. This coincides with the shift in the pseudocritical region to higher temperatures with an increase in pressure, as shown in Fig. 5.6. In addition, at higher pressures, the variation in the heat transfer coefficient with temperature is smaller than at pressures near the critical point. This is due to the variation in thermophysical properties (specific heat) at maximum near the critical point, which decreases as the pressure increases from the critical pressure (7.353 MPa) (Fig. 5.7).

The Pitla correlation has been compared with measurements of a purposely built test rig (Pitla et al. 2001a). For the experimental measurements, the tube inner diameter was 4.72 mm, the CO_2 inlet and outlet temperature ranges were from 101 to 134 °C and 20 to 34 °C, respectively, inlet pressure varied from 94 to 134 bar, and CO_2 mass flow rates from 0.0196 to 0.0387 kg/s. It was found that 85% of the heat transfer coefficient values predicted by the correlation were within 20% accuracy when compared with the corresponding measurements. In addition, the

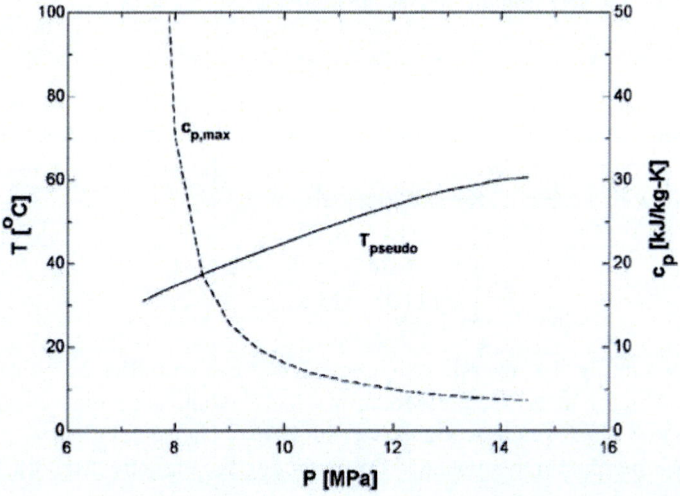

Fig. 5.7 Pseudocritical temperature and maximum isobaric specific heat of CO_2 (Liao and Zhao 2002)

correlation was also compared with three other existing correlations from the literature (Gnielinski 1976; Krasnoshechekov et al. 1970; Baskov et al. 1997) which found that the Pitla correlation was more accurate, particularly in the pseudocritical region when the pressure is relatively higher than the critical pressure.

Experimental investigations were carried out to obtain the heat transfer and pressure drop characteristics during the CO_2 gas cooling process in a horizontal tube with an inner diameter of 7.73 mm (Yoon et al. 2003). The tested CO_2 mass fluxes were fixed at 225, 337, and 450 kg m^{-2} s^{-1}, while the CO_2 pressures were controlled between 7.5 and 8.8 MPa which were close to the critical pressure. Accordingly, a simple formula for the heat transfer coefficient calculation was correlated

$$\mathrm{Nu}_{\mathrm{bulk}} = a\,\mathrm{Re}_{\mathrm{bulk}}^{b}\,Pr_{\mathrm{bulk}}^{c}\left(\frac{\rho_{\mathrm{pc}}}{\rho_{\mathrm{bulk}}}\right)^{n} \tag{5.12}$$

where $a = 0.14$, $b = 0.69$, $c = 0.66$, $n = 0$ for $T_{\mathrm{bulk}} > T_{\mathrm{pc}}$.

$a = 0.013$, $b = 1.0$, $c = -0.05$, $n = 1.6$ for $T_{\mathrm{bulk}} \leq T_{\mathrm{pc}}$.

Compared to the experimental data, the correlation can have an absolute average deviation of 12.7%.

5.3.2 *Pressure Drop*

Since the flow of carbon dioxide in the supercritical gas cooling process is somewhat similar to that of a conventional single-phase flow, it seems reasonable to apply the single-phase pressure drop correlation in calculating the pressure drop during the cooling process. The frictional pressure drop for a fully developed turbulent single-phase flow in a smooth tube is calculated as

$$\Delta P = f\frac{G^2 L}{2\rho D_i} \tag{5.13}$$

A number of equations have been developed for the friction factor, f. However, Blasius' equation (Incropera and DeWitt 1996) is the most widely used for the turbulent flow in a smooth tube and is calculated as

$$f = \begin{cases} 0.316\,\mathrm{Re}^{-\frac{1}{4}}\mathrm{for Re} \leq 2\times10^4 \\ 0.184\,\mathrm{Re}^{-\frac{1}{5}}\mathrm{for Re} > 2\times10^4 \end{cases} \tag{5.14}$$

As shown in Fig. 5.8, the pressure drop calculations with Blasius' equation were compared with the measured pressure drop data to obtain an absolute average deviation of about 4.9% (Incropera and DeWitt 1996). The calculation with Blasius' equation is, therefore, recommended for predicting the pressure drop of CO_2 in the supercritical cooling process.

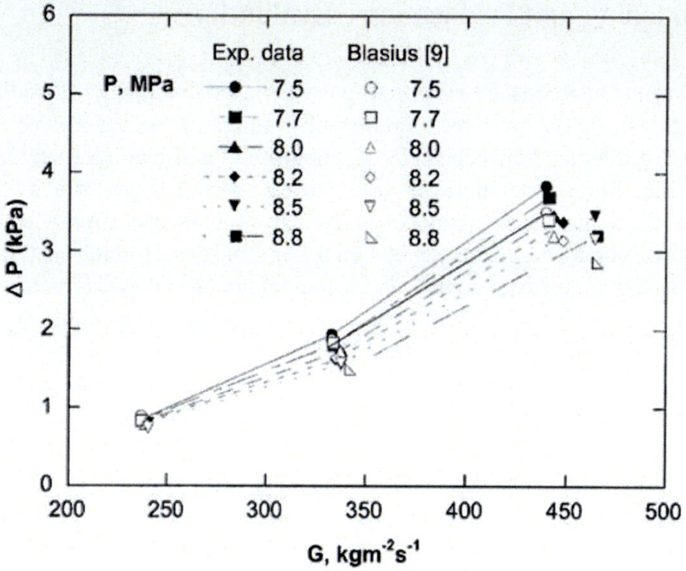

Fig. 5.8 Comparison of the measured pressure drop data with those predicted by Blasius' equation (Incropera and DeWitt 1996)

5.3.3 Calculations of Smaller-Diameter Tubes

For smaller-diameter tubes such as microchannels, different correlations might be applied for the calculations of heat transfer and pressure drop during CO_2 supercritical cooling processes. The heat transfers during CO_2 supercritical cooling processes in 0.8 mm microchannel tubes were measured and correlated (Pettersen et al. 2000). It was found that the standard single-phase correlations such as the widely used Dittus–Boelter model and the Gnielinski correlation (Gnielinski 1976) gave good correspondence between measured and calculated heat transfer coefficients. Meanwhile, the Colebrook and White correlation reproduced the pressure drop data well.

In addition, the heat transfer coefficients for CO_2 supercritical cooling processes in horizontal micro/mini tubes were measured (Liao and Zhao 2002). Test tubes were stainless steel tubes with inside diameters of 0.5, 0.7, 1.1, 1.4, 1.55, and 2.16 mm, respectively. The pressures and temperatures measured ranged from 7.4 to 12 MPa and 20 to 110 °C, respectively. The buoyancy force significantly affected both supercritical CO_2 flow and heat transfer. However, the buoyancy effect became smaller as the tube diameter decreased. They reported that the existing correlations for larger tubes deviated notably from their test data for the micro/mini tubes. Based on the test data, they developed a correlation for the axially averaged Nusselt number with a mean relative error of 9.8%.

5.4 Modelling and Performance Evaluation

IT is known that at a constant evaporating temperature the maximum cooling COP increases greatly with lower refrigerant temperatures at the gas cooler exit. The temperature difference between the refrigerant outlet and incoming ambient air is known as the approach temperature (AT) for an air-cooled gas cooler. The minimization of the approach temperature will greatly affect the system efficiency (Fang et al. 2001), this being mainly dependent on the optimal design of the heat exchanger. Considering that circuit arrangement and structural parameters will affect the optimal design of the heat exchanger, an efficient and economical option would be to utilize the simulation technique for the optimal design.

In CO_2 transcritical cycles, finned-tube gas coolers are not as popular as aluminium minichannel heat exchangers, which have the advantages of being lightweight and compact, with a lower risk of high-pressure stresses, and are already widely used in automobile air-conditioning. Therefore, a great deal of research and development effort has been put into minichannel heat exchangers (Yin et al. 2001; Ortiz et al. 2003; Pettersen et al. 1998). However, because of the lower cost, the finned-tube coils are still believed to be competent types of gas coolers. Theoretically, three modelling methods could be used in the performance analysis of such gas coolers: Effectiveness-NTU or LMTD, i.e. lumped method, tube-in-tube, and distributed method. Since there is a rapid change of CO_2 thermophysical properties with temperatures during an isobaric gas cooling process, it is not practical to use the overall Effectiveness-NTU or LMTD method to simulate gas coolers, particularly if the property profiles (such as temperature) need to be predicted (Kim et al. 2004b). The tube-in-tube method developed from the research of Domanski (1989), Domanski and Yashar (2007) was utilized in the simulation of a gas cooler by Chang and Kim (2007). By means of the model simulation, the effects of coil structural parameters on the performance of the gas cooler were investigated. It was found from the simulation results that the approach temperature can be reduced with an increased heat exchanger front area. Although the model demonstrates significant improvement with this method, a more detailed modelling strategy in the distributed method is still expected to further enhance simulation accuracy, and therefore, to obtain more reliable conclusions.

5.4.1 Distributed Method

The distributed method was used in developing the simulation model for finned-tube air-cooled CO_2 gas coolers (Ge and Cropper 2009). A diagram with sub-elements of the coil in a three-dimensional (3D) space for the model is schematically drawn in Fig. 5.9. Tubes are arranged parallel to the i direction, j is specified in the longitudinal direction, while k is in the transverse direction. Air is flowing parallel to j direction and refrigerant is assumed to be in approximate counter-cross direction to air for this sample. The number selection of small elements in the i direction is arbitrary from one to infinity. The larger this value is, the more accurate the simulation will be, but expensive computing time will be sacrificed. The coordinate of each divided element in the 3D space can then be determined. The coordinate value i represents the number of sub-elements for each tube selected by the model; j corresponds to tube row numbers in longitudinal paths starting from the air inlet, and k equals the tube numbers in the transverse path originating from the bottom. Therefore, the state point of either refrigerant or air at each specified sub-element in the 3D space can be positioned with its corresponding coordinate values $i, j,$ and k, which vary according to the circuit number and tube number. The tube number starts from the refrigerant inlet to the refrigerant outlet for each circuit. The solving routine first starts from the circuit loop if there is more than one circuit for the coil. For each circuit, the simulation will run through each numbered tube starting from the refrigerant inlet and then the element loop for each pipe. The whole modelling work depends on setting up the conservation equations for each sub-element and an efficient routine to solve these equations. The solutions for one sub-element can be used as the inputs for the next sub-division. The air-side parameters for each element, which are unknown initially, will be assumed first. These parameters will be updated by the next iteration. The total heating load of the gas cooler is calculated at the end of each iteration. The iteration will carry on until all the loops are cycled and the total heating loads for two continuous iterations are almost unchanged.

5.4.1.1 Refrigerant Side Conservation Equations

Before setting up the refrigerant side conservation equations for each element, the following assumptions are proposed:

- The system is in a steady state.
- No heat conduction in the direction of the tube axis and nearby fins.
- Air is in a homogeneous distribution, that is, the air-facing velocity to each element is the same.
- No contact heat resistance between the fin and the tube.
- At any point in the flowing direction, the refrigerant is in thermal equilibrium condition.

Mass equation

$$\frac{\mathrm{d}}{\mathrm{d}z}(\dot{m}_r) = 0 \tag{5.15}$$

Momentum equation

$$\frac{1}{A_i}\frac{\mathrm{d}}{\mathrm{d}z}(\dot{m}_r u) = -\frac{\mathrm{d}P}{\mathrm{d}z} - \frac{\tau_{\mathrm{wi}} s_{\mathrm{wi}}}{A_{\mathrm{wi}}} \tag{5.16}$$

Energy equation

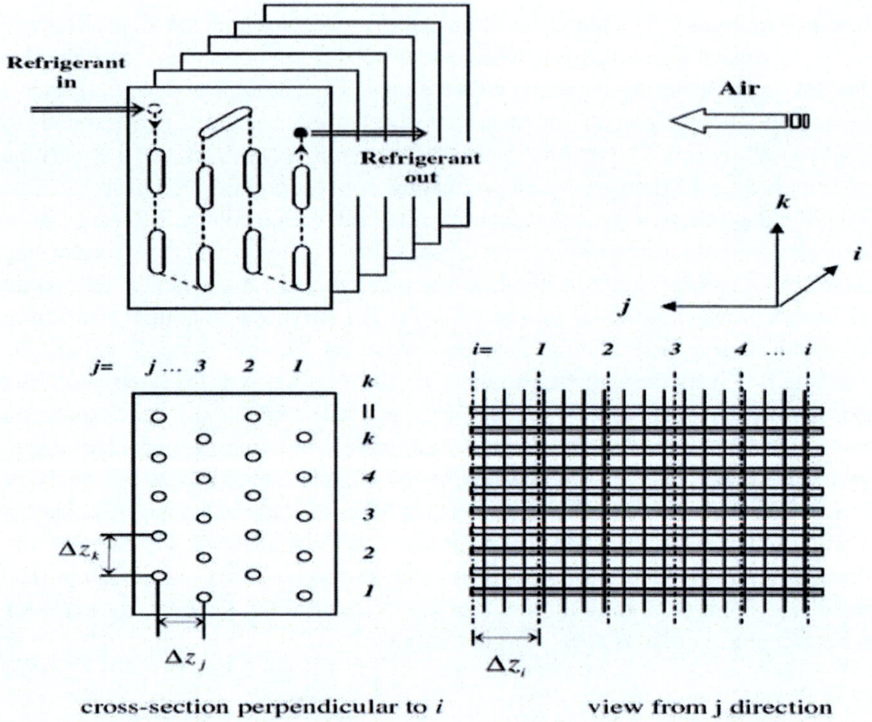

Fig. 5.9 Three-dimensional coordinate of sub-elements in the coil for the gas cooler model

$$\frac{d}{dz}(\dot{m}_r h) = -(\pi\, d_o)\dot{q} \tag{5.17}$$

The above equations can be easily discretized as below for a sub-element shown in Fig. 5.9 with a coordinate from (*i, j, k*) to (*i* + *1*, *j*, *k*). The dimensions of the sub-element at (*i, j, k*) directing to *i, j, and k* are $\Delta z_{i,}$ $\Delta z_{j,}$ and Δz_k, respectively.

Mass equation

$$\dot{m}_r\big|_{(i+1,j,k)} - \dot{m}_r\big|_{(i,j,k)} = 0 \tag{5.18}$$

Momentum equation

$$\frac{1}{A_i}[(\dot{m}_r u)\big|_{(i+1,j,k)} - (\dot{m}_r u)\big|_{(i,j,k)}] = -\Delta P - \Delta P_f \tag{5.19}$$

where

$$\Delta P_f = f\frac{G^2 \Delta z_i}{2\rho d_i} \tag{5.20}$$

Energy equation

$$(\dot{m}_r h)|_{(i+1,j,k)} - (\dot{m}_r h)|_{(i,j,k)} = -(\pi d_0)\dot{q} \times \Delta z_i \tag{5.21}$$

The conservation equations can also be applied to the air-side calculation. The pressure drop calculation is used instead of the momentum equation, and for this side, the heat transfer calculation is included in the energy equation. In addition, there is a heat balance between the air and refrigerant sides for each element.

5.4.1.2 Air-Side Heat Transfer

Effectiveness-NTU method is used in the calculation of heat transfer for airside in one grid section.

$$\dot{Q}_a = \varepsilon C_{\min}[T_r(i, j, k) - T_a(i, j, k)] \tag{5.22}$$

where the effectiveness ε is calculated as

$$\begin{aligned} &\varepsilon = 1 - \exp(-\gamma \frac{C_{\max}}{C_{\min}}) \text{ for } C_{\max} = C_h \\ &\text{where } \gamma = 1 - \exp(-\frac{UA}{C_{\max}}) \end{aligned} \tag{5.23}$$

and

$$\begin{aligned} &\varepsilon = \frac{C_{\max}}{C_{\min}}(1 - \exp(-\gamma \frac{C_{\min}}{C_{\max}})) \text{ for } C_{\min} = C_h \\ &\text{where } \gamma = 1 - \exp(-\frac{UA}{C_{\min}}) \end{aligned} \tag{5.24}$$

The product UA (overall heat-transfer coefficient times area) can be calculated as

$$\text{UA} = (\frac{1}{\alpha_a \eta_0 A_0} + \sum R_i + \frac{1}{\alpha_r A_r})^{-1} \tag{5.25}$$

where ΣR_i is the sum of heat conduction resistances through the pipe wall and fin.

The heat transfer from the airside can be calculated as

$$\begin{aligned} \dot{Q}_a &= \dot{m}_a(i, j, k) \times Cp_a(i, j, k) \times [T_a(i, j+1, k) \\ &\quad - T_a(i, j, k)] \\ &= UA(i, j, k) \times [T_r(i, j, k) - T_a(i, j, k)] \end{aligned} \tag{5.26}$$

The parameters at grid points $(i + 1, j, k)$ for refrigerant and $(i, j + 1, k)$ for air can be obtained when Eqs. (5.18)–(5.26) are solved together.

The accurate model prediction also relies on the precise calculations of fluid properties, heat transfer coefficients and pressure drops on both refrigerant and air sides. The CO_2 refrigerant properties are calculated using subroutines from the National Institute of Standards and Technology software package REFPROP (Lemmon et al. 2007). For calculating the refrigerant heat transfer coefficient, the correlation from Pitla et al. is utilized (Pitla et al. 2002). The friction pressure drop is calculated in Eq. (5.20) and the Blasius equation (Incropera and DeWitt 1996) is used to calculate the friction factor f. The air-side heat transfer and friction coefficients are computed using the correlations by Wang et al. (1999, 2001).

5.4.2 Model Validations

To develop a performance database for the component design in CO_2 transcritical cycles, a specially designed test facility was set up by Hwang et al. (2005). The test system was composed of an air duct and two environmental chambers that housed an evaporator, a gas cooler, an expansion valve, and a compressor. By means of this test rig, a set of parametric measurements at various inlet air temperatures and velocities, refrigerant inlet temperatures, mass flow rates, and operating pressures were carried out on a specified CO_2 gas cooler. The side view of the circuit arrangement for the tested gas cooler is shown in Fig. 5.10. The airflow is from right to left, and the refrigerant inlet is at the upper left numbered "0", while the refrigerant outlet is at the lower right numbered "54" for the heat exchanger. The dashed lines in the Figure indicate the U-bends of the rear side noted with odd numbers, while the solid lines signify the U-bends of the front side noted with even numbers. To measure the variation of refrigerant temperature along the heat exchanger pipes, number of thermocouples were attached on the outside surfaces of the front side U-bend pipes and at the refrigerant inlet and outlet as well. These thermocouples were well-insulated to get a more accurate measurement. The structural specification of the gas cooler is listed in Table 5.1.

The test conditions, 36 in total, are listed in Table 5.2. Each test condition contains the measurements of air inlet temperature, air velocity, refrigerant inlet temperature, refrigerant inlet pressure, and refrigerant mass flow rate. These measurements and the coil structural parameters will be used as model inputs and parameters, respectively. Therefore, the predicted refrigerant temperature profile at each test condition is compared with the corresponding test result in order to validate the model. To save space, comparison results for twelve test conditions with numbers 1–3, 10–12, 19–21, and 25–27, listed in Table 5.2, are selected and shown in Figs. 5.11, 5.12, 5.13 and 5.14, respectively. It can be seen from both simulation and test results that a sharp refrigerant temperature decrease occurs in the third pipe row ($j = 3$), pipes numbered from 0 to 18 in Fig. 5.11. The temperature changing rates in the second ($j = 2$) and first rows ($j = 1$) are gradually reduced. In addition, at constant refrigerant pressures and mass flow rates, similar refrigerant inlet temperatures and unchanged air inlet temperatures, the refrigerant temperature at any specified location is always

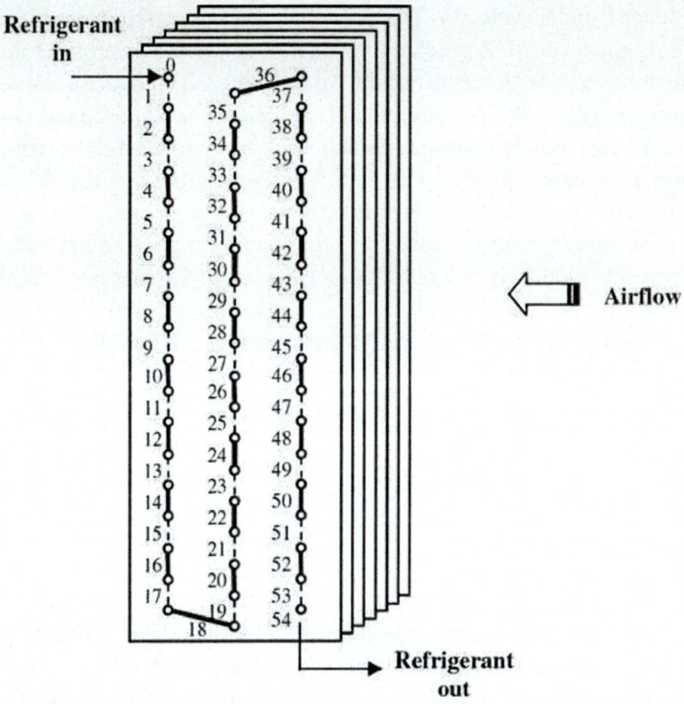

Fig. 5.10 Tested gas cooler with numbered pipes

Table 5.1 Specification of the tested gas cooler

Dimension	
W × H × D (m)	0.61 × 0.46 × 0.05
Front area (m^2)	0.281
Fin	
Shape	Raised lance
Fin pitch (mm)	1.5
Thickness (mm)	0.13
Tube	
Number of tubes row	3
Number of tubes per row	18
Tube outside diameter (mm)	7.9
Tube inside diameter (mm)	7.5
Tube shape	Smooth

lower for higher front air velocity. This is because, at higher front air velocities, the heat transfer is enhanced. The predicted refrigerant temperature profile for each test condition matches well with the test result. To facilitate the comparison, the constant inlet air temperature line is also presented in each plot. For all of the test conditions, the refrigerant temperatures at the gas cooler outlet are predicted and compared with those of the test results, as shown in Fig. 5.15. The temperature discrepancies between the simulation and test results for refrigerant outlet temperatures are mostly within ±2 °C when air front velocity is above 1 m/s. Larger errors are predominantly caused when the air front velocity is at 1 m/s. The correlation of air-side heat transfer coefficient at lower air velocities, therefore, needs to be further revised. The simulation can thus be concluded to fairly represent the test results and the model is, therefore, validated.

5.5 Microchannel CO_2 Gas Coolers

CO_2 refrigeration, air-conditioning, and heat pump systems can be applied to commercial, residential, and industrial purposes. Successful application of CO_2-based technology depends on the development of efficient and compact components with low weight, good reliability, and low cost. Since 2008, in the EU, the focus on the greenhouse effect of fluorinated compounds has led to a proposed gradual phase-out of refrigerant R134a in mobile air-conditioning. Subsequently, as one of the most significant applications, CO_2 air-conditioning has been widely applied in automotive industries in which the mass and space requirements are of particular importance in mobile systems.

The first automobile air-conditioning gas cooler prototypes were developed and manufactured in 1990–1991 (Lorentzen and Pettersen 1992). At that time, mechanically expanded round-tube units were commonly used in European cars, and a gas cooler was designed based on OD/ID 4.9 mm/3.4 mm aluminium tubes and plain aluminium fins. Core dimensions were based on the 1990 cross-flow condenser of a European passenger car (Lorentzen and Pettersen 1992). The tube configuration and circuiting of the CO_2 unit are shown in Fig. 5.16. Core depths of the CO_2 and baseline unit were 34 mm.

A problem in this first design was 'thermal short circuiting' due to conduction through the fins from hot tubes to colder tubes. The temperature gradient in gas cooling made this more important than with a condensing refrigerant. Thermocouples mounted on the tubes indicated that the refrigerant temperature actually increased towards the gas-cooler outlet as a consequence of conduction from the hot inlet tubes. The gas-cooler fins were then modified by a split between the second and third tube rows (in the direction of airflow). In addition, the refrigerant inlet was moved from the centre row to the rear row. Recorded tube-wall temperatures before and after this modification are shown in Fig. 5.17. The temperature approach was reduced from 12.2 K to 3.7 K, at an air-face velocity of 2.5 m s^{-1}. A reduction in air-face velocity to 1.0 m s^{-1} (and a reduction in compressor speed from 1600 to 700 revmin^{-1},

Table 5.2 Test conditions

Test conditions	Air inlet air temperatures (°C)	Air velocity (m/s)	Refrigerant inlet temperature (°C)	Refrigerant inlet pressure (MPa)	Refrigerant flow rate (kg/s)	Tested refrigerant outlet temperature (°C)	Simulated refrigerant outlet temperature (°C)
1	29.4	1	118.1	9	0.038	40.4	38.0
2	29.4	2	109.5	9	0.038	33.5	33.5
3	29.4	3	113.5	9	0.038	31.3	31.5
4	29.4	1	124	10	0.038	41.5	36.9
5	29.4	2	118	10	0.038	32.3	31.2
6	29.4	3	117.1	10	0.038	31.1	30.3
7	29.4	1	128.8	11	0.038	40.4	34.3
8	29.4	2	123.5	11	0.038	31.7	30.4
9	29.4	3	123.1	11	0.038	30.9	29.9
10	35	1	121.3	9	0.038	43.1	40.6
11	35	2	119.4	9	0.038	39.8	38.8
12	35	3	118.8	9	0.038	38.2	37.9
13	35	1	127.7	10	0.038	45.5	41.9
14	35	2	122.6	10	0.038	38.7	37.9
15	35	3	122.2	10	0.038	37.2	36.6
16	35	1	133.3	11	0.038	46.0	40.9
17	35	2	128.9	11	0.038	38.0	36.6
18	35	3	128.4	11	0.038	36.7	35.6
19	29.4	1	94.8	9	0.076	41.1	41.1
20	29.4	2	90.8	9	0.076	38.4	38.8

(continued)

Table 5.2 (continued)

Test conditions	Air inlet air temperatures (°C)	Air velocity (m/s)	Refrigerant inlet temperature (°C)	Refrigerant inlet pressure (MPa)	Refrigerant flow rate (kg/s)	Tested refrigerant outlet temperature (°C)	Simulated refrigerant outlet temperature (°C)
21	29.4	3	86.9	9	0.076	37.2	37.8
22	29.4	1	103.3	10	0.076	45.8	44.9
23	29.4	2	94.8	10	0.076	39.1	40.4
24	29.4	3	90.7	10	0.076	35.3	37.5
25	29.4	1	110.6	11	0.076	49.3	47.0
26	29.4	2	100.7	11	0.076	38.4	39.5
27	29.4	3	97.1	11	0.076	33.9	35.6
28	35	1	92.5	不	0.076	43.8	43.3
29	35	2	90	9	0.076	40.2	40.9
30	35	3	88.4	9	0.076	39.4	40.0
31	35	1	104.1	10	0.076	48.0	47.2
32	35	2	98.4	10	0.076	43.4	43.6
33	35	3	93.9	10	0.076	41.1	42.0
34	35	1	109.6	11	0.076	51.5	49.7
35	35	2	101.9	11	0.076	43.6	44.3
36	35	3	98.4	11	0.076	40.5	41.6

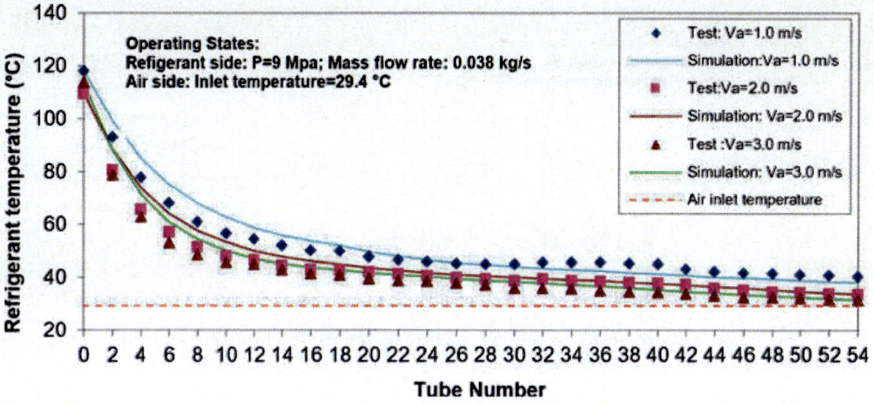

Fig. 5.11 Comparison of simulation with test results of test condition Nos. 1–3 for refrigerant temperature profile

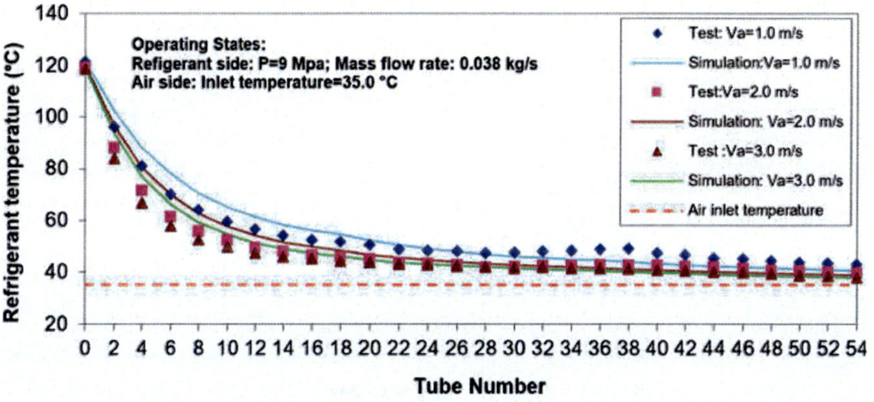

Fig. 5.12 Comparison of simulation with test results of test condition Nos. 10–12 for refrigerant temperature profile

i.e. to 'idle' conditions) typically increased the temperature approach by 10–20% compared to normal driving conditions. The change in gas-cooler refrigerant inlet temperature in Fig. 5.17 is a result of the differences in evaporating pressure and compressor inlet temperature.

Owing to the high-pressure level, large pressure drops can be tolerated in the gas cooler. Thus, heat exchangers can have refrigerant mass fluxes typically ranging from 600 to 1200 $kgm^{-2}\ s^{-1}$, with even higher numbers reported for water-heating heat exchangers. The high working pressure and favourable heat transfer properties of CO_2 enable reduced tube diameters and small refrigerant-side surface areas. Since these reductions may give room for more air-side surface per unit core volume, their compactness can be increased. Table 5.3 gives examples of estimated heat transfer and pressure drop data for supercritical CO_2 flow in compact air-conditioning gas coolers,

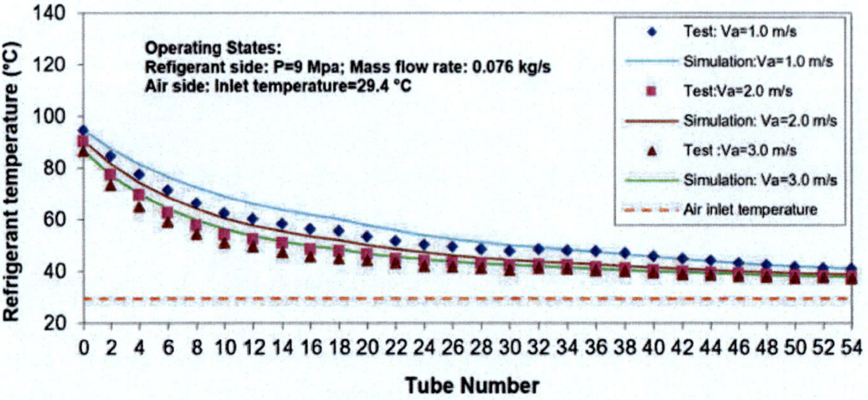

Fig. 5.13 Comparison of simulation with test results of test condition Nos. 19–21 for refrigerant temperature profile

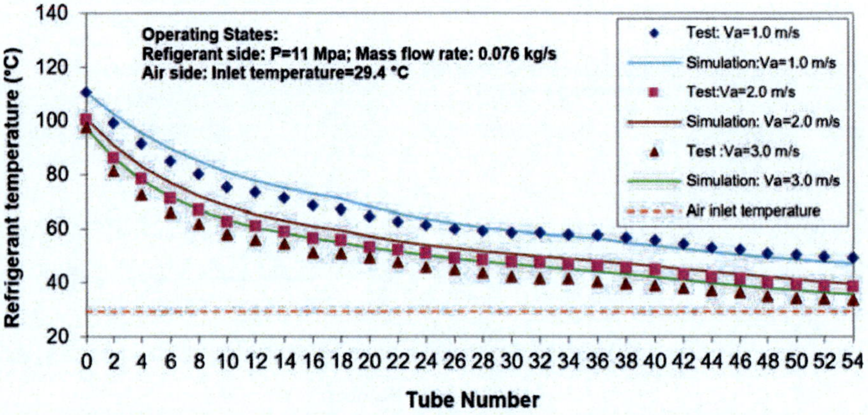

Fig. 5.14 Comparison of simulation with test results of test condition Nos. 25–27 for refrigerant temperature profile

with tube diameters of 2.0 and 0.8 mm (Pettersen et al. 2000). These dimensions are representative of round-tube and microchannel heat exchangers, respectively. As may be observed, the heat transfer coefficients are quite high for single-phase flow. Although the pressure gradient is higher for microchannel flow, the shorter circuits usually give lower overall pressure drops in this type of heat exchanger. Supercritical flow of CO_2 in microchannels is usually turbulent, although the transition regime may be encountered near the gas cooler outlet at low temperatures. Even though small-diameter round-tube heat exchangers can achieve low weight and compact design for a high-pressure fluid like CO_2, the added performance and compactness of brazed microchannel heat exchangers make these very attractive, especially in transport applications.

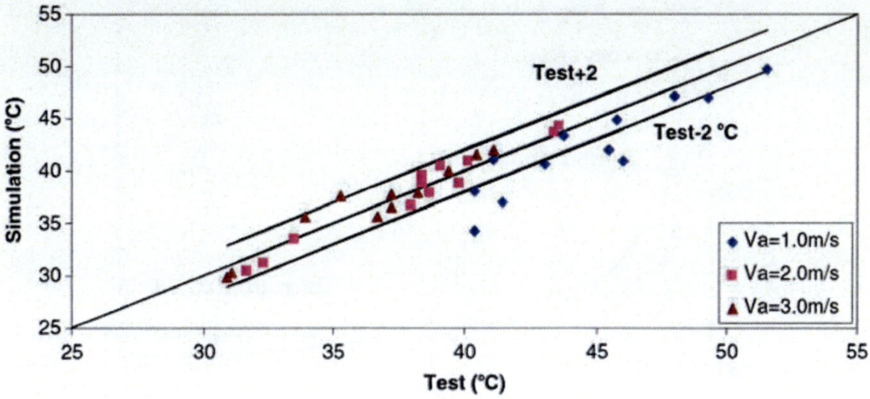

Fig. 5.15 Comparison of simulation with test results of all test conditions for refrigerant temperatures at gas cooler outlet

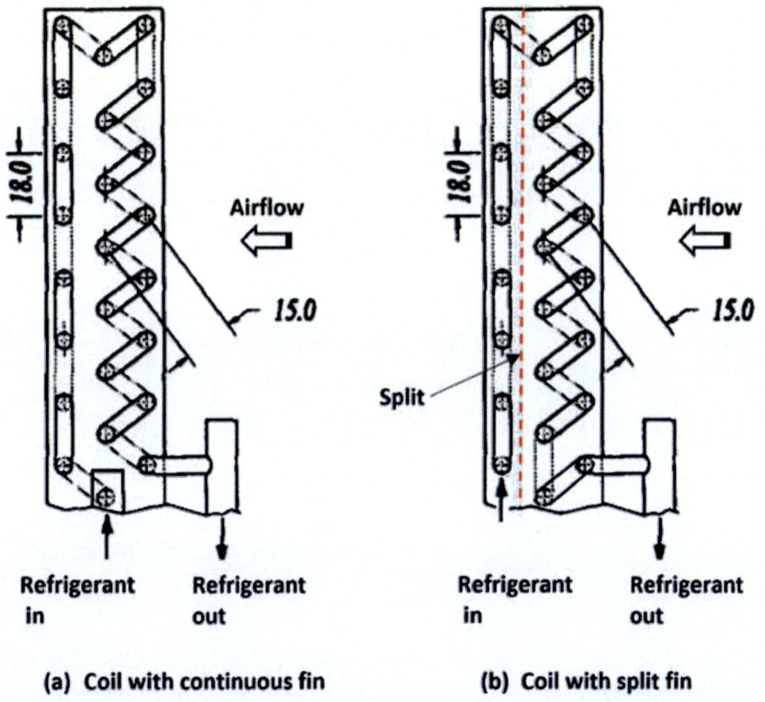

Fig. 5.16 Tube configuration and refrigerant flow in one (upper) of three equal circuits in a gas-cooler prototype (side view). Dimensions in mm

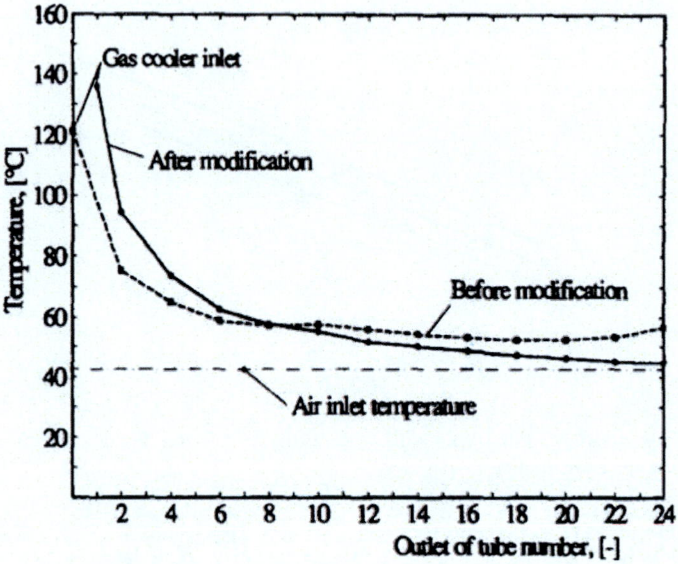

Fig. 5.17 Gas-cooler tube-wall temperatures from refrigerant inlet to outlet. The temperature profiles before modification (dashed) and after introducing a split in the fin material between the second and third tube row (full line) are shown. Data recorded at 43 °C air inlet temperature and 2.5 m s^{-1} air-face velocity

Table 5.3 Estimated heat transfer and pressure drop data for supercritical CO_2 flow at 100 bar and 45 °C (Pettersen et al. 2000)

	Diameter 2.0 mm	Diameter 0.8 mm
Mass flux (kgm-2 s-I)	1000	800
Reynolds number	54,400	17,400
Pressure drop gradient (barm-I)	0.13	0.26
Nusselt number	215	87
Heat transfer coefficient (Wm-2 K-1)	6900	7000

[a] Assuming a tube roughness of 0.0015 mm (drawn tube)
[b] Based on the Dinus-Boelter correlation for cooled flow ($Nu = 0.023Re^{0.8}Pr^{0.3}$)

To handle the high pressures associated with the CO_2 cycle, many CO_2 systems employ heat exchangers with flat multiport (microchannel) tubes as shown in Fig. 5.18. This technology, with its folded louvered fins, provides additional benefits as a by-product. Compared to conventional flat-fin/round-tube designs, microchannel heat exchangers increase the refrigerant-side area by about a factor of three and have far less air-side pressure drops due to the streamlined profile presented by the tubes. The flat tubes enable higher face velocities that increase the air-side heat transfer coefficient.

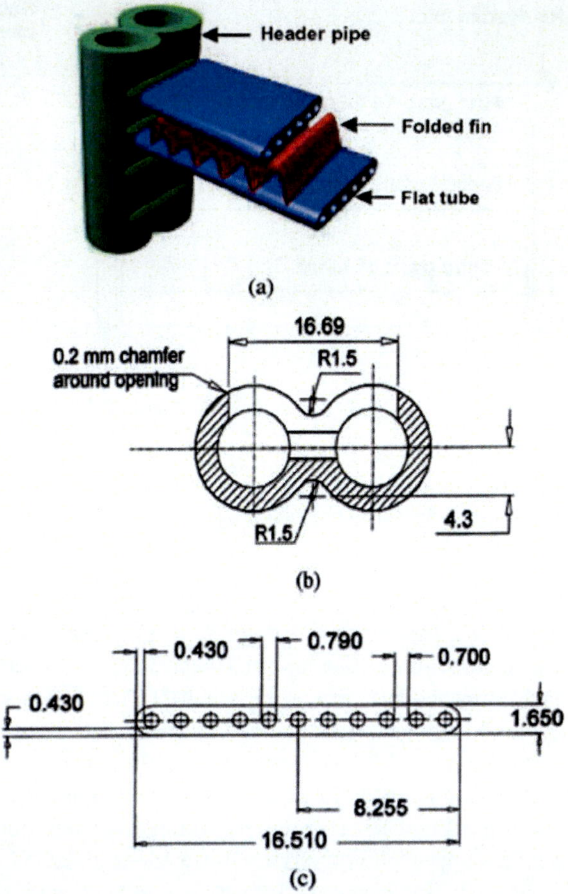

Fig. 5.18 A prototype microchannel CO_2 gas cooler for a car air-conditioning system (Pettersen et al. 1998). **a** Geometry of heat exchanger, **b** cross section of header pipe, **c** flat microchannel tube

One issue in compact gas cooler design is internal conduction due to large temperature differences across small lengths. As pointed out by Pettersen et al. (1998) internal conduction in fins, tubes, and manifolds may lead to performance reduction. Solutions to avoid these problems include splitting of fins, use of several heat exchanger sections, and careful design of manifold geometries.

As indicated in Fig. 5.3, a CO_2 transcritical cycle is so sensitive to refrigerant exit conditions that a counter-flow configuration is important for the gas cooler to exploit the large refrigerant-side temperature glide. Moreover, the steep refrigerant temperature glide allows for ideal cycle efficiency to be achieved at finite air flow rates, in contrast to the infinite air flow required to achieve ideal efficiency in the subcritical cycle.

Yin et al. (2001) validated a gas cooler simulation model using measured inlet data for a diverse set of 48 operating conditions, predicting refrigerant outlet temperature within ± 0.5 °C for most of the experimental data. They proposed a multi-slab

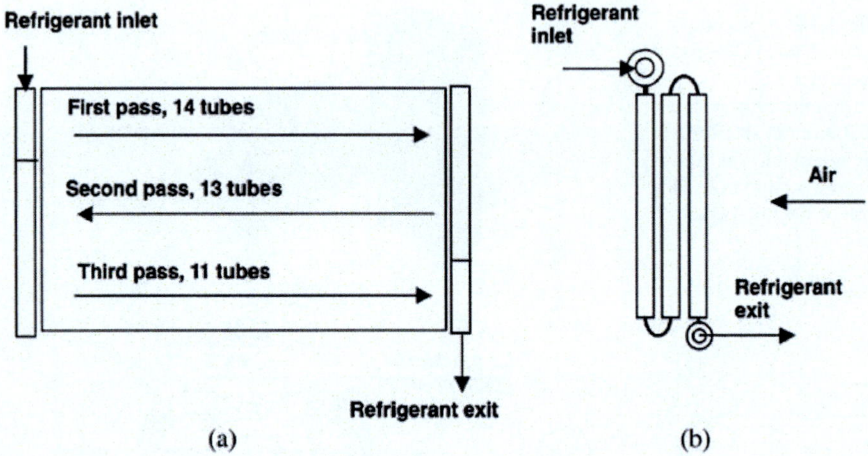

Fig. 5.19 Gas cooler design for a CO_2 air-conditioning system (Yin et al. 2001). **a** One-slab three-pass design, **b** three-slab one-pass design

gas cooler design (Fig. 5.19) and reported that the new design offered better performance than the commonly used multi-pass design (Fig. 5.19a). For the given heat exchanger volume, they reported that a newly designed cross-counter flow gas cooler could improve system capacity and COP by 3–4 and 5%, respectively, compared to the old design (Fig. 5.19a). The model was used to design the next-generation prototype gas cooler shown in Fig. 5.19b, where a multi-slab overall counter flow configuration concentrates the cool air stream on the exiting refrigerant, because the transcritical cycle is so sensitive to this exit condition. The new gas cooler design achieves approach temperature differences of <2 °C at most operating conditions, because air flowing over the first slab undergoes only a small temperature change, and this ΔT is what places an upper bound on the approach temperature difference. The flat tubes are vertical in this prototype to facilitate condensate drainage and defrosting in heating mode. Finally, the refrigerant flows in a single pass from the inlet to the outlet, with no intermediate headers, to accommodate reversibility and facilitate refrigerant distribution in heating mode. It is clear that flat tubes must be oriented vertically for any air-source heat pump, for reasons of defrost and condensate drainage, because both indoor and outdoor heat exchangers must function as evaporators as well as gas coolers.

References

Baskov VL, Kuraeva IV, Protopopov VS (1997) Heat transfer with the turbulent flow of a liquid at supercritical pressure in tubes under cooling conditions. Teplofiz Vysok Temp 15(1):96–102

Chang YS, Kim MS (2007) Modelling and performance simulation of a gas cooler for CO_2 heat pump system. HVAC&R Res 13(3):445–456

Domanski PA (1989) EVSIM-An evaporator simulation model accounting for refrigerant and one-dimensional air, NISTIR-89-4133. NIST, Washington, DC

Domanski PA, Yashar D (2007) Optimization of finned-tube condensers using an intelligent system. Int J Refrig 30:482–488

Fang X, Bullard C, Hrnjak P (2001) Heat transfer and pressure drop of gas coolers. ASHRAE Trans 107(1):255–266

Ge YT, Cropper RT (2009) Simulation and performance evaluation of finned-tube CO_2 gas coolers for refrigeration systems. Appl Therm Eng 29:957–965

Gnielinski V (1976) New equation for heat and mass transfer in turbulent pipe and channel flow. Int Chem Eng 16:359–368

Hwang Y, Jin DDH, Adermacher R, Hutchins JW (2005) Performance measurement of CO_2 heat exchangers. ASHRAE Trans 306–316

Incropera FP, DeWitt DP (1996) Introduction to heat transfer, 3rd edn. Wiley, New York

Inokuty H (1928) Graphical method of ®nding compression pressure of CO_2 refrigerating machine for maximum coeffcient of performance. In: Proceedings of the 5th international congress refrigeration, Rome, pp 185–192

Kim M-H, Pettersen J, Bullard CW (2004a) Fundamental process and system design issues in CO_2 vapor compression systems. Prog Energy Combust Sci 30:119–174

Kim MH, Pettersen J, Bullard CW (2004b) Fundamental process and system design issues in CO_2 vapour compression systems. Prog Energy Combust Sci 30:119–174

Krasnoshechekov EA, Kuraeva IV, Protopopov VS (1970) Local heat transfer of carbon dioxide at supercritical pressure under cooling conditions. Teplofiz Vys Temp 7(5):922–930

Lemmon EW, Huber ML, McLinden M (2007) NIST thermodynamic and transport properties of refrigerants and refrigerant mixtures- REFPROP, v8.0. NIST Standard Reference Database 23

Liao S, Zhao T (2002) Measurements of heat transfer coefficients from supercritical carbon dioxide flowing in horizontal mini/meso channels. J Heat Transf 124:413–420

Liao SM, Zhao TS, Jakobsen A (2000) A correlation of optimal heat rejection pressures in transcritical carbon dioxide cycles. Appl Therm Eng 20:831–841

Lorentzen G, Pettersen J (1992) New possibilities for non-CFC refrigeration. In: Proceedings of IIR International Symposium on Refrigeration, Energy and Environment, Trondheim, Norway, 22–24 June 1992, pp 147–163

Ortiz TM, Li D, Groll EA (2003) Evaluation of the performance potential of CO_2 as a refrigerant in air-to-air air conditioners and heat pumps:system modeling and analysis. Final report, ARTI-21CR/610-10030, December 2003

Pettersen J, Hafner A, Skaugen G, Rekstad H (1998) Development of compact heat exchangers for CO_2 air-conditioning systems. Int J Refrig 21(3):180–193

Pettersen J, Rieberer R, Leister A (2000) Heat transfer and pressure drop characteristics of super-critical carbon dioxide in microchannel tubes under cooling. In: Groll EA, Robinson DM (eds) The Fourth IIR-Gustav Lorentzen conference on natural working fluids, West Lafayette, IN. pp 99–106

Pettersen J, Skaugen G (1994) Operation of trans-critical CO_2 vapour compression circuits in vehicle air conditioning. In: Proceedings of the new applications of natural working fluids in refrigeration and air conditioning, Hannover, Germany, pp 495–505

Pitla SS, Groll EA, Ramadhyani S (2001b) Convective heat transfer from in-tube cooling of turbulent supercritical carbon dioxide: part 2—experimental data and numerical predictions. Int J HVAC&R Res 7(4):367–382

Pitla SS, Groll EA, Ramadhyani S (2002) New correlation to predict the heat transfer coefficient during in-tube cooling of turbulent supercritical CO_2. Int J Refrig 25:887–895

Pitla SS, Ramadhyani S, Groll EA (2001a) Convective heat transfer from in-tube flow of turbulent supercritical carbon dioxide: part 1—numerical analysis. Int J HVAC&R Res 7(4):345–366

Pitla SS, Robinson DM, Groll EA, Ramadhyani S (1998) Heat transfer from supercritical carbon dioxide in tube flow: a critical review. Int J HVAC&R Res 4(3):281–301

Wang CC, Jang JY, Chiou NF (1999) A heat transfer and friction correlation for wavy fin-and-tube heat exchangers. Int J Heat Mass Transf 42:1919–1924

Wang CC, Lee WS, Sheu WJ (2001) A comparative study of compact enhanced finand-tube heat exchangers. Int J Heat Mass Transf 44:3565–3573

Yin JM, Bullard CW, Hrnjak PS (2001) R-744 gas cooler model development and validation. Int J Refrig 24:692–701

Yoon SH, Kim JH, Hwang YW, Kim MS, Min K, Kim Y (2003) Heat transfer and pressure drop characteristics during the in-tube cooling process of carbon dioxide in the supercritical region. Int J Refrig 26:857–864

Chapter 6
CO_2 Evaporation Process Modeling and Evaporator Design

Lixin Cheng, Guodong Xia, and Qinling Li

Nomenclature

A	Cross-sectional area of flow channel, m^2
A_L	Cross-sectional area occupied by liquid-phase, m^2
A_{LD}	Dimensionless cross-sectional area occupied by liquid-phase
A_V	Cross-sectional area occupied by vapor phase, m^2
A_{VD}	Dimensionless cross-sectional area occupied by vapor phase
CO	Confinement number
D	Internal tube diameter, m
Fr_L	Liquid Froude number $[G^2/(\rho_L{}^2 g D_{eq})]$
$Fr_{V,Mori,}$	Vapor Froude number $[G^2/(\rho_v(\rho_L - \rho_v)gD_{eq})]$ defined by Mori et al. (Koyama et al. 2004)
f	Friction factor
G	Total vapor and liquid two-phase mass flux, kg/m^2s
g	Gravitational acceleration, 9.81 m/s^2
h	Heat transfer coefficient, W/m^2K
h_L	Vertical height of liquid, m
h_{LD}	Dimensionless vertical height of liquid
h_{LV}	Latent heat of vaporization, J/kg
k	Thermal conductivity, W/mK
L	Channel length, m
M	Molecular weight, kg/kmol
N	Channel number

L. Cheng · G. Xia (✉)
College of Environment and Energy Engineering, Beijing University of Technology, Beijing 100124, Chaoyang District, China
e-mail: xgd@bjut.edu.cn

L. Cheng · Q. Li
Department of Engineering and Mathematics, Sheffield Hallam University, Sheffield S11WB, UK

X.-R. Zhang and T. M. Eikevik (eds.), *CO_2 Refrigeration Cycle and Systems*, Lecture Notes in Energy 96, https://doi.org/10.1007/978-3-031-22512-3_6

P_i	Perimeter of interface, m
P_{iD}	Dimensionless perimeter of interface
P_L	Perimeter of tube wetted by liquid, m
P_{LD}	Dimensionless perimeter of tube wetted by liquid
Pr	Prandtl number $[c_p\mu/k]$
P_V	Perimeter of tube in contact with vapor, m
P_{VD}	Dimensionless perimeter of tube in contact with vapor
p	Pressure, bar
p_r	Reduced pressure $[p/p_{crit}]$
q	Heat flux, W/m^2
Re	Homogeneous Reynolds number $\{(GD_{eq}/\mu_V)[x + \rho_V/\rho_L(1\text{-}x)]\}$
Re_{LO}	Reynolds number considering the total vapor–liquid flow as liquid flow $[GD_{eq}/(\mu_L)]$
Re_M	Reynolds number $[GD_{eq}/(\mu_H)]$ defined in mist flow.
Re_V	Vapor phase Reynolds number $[GxD_{eq}/(\mu_V\varepsilon)]$
Re_δ	Liquid film Reynolds number $[4G(1\text{-}x)\delta/(\mu_L\ (1-\varepsilon))]$
S	Nucleate boiling suppression factor
T	Temperature, °C
t	Thickness of fin, m
u	Mean velocity, m/s
We_L	Liquid Weber number $[G^2D_{eq}/(\rho_L\sigma)]$ defined by Eq. (6.17); $[\rho_L u_L{}^2 D_{eq}/\sigma]$ defined by Eq. (6.51)
We_V	Vapor Weber number $[G^2D_{eq}/(\rho_v\sigma)]$ defined by Eq. (6.22)
x	Vapor quality
Y	Correction factor

Greek Symbols

Δp	Pressure drop, Pa
δ	Liquid film thickness, m
ε	Cross-sectional vapor void fraction
ε_{IA}	Vapor void fraction at $x = x_{IA}$
η	Fin efficiency
μ	Dynamic viscosity, Ns/m^2
θ_{dry}	Dry angle of tube perimeter, rad
$\theta^*{}_{dry}$	Dimensionless dry angle $[\theta_{dry}/(2\pi)]$
θ_{strat}	Stratified flow angle of tube perimeter, rad
$\theta^*{}_{strat}$	Dimensionless stratified flow angle $[\theta_{strat}/(2\pi)]$
θ_{wet}	Wet angle of the tube perimeter, rad
ρ	Density, kg/m^3
σ	Surface tension, N/m

Subscripts

A	Annular flow
B	Bubbly flow
B	Base
cb	Convective boiling
crit	Critical
de	Dryout completion
di	Dryout inception
dry	Dry
dryout	Dryout region
eq	Equivalent
f	Frictional
g	Gas
H	Homogeneous
h	Hydraulic
I	Intermittent flow
IA	Intermittent flow to annular flow transition
i	Liquid-vapor interface
iD	Interface in cross section
in	Tube inlet
L	Liquid
LD	Liquid in cross section of the tube
LO	Considering the total vapor–liquid flow as liquid flow
LV	Liquid-vapor
M	Mist flow
m	Momentum
nb	Nucleate boiling
pc	Pseudocritical
out	Tube outlet
Slug	Slug flow
SW	Stratified-wavy flow
sat	Saturation
si	Silicon
static	Static
strat	Stratified flow
strat($x \geq xIA$)	Stratified flow at $x \geq xIA$
strat($x < xIA$)	Stratified flow at $x < xIA$
total	Total
tp	Two-phase flow
V	Vapor
VD	Vapor in cross section of the tube
W	Wall
wavy	Wavy flow
wet	Wet perimeter

This chapter addresses fundamental issues related to CO_2 evaporation processes such as two-phase flow patterns, evaporation heat transfer and two-phase pressure drops in horizontal macro-scale and micro-scale tubes and evaporator simulation and design. The oil effects on the evaporation heat transfer and two-phase frictional pressure drops are presented as well. First, the thermal physical and transport properties of CO_2 are presented and compared to a traditional refrigerant R134a. Then, some existing experimental data of CO_2 are analyzed. Next, emphases are given to the Cheng et al. two-phase flow pattern map for CO_2 evaporation, flow pattern based evaporation heat transfer and pressure drop models specifically for CO_2. Furthermore, CO_2 evaporator simulations in the literature are reviewed. It shows that the Cheng et al. flow pattern map and models favorably predict the flow patterns, heat transfer and pressure drops in the evaporators in the thermal systems. The future research needs in CO_2 two-phase flow and evaporation heat transfer have been identified according to the analysis. In particular, new thermal and energy systems using CO_2 should be targeted. In the end, some practical designs of CO_2 evaporators are presented.

6.1 Introduction

As a natural working fluid, CO_2 (R744) has no ozone depletion potential (ODP = 0) and a negligible direct global warming potential (GWP = 1) when used as a refrigerant. CO_2 has been receiving renewed interest as an efficient and environmentally safe refrigerant in a number of applications, including automotive air-conditioning, residential heat pumps and as a primary and secondary refrigerant in refrigeration systems at low temperatures in recent years (Lorentzen 1994, 1995; Thome 1996; Thome and Ribatski 2005; Thome et al. 2008; Cheng et al. 2008a; Kim et al. 2000; Sawalha 2005). It is also an excellent coolant which has potential applications in various engineering applications such as cooling for electronic chips and other high heat flux removal (Cheng and Thome 2009; Cheng 2013, 2014; Cheng and Xia 2017). CO_2 has a low critical temperature ($T_{cr} = 31.1$ °C) and a high critical pressure ($p_{cr} = 7.38$ MPa), therefore, CO_2 is utilized at much higher operating pressures in air-conditioning and heat pump systems compared to other conventional refrigerants.

CO_2 has positive attributes as a secondary refrigerant at low temperatures in commercial refrigeration used in supermarkets, shops and large kitchens etc., in indirect and low temperature cascade systems and as a primary refrigerant in all CO_2 centralized refrigeration systems. Figure 6.1 shows a schematic of an ammonia-CO_2 secondary loop of an indirect refrigeration system which operates at low temperatures. In this refrigeration system, CO_2 evaporation and condensation processes are nearly at the same pressures. In particular, this system does not involve any oil effect for the CO_2 secondary loop. The main benefits of using CO_2 in indirect system arrangements include the simplicity of the system and the possibility of using components for other refrigerants to build the CO_2 circuit. In recent years, other arrangements, such as cascade and multistage systems have been used commercially (Kim et al. 2000; Sawalha 2005). Advantages of CO_2 cascade systems include greatly reduced low temperature compressor size, the absence of a liquid pump and fewer stages of heat transfer. Furthermore, two-stage and multistage CO_2 centralized refrigerant systems are also used in supermarket refrigeration. These systems are most suitable for cold climates or where heat sinks are available (Sawalha 2005). The operation of such refrigeration systems is mostly in the sub-critical region. If the ambient temperature is higher than the critical temperature, the system will be supercritical and this generally should be connected to another thermal system such as hot water or space heating heat pumps for efficient energy use. Due to its excellent safety characteristics (nonflammable, non-explosive, inexpensive and relatively nontoxic), CO_2 is an ideal refrigerant to be used in the refrigerated spaces. Due to the high working pressure, CO_2 as a phase change secondary refrigerant has a high volumetric refrigeration capacity, which equates to approximately five times or more than that of R-22 and NH_3. As the effects of good thermophysical properties, favorable evaporation heat transfer and two-phase flow characteristics of CO_2, smaller pipe dimensions can be used in its refrigeration systems. The use of CO_2 in these refrigeration systems requires the understanding and prediction of convective boiling and condensation heat transfer, two-phase flow patterns and pressure drops at low temperatures for achieving more accurate designs of evaporators and more energy-efficient cycles using CO_2.

CO_2 has favorable thermal performance when used in air-conditioning and heat pump systems as compared to conventional refrigerants. For example, in the CO_2 automobile air-conditioning system as shown in Fig. 6.2 for ambient air temperatures, the heat transfer process on the high pressure side of a CO_2 cycle is not a condensation process as in conventional systems but a supercritical gas cooling process (Sawalha 2005). Figure 6.3 shows a comparison of pressure-enthalpy diagrams of

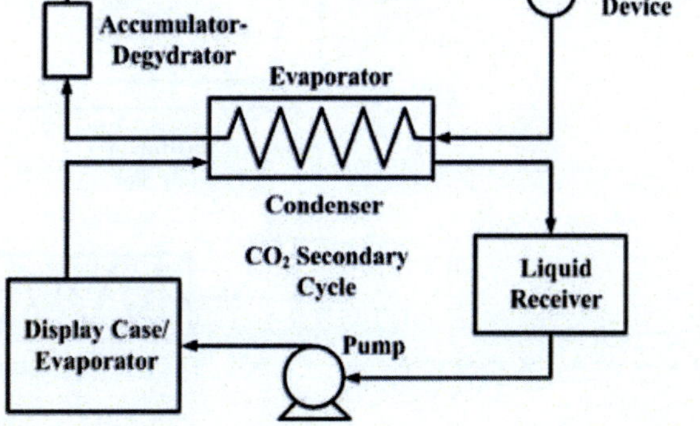

Fig. 6.1 Schematic of ammonia-CO_2 in secondary loop of an indirect refrigeration system

CO_2 and R134a in automobile air-conditioning systems. Hence, the conventional condenser is replaced by a gas cooler that rejects heat under variable temperatures at high pressure. To improve the performance of the CO_2 system, an internal heat exchanger is used to subcool the refrigerant between gas cooler and the expansion device and to superheat the refrigerant in the suction line of the compressor. For the evaporation processes in the evaporator, CO_2 evaporates at a much higher pressure than conventional refrigerant R134a. The physical and transport properties of CO_2 are quite different from those of conventional refrigerants at the same saturation temperatures. The physical properties have a significant effect on the evaporation processes including the two-phase flow patterns, evaporation heat transfer characteristics and two-phase frictional pressure drops in the compact heat exchangers with micro-channels. Therefore, it is essential to understand the two-phase flow and evaporation processes for properly designing the evaporators which use micro-scale channels for automobile air-conditioning and heat pump systems.

Both macro- and micro-scale channels are used in the CO_2 refrigeration, air-conditioning and heat pump systems. For example, in the automobile air-conditioning systems, micro-scale channels with diameters of 0.6 to 1 mm are generally used in evaporators, internal heat exchangers and gas coolers while macro- and micro-channels are used for CO_2 refrigeration and heat pump systems. Due to the significant differences of two-phase flow and evaporation heat transfer phenomena in micro-scale channels as compared to conventional size channels or macro-scale channels, emphasis has been put on the characteristics of two-phase flow and heat transfer in small and micro-scale flow passages due to the rapid development of micro-scale devices in recent years (Cheng and Thome 2009; Cheng 2013, 2014, 2016a, b; Cheng and Xia 2017; Kandlikar 2002; Cheng et al. 2007a; Thome 2004, 2006; Cheng and Mewes 2006; Kew and Cornwell 1997). Experimental studies and models on CO_2

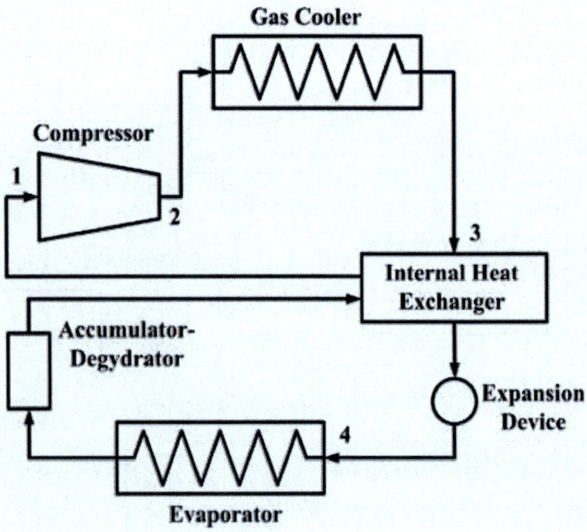

Fig. 6.2 Schematic of CO_2 automobile air-conditioning system

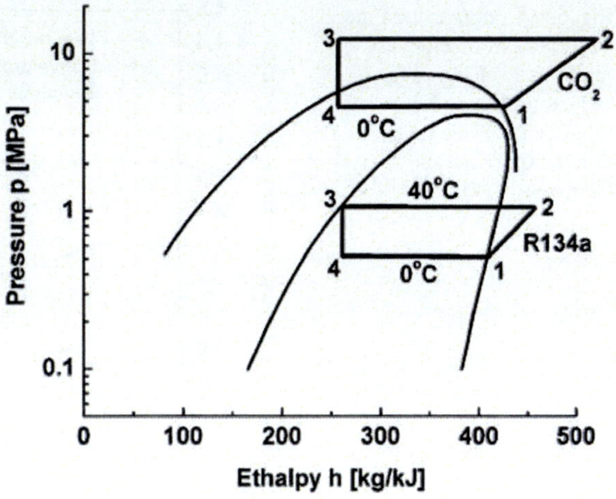

Fig. 6.3 Pressure-enthalpy diagrams of CO_2 and R134a in automobile air-conditioning systems

flow boiling and two-phase flow characteristics have been conducted in both macro- and micro- channels as well.

One very important issue should be the clarification of the distinction between micro-scale channels and macro-scale channels. However, a universal agreement is not clearly established in the literature. Instead, there are various definitions on this issue (Cheng and Mewes 2006; Cheng 2016a, b; Kew and Cornwell 1997; Mehendale et al. 2000; Cheng et al. 2008c). Here, just to show two examples, based on engineering practice and application areas such as refrigeration industry in the small tonnage units, compact evaporators employed in automotive, aerospace, air separation and cryogenic industries, cooling elements in the field of microelectronics and micro-electro-mechanical-systems (MEMS), Kandlikar (2002) defined the following ranges of hydraulic diameters D_h which are attributed to different classifications:

- Conventional channels: $D_h > 3$ mm.
- Minichannels: $D_h = 200$ μm–3 mm.
- Microchannels: $D_h = 10$–200 μm.

According to this definition, the distinction between small and conventional size channels is 3 mm and the distinction between mini and micro-channels is 200 μm.

Kew and Cornwell (1997) earlier proposed the Confinement number Co for the distinction of macro- and micro-scale channels, as

$$Co = \frac{1}{D_h}\sqrt{\frac{4\sigma}{g(\rho_L - \rho_G)}} \tag{6.1}$$

which is based on the definition of the Laplace constant (Cheng and Mewes 2006; Cheng et al. 2008c). Other different definitions are also proposed in the reviews (Cheng and Xia 2017; Cheng and Mewes 2006; Cheng et al. 2008c). Obviously, there is no agreement on the definition of a micro-scale channel so far. Figure 6.4 shows

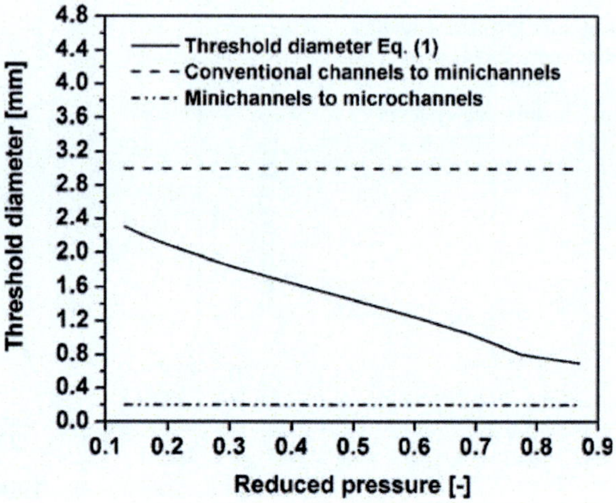

Fig. 6.4 Comparison of the definitions of macro- and micro-scale channels for CO_2 according to Kandlikar (2002) and Confinement number *Co* (Kew and Cornwell 1997)

the comparable definitions macro- and micro-scale channels for CO_2 according to Kandlikar (Kandlikar 2002) and the Confinement number *Co* (Eq. 6.1), which shows the big difference among these criteria. In this chapter, the distinction between macro- and micro-scale channels by the threshold diameter of 3 mm is adopted due to the lack of a well-established theory but is in line with those recommended by Kandlikar (2002) and also for the practical use in the CO_2 air-conditioning, heat pump and refrigeration systems.

From a predictive standpoint, many features of the existing flow pattern maps, evaporation heat transfer and two-phase pressure drop correlations require refinement to attain the desired level of accuracy for refrigerant heat exchangers (evaporators, internal heat exchangers, gas coolers and condensers) design as pointed out by Thome (1996). Therefore, this chapter addresses all the issues related to CO_2 two-phase flow, flow patterns and evaporation heat transfer without and with oil effect. Emphasis is given to the CO_2 two-phase flow patterns, evaporation heat transfer and pressure drop models for CO_2 evaporation inside horizontal tubes. The future research needs in the CO_2 two-phase flow and evaporation heat transfer are identified according to the analysis of the existing studies. Finally, design of CO_2 evaporators is discussed.

6.2 CO_2 Evaporation Heat Transfer and Two-Phase Flow Characteristics Inside Tubes

6.2.1 Thermal Physical and Transport Properties of CO_2

Thermal physical and transport properties of CO_2 have a significant effect on the flow patterns, two-phase flow and evaporation heat transfer characteristics and the corresponding prediction models in the evaporator tubes. CO_2 has higher liquid and vapor thermal conductivity, a lower vapor–liquid density ratio (lower liquid and higher vapor densities), a very low surface tension, and a lower liquid–vapor viscosity ratio (lower liquid and higher vapor viscosity) than conventional refrigerants. Figures 6.5, 6.6, 6.7, 6.8, 6.9, 6.10, 6.11, 6.12 and 6.13 shows the comparisons of the CO_2 and R134a physical properties obtained using REFPROP Version 7.0. As a result, the flow boiling heat transfer, two-phase flow pattern and pressure drop characteristics are quite different from those of conventional low pressure refrigerants. Previous experimental studies have shown that CO_2 has higher flow boiling heat transfer coefficients and lower pressure drops than those of conventional refrigerants at the same saturation temperatures. The available flow boiling heat transfer correlations developed for conventional low pressure refrigerants generally significantly under-predict the experimental data of CO_2. In addition, dryout may occur much earlier (at moderate vapor quality) in CO_2 flow boiling, particularly at high mass flux and high temperature conditions. Significant deviations for the flow patterns of CO_2 compared to the flow pattern maps that were developed for other fluids at lower pressures have been observed as well. Two-phase pressure drops of CO_2 are also much lower than conventional low pressure refrigerants. Furthermore, lubricant oil has a great effect on heat transfer and pressure drop, which should be clarified for both flow boiling, supercritical gas heat transfer and pressure drops. Therefore, it is very important to understand and to predict the flow patterns, pressure drop and heat transfer in flow boiling without and with the oil effect in evaporators and oil–gas two-phase flow in the gas coolers.

Physical properties have a significant effect on two-phase flow and heat transfer characteristics. The physical and transport properties of CO_2 are quite different from those of conventional refrigerants when compared at the same saturation temperature. Comparisons of the physical properties of CO_2 and R134a were obtained using REFPROP.NIST Ver 7.0 (2002) are shown in Figs. 6.5, 6.6, 6.7, 6.8, 6.9, 6.10, 6.11, 6.12 and 6.13. CO_2 has a much higher saturation pressure than R134a at the same

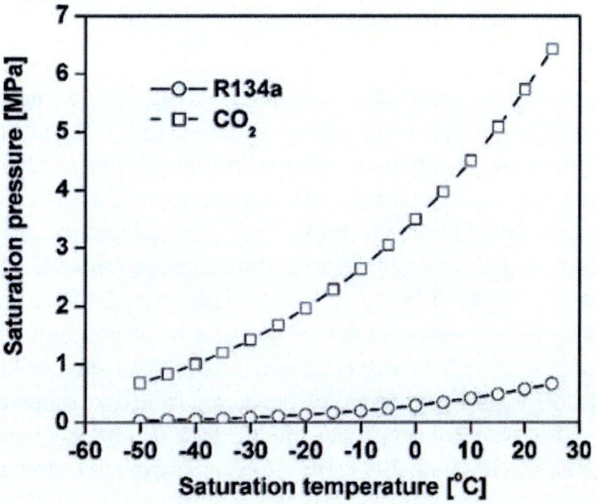

Fig. 6.5 Comparison of saturation pressures of CO_2 and R134a (REFPROP 2002)

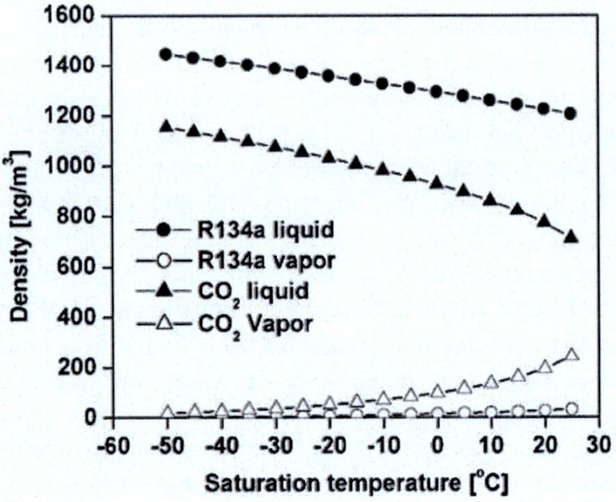

Fig. 6.6 Comparison of liquid and vapor densities of CO_2 and R134a (REFPROP 2002)

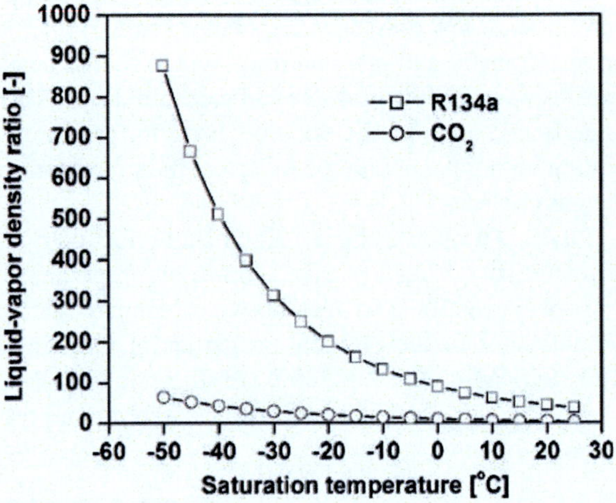

Fig. 6.7 Comparison of liquid–vapor density ratios of CO_2 and R134a (REFPROP 2002)

saturation temperature. Furthermore, CO_2 has a much lower vapor–liquid density ratio (lower liquid and higher vapor densities), higher liquid and vapor specific heats, a lower liquid–vapor viscosity ratio (lower liquid and higher vapor viscosity), a higher latent heat (only near the critical point, the CO_2 latent heats are lower than R134a), much higher liquid and vapor thermal conductivity and much lower surface tensions than R134a and other low pressure refrigerants. The different physical properties result in quite different evaporation heat transfer, two-phase flow pattern and pressure drop behaviors as compared to those of conventional low pressure refrigerants. In the next section, these behaviors and their mechanisms are described and explained according to the thermal physical and transport properties.

The physical properties of CO_2 may be obtained from several software packages (REFPROP 1998, 2002; EES 2005). However, it should be pointed out that there are

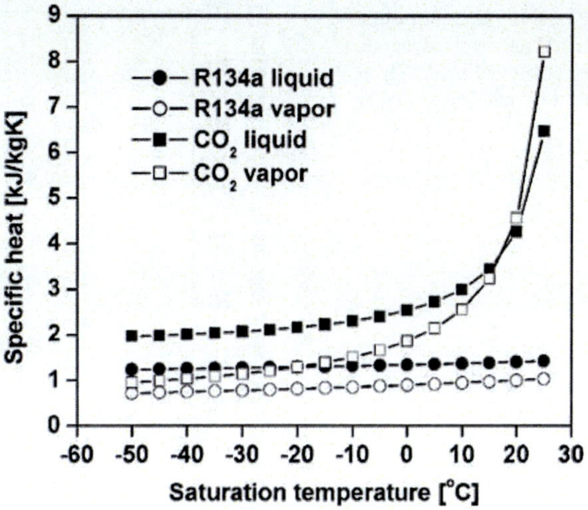

Fig. 6.8 Comparison of liquid and vapor specific heats of CO_2 and R134a (REFPROP 2002)

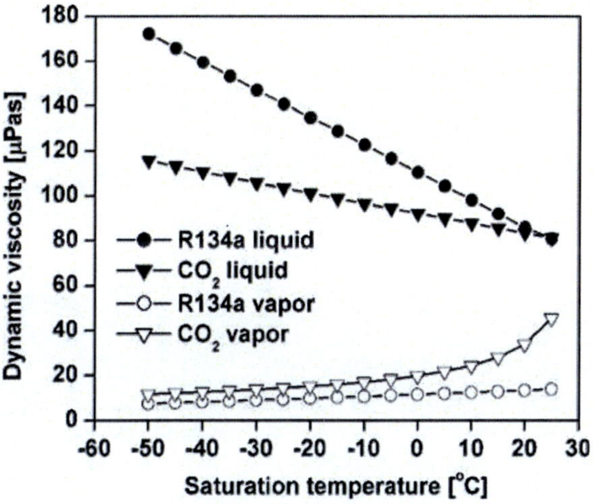

Fig. 6.9 Comparison of liquid and vapor dynamic viscosities of CO_2 and R134a (REFPROP 2002)

some differences among these software packages. Utilization of different software packages for physical properties has some effect on reducing experimental results and implementation of prediction methods. For example, the CO_2 flow pattern map, evaporation heat transfer and two-phase pressure drop models in this chapter were developed using the physical properties from REFPROP.NIST version 6.01 (1998). When using other software packages, the results differ but not significantly.

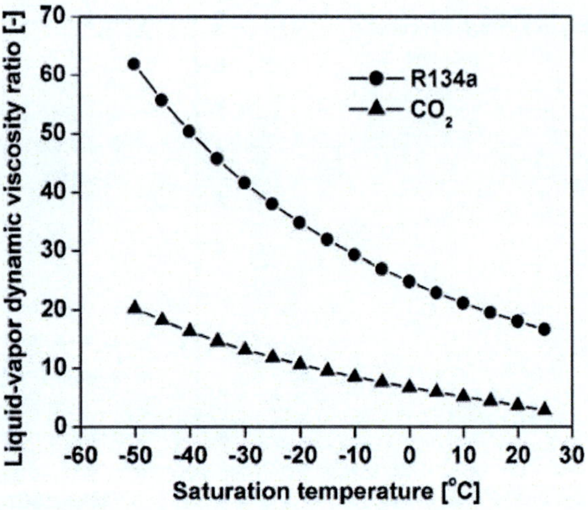

Fig. 6.10 Comparison of liquid and vapor dynamic viscosity ratios of CO_2 and R134a (REFPROP 2002)

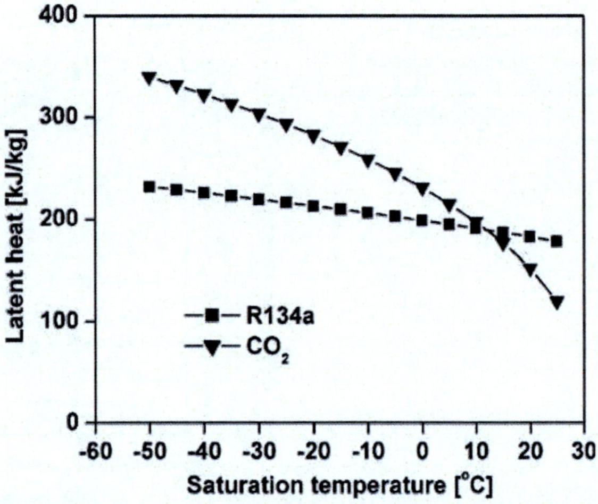

Fig. 6.11 Comparison of latent heats of CO_2 and R134a (REFPROP 2002)

6.2.2 *Analysis of Experimental Data of CO_2 Evaporation Inside Tubes*

Thome and Ribatski (2005) presented an overall review on two-phase flow and flow boiling of CO_2 in macro- and micro-channels in 2005. Since then, a number of experimental and modeling work has been conducted. In particular, Cheng et al. (2006a, b, 2008b, d) did a comprehensive literature review on the relevant topics, collected the experimental results from different studies and critical analyzed the results in developing new flow pattern map and models for evaporation heat transfer and frictional pressure drops of CO_2 inside horizontal tubes. Based on the database setup, Cheng et al. have developed a general flow pattern map covering all flow patterns,

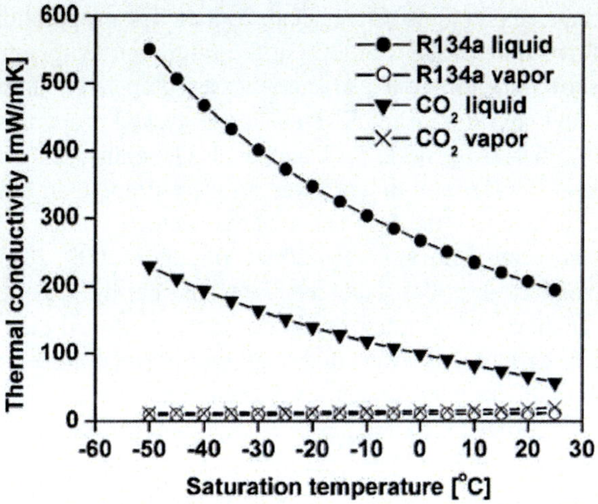

Fig. 6.12 Comparison of liquid and vapor thermal conductivities of CO_2 and R134a (REFPROP 2002)

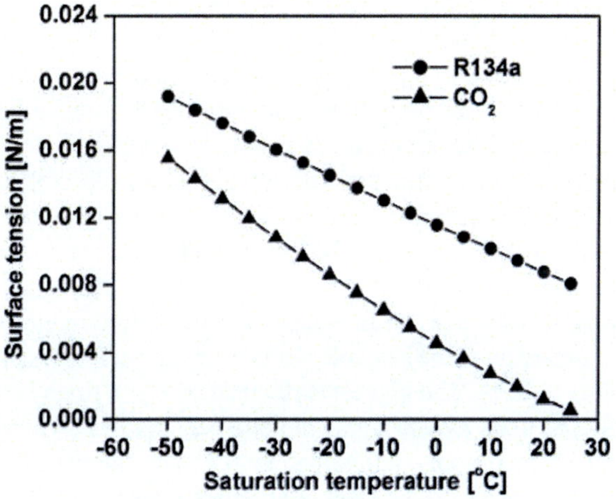

Fig. 6.13 Comparison of surface tensions of CO_2 and R134a (REFPROP 2002)

flow pattern based evaporation heat transfer and two-phase frictional pressure drop models for CO_2 evaporating inside horizontal models. Over the past 10 years, these prediction methods for flow patterns, heat transfer coefficients and pressure drops have been proved to be able to favorably capture the experimental data by various researchers. Here we do not do another comprehensive review but present a systematic knowledge on this topic. Behaviors of flow boiling heat transfer, two-phase flow patterns and two-phase pressure drops without oil effect are briefly summarized according to the available studies. The CO_2 flow map, flow pattern based flow boiling heat transfer model and phenomenological two-phase frictional pressure drop model are mainly presented in the following sections.

According to the available studies in the literature, quite different evaporation heat transfer and two-phase flow behaviors of CO_2 have been shown for high and low reduced pressures (Cheng et al. 2006a). The evaporation heat transfer and two-phase flow characteristics of CO_2 at the saturation temperatures ranging from 0 to 25 °C show quite different characteristics as compared to those of conventional refrigerants due to the significant differences in physical properties shown in Figs. 6.5, 6.6, 6.7, 6.8, 6.9, 6.10, 6.11, 6.12 and 6.13. Generally, CO_2 has much higher evaporation heat transfer and much lower pressure drops than other low pressure refrigerants. One feature is the dominance of the nucleate boiling at low/moderate vapor qualities prior to dryout (Pettersen 2004; Yun et al. 2003, 2005a; Hihara and Tanaka 2000; Yoon et al. 2004). Another feature is that the dryout in CO_2 flow boiling occurs much earlier (at relatively lower vapor qualities) than in conventional refrigerants. Furthermore, the effect of the saturation temperature on the evaporation heat transfer coefficients is noticeable. At higher saturation temperatures, the nucleate boiling is more pronounced and plays an important role at low vapor quality.

It should be pointed out here that the experimental data from the different independent studies show somewhat quite different evaporation heat transfer trends at similar test conditions. Just to show several examples here, Fig. 6.14 depicts two opposite evaporation heat transfer behaviors with the saturation temperature in the studies of Pettersen (2004) and Yoon et al. (2004). Heat transfer coefficients increase with increasing saturation temperature in the study of Pettersen while they decrease in the study of Yoon et al. The only big difference between the two studies is the diameter of the test channels as indicated in Fig. 6.14. Figure 6.15 shows the comparison of the experimental data of Yun et al. (2005b) for two diameters of 1.53 and 1.54 mm at the same test conditions. According to their results, the evaporation heat transfer coefficients can be higher up to 80% with a very little change of hydraulic diameter from 1.53 to 1.54 mm at the same test conditions. No explanation of why there is such a big difference even was offered in their paper. Figure 6.16 shows the comparison of the heat transfer coefficients of Pettersen (2004) with those of Koyama et al. (2001). The biggest difference between them is that in Koyama et al. the heat flux is 32.06 kW/m^2 while in Pettersen it is 10 kW/m^2. The heat transfer coefficients fall off at a vapor quality of about 0.7 in the study of Pettersen while the heat transfer coefficients increase even at qualities larger than 0.7 in the study of Koyama et al. It is difficult to explain why the heat transfer coefficients fall off at the lower heat flux in one study while they still increase at the higher heat flux in the other study. This could be an effect of the heating methods or because of multi-channel vs. single channel test setups. Figure 6.17 shows the heat transfer data of Hihara (2006) at a mass velocity of 360 kg/m^{2}s, a saturation temperature of 15 °C and a heat flux of 18 kW/m^2 with two different tube diameters, 4 and 6 mm. It shows that the heat transfer coefficients of the 4 mm tube are twice those of the 6 mm tube. In addition, the trends of the heat transfer coefficients are totally different. As both diameters are in the range of macro-scale, it is surprising that the diameter has such a big effect on the heat transfer values and trends. Hence, in summary, there is still not a clear view of why CO_2 data do not conform to conventional trends and also differ widely from one study to another (Fig. 6.18).

However, the available studies have shown different heat transfer behaviors at lower saturation temperatures from those at higher saturation temperatures. In fact,

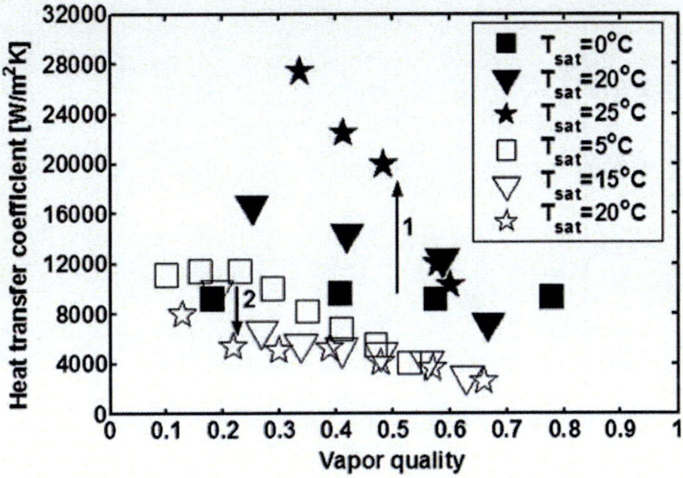

Fig. 6.14 The experimental heat transfer coefficients in two different studies showing two opposite trends with the increase of saturation temperature. Arrow 1 showing the trend of the experimental flow boiling heat transfer coefficients (solid symbols) of Pettersen (2004): D_h = 0.8 mm, G = 280 kg/m^{2}s and q = 10 W/m^2 at 0, 20 and 25 °C. Arrow 2 showing the trend of the experimental flow boiling heat transfer coefficients (hollow symbols) of Yoon et al. (2004): D_h = 7.53 mm, G = 318 kg/m^{2}s and q = 16.4 W/m^2 at 5, 15 and 20 °C

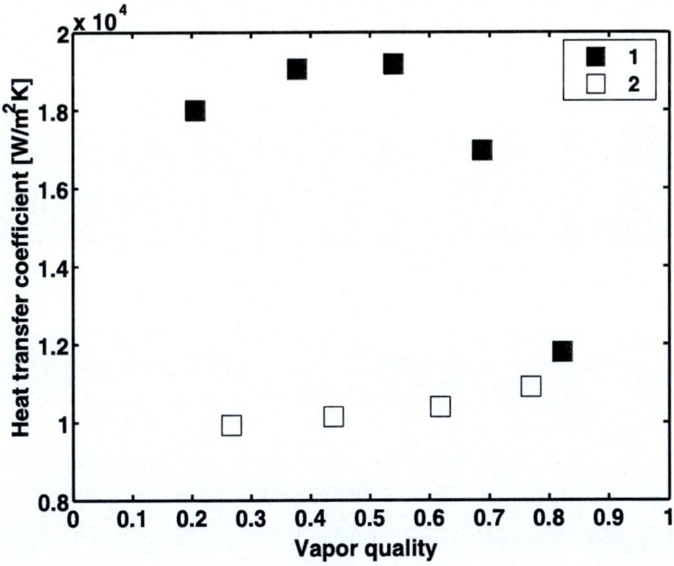

Fig. 6.15 The experimental flow boiling heat transfer coefficients in the same study showing different results with a very little change in hydraulic diameters from 1.53 to 1.54 mm. Solid symbols showing the experimental flow boiling heat transfer coefficients of Yun et al. (2005b): D_h = 1.53 mm, G = 300 kg/m^{2}s, T_{sat} = 5 °C and q = 20 W/m^2. Hollow symbols showing the experimental flow boiling heat transfer coefficients of Yun et al. (2005b): D_h = 1.54 mm, G = 300 kg/m^{2}s, T_{sat} = 5 °C and q = 20 W/m^2

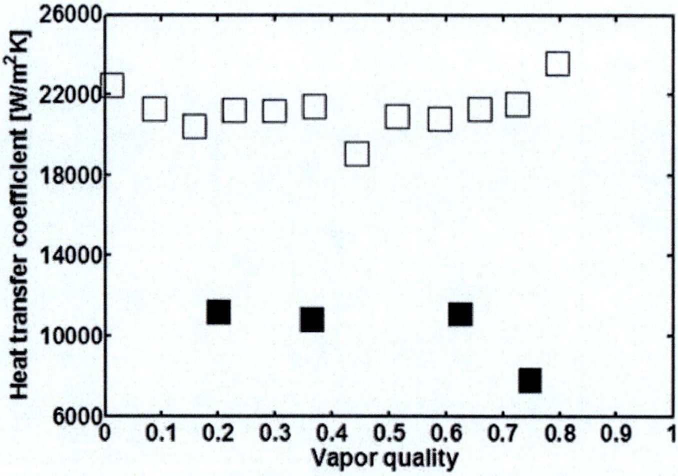

Fig. 6.16 The experimental flow boiling heat transfer coefficients in two different studies showing opposite flow boiling heat transfer coefficient trends. Solid symbols showing the experimental flow boiling heat transfer coefficients of Pettersen (2004): $D_h = 0.8$ mm, $G = 190$ kg/m^{2}s, $T_{sat} = 0$ °C and $q = 10$ W/m^2. Hollow symbols showing the experimental flow boiling heat transfer coefficients of Koyama et al. (2001): $D_h = 1.8$ mm, $G = 260$ kg/m^{2}s, $T_{sat} = 0.26$ °C and $q = 32.06$ W/m^2

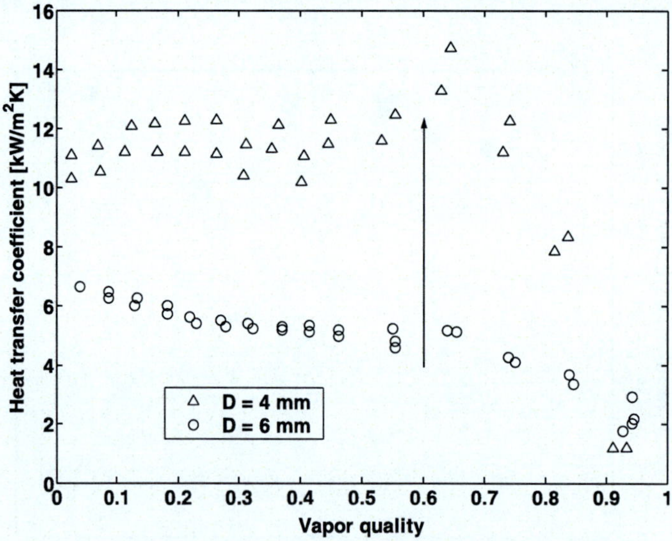

Fig. 6.17 Experimental heat transfer data of Hihara (2006) ($G = 360$ kg/m^{2}s, $T_{sat} = 15$ °C and $q = 18$ kW/m^2)

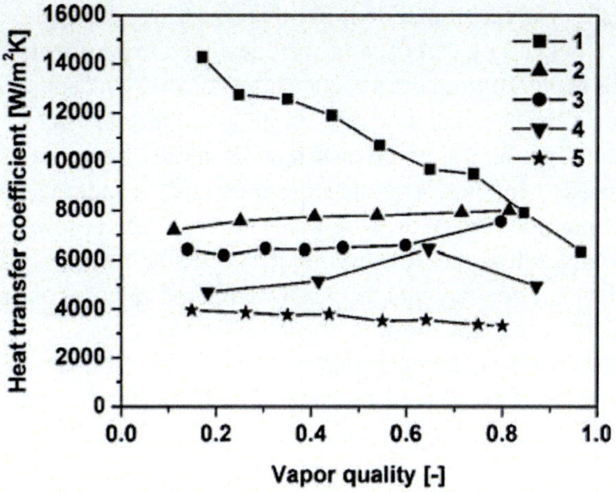

Fig. 6.18 Comparison of different experimental data of CO_2 at experimental conditions: (1) Bredesen et al. (1997): $T_{sat} = 5$ °C, D = 7 mm, G = 200 kg/m^{2}s, q = 6 kW/m^2; (2) Park and Hrnjak (2005): $T_{sat} = -30$ °C, D = 6.1 mm, G = 200 kg/m^{2}s, q = 15 kW/m^2; (3) Bredesen et al. (1997): $T_{sat} = -30$ °C, D = 7 mm, G = 200 kg/m^{2}s, q = 6 kW/m^2; (4) Zhao and Bansal (2007): $T_{sat} = -28.7$ °C, D = 4.57 mm, G = 196.8 kg/m^{2}s, q = 17 kW/m^2; and (5) Knudsen and Jensen (1997): $T_{sat} = -28$ °C, D = 10.08 mm, G = 80 kg/m^{2}s, q = 8 kW/m^2

at low evaporation temperatures down to −40 °C, the CO_2 reduced pressures (e.g. the reduced pressure $p_r = 0.136$ at −40 °C) are still much higher than those of conventional refrigerants such as R134a (e.g. the reduced pressure $p_r = 0.0126$ at −40 °C). The physical properties at the lower temperatures are much different from those of R134a as shown in Figs. 6.5, 6.6, 6.7, 6.8, 6.9, 6.10, 6.11, 6.12 and 6.13, showing a similar trend as those of CO_2 at higher temperatures as indicated.

It is difficult to explain the experimental results in some studies at low temperatures. So far, there are several studies of CO_2 at low temperatures in the literature but still very limited information is available. Bredensen et al. (1997) performed the boiling heat transfer experiments with CO_2 at temperatures of −10 and −25 °C. The experimental results show the heat transfer coefficient increases with vapor quality until dryout, which is opposite to the trend of their data at 0 °C. Knudsen and Jensen (1997) measured flow boiling heat transfer coefficients of CO_2 in a horizontal tube of diameter 10.06 mm at the saturation temperatures of −28 and −30 °C. Their boiling heat transfer coefficients are much lower than others' data. Zhao and Bansal (2007) presented experimental heat transfer data at −30 °C. Park and Hrnjak (2005) showed the heat transfer coefficients in a 6.1 mm inner diameter tube at −30 and −15 °C for various mass fluxes and heat fluxes. Figure 6.22 shows the comparison of the experimental heat transfer data in these studies. Quite big differences among these data are found. It is difficult to explain why there are such big differences although the test conditions are similar. Zhao and Bansal also found that the Liu and Winterton (1991) correlation predicted their data rather well while it does not predict other data. In fact, there are only a few data points in their study. Considering the big differences among the available data, it is recommended that more accurate experimental data at low temperatures are needed by careful and well designed experiments. Empirical heat transfer methods do not capture the parametric trends in dryout and mist flow regimes and cannot explain the physical mechanisms although they predict some data well in some cases. Therefore, an improved heat transfer model based on flow regimes for flow boiling is needed, but first accurate experimental data under wide test conditions are needed. Furthermore, no two-phase pressure drop data at low temperatures are available so far. Regarding the flow boiling heat transfer mechanisms, high reduced pressures and low surface tensions for CO_2 compared to conventional refrigerants have major effects on nucleate boiling heat transfer characteristics. Previous studies have suggested a clear dominance of nucleate boiling heat transfer even at very high mass flux. Therefore, CO_2 has much higher heat transfer coefficients than those of conventional refrigerants at the same saturation temperature and the available heat transfer correlations generally underpredict the experimental data of CO_2. In addition, previous experimental studies have demonstrated that dryout trends occur earlier at moderate vapor qualities in CO_2, particularly at high mass flux and high temperature conditions. However, it is difficult to explain the available boiling data at low temperatures according to these mechanisms although nearly all these studies pointed to nucleate boiling dominant mechanism with respect to their data. From the physical properties at low temperatures, it seems that these heat transfer behaviors should be similar to those at high saturation temperatures but they are not indeed. Thus, understanding the two-phase flow and heat transfer characteristics of CO_2 at low temperatures is essential.

Furthermore, for flow boiling in enhanced tubes, Koyama et al. (2004a) conducted experiments on flow boiling in a smooth copper tube and in a micro-fin copper tube at 5.3 °C. From their results, the heat transfer coefficients are only slightly higher than in the micro-fin tubes with a slight pressure drop increase as well. In this case, micro-fin tubes are not appropriate for CO_2. Cho and Kim (2007) conducted experimental studies of CO_2 for micro-channels and their data show that the average evaporation heat transfer coefficients are 150 to 210% higher than those of smooth tubes. The increase of pressure drop was much lower than the heat transfer increase. So far, only limited studies of CO_2 flow boiling in micro-fin tubes are available. Whether they significantly enhance CO_2 flow boiling heat transfer or not is still unclear due to the lack of such information. Furthermore, (Koyama et al. 2005; Siegismund and Kauffeld 2004) conducted experimental studies of flow boiling of CO_2-oil mixture in micro-fin tubes. Similar conclusions to those in smooth tubes were obtained.

According to the overall review of evaporation heat transfer and two-phase flow of CO_2 in the literature conducted by Thome and Ribatski (2005), none of the available prediction methods was able to predict the experimental data of CO_2 well. Therefore, they suggested that a new evaporation heat transfer prediction method should be developed and the evaporation heat transfer model should include the CO_2 effects on the annular to dryout and dryout to mist flow transitions in order to more accurately predict heat transfer coefficients at moderate/high vapor qualities. In response, Cheng et al. (2006a, b) proposed a new flow pattern map and a new evaporation heat transfer model based on the flow patterns for CO_2 evaporating inside horizontal tubes. The flow map and flow pattern based mechanistic heat transfer model were developed by modifying the methods of Wojtan et al. (2005a, b), which is an updated version of the Kattan–Thome–Favrat (1998a, b, c) flow pattern map and evaporation heat transfer model which were developed for five conventional refrigerants. Cheng et al. related the flow patterns to the corresponding evaporation heat transfer mechanisms for CO_2, thus, different from the numerous empirical models, such as the correlations of Chen (1966), Shah (1982), Gungor and Winterton (1986), Kandlikar (1990) and Liu and Winterton (1991), etc., which do not include flow pattern information. In fact, some of these correlations predicted the data well to some extent but fail to capture the parametric trends, or ignore the dryout and mist flow regimes which are typical working conditions for CO_2 evaporating in horizontal channels. The Cheng et al. CO_2 evaporation heat transfer model is applicable to a wide range of conditions: tube diameters (equivalent diameters defined by Eq. (6.2) is used for non-circular channels) from 0.8 to 10 mm, mass fluxes from 80 to 570 kg/m^{2}s, heat fluxes from 5 to 32 kW/m^2, saturation temperatures from –28 to 25 °C (the corresponding reduced pressures are from 0.21 to 0.87). The model reasonably predicts the database and it covers channel diameters found in most CO_2 evaporation applications. However, their model is limited by its parameter ranges from being applicable to some important applications, for example, the mass velocity ranges from 50 to 1500 kg/m^{2}s in CO_2 automobile air-conditioning systems and other thermal systems. In addition, the heat fluxes in some applications go beyond the maximum value in the Cheng et al. evaporation heat transfer model. Furthermore, the model does not extrapolate well to these conditions. In addition, the heat transfer model does not include heat transfer methods for CO_2 in mist flow and bubbly flow regimes due to the lack of the experimental data in these regimes, which were not available at that time. Therefore, it is necessary to update the heat transfer model for CO_2 to cover a wider range of conditions and these flow regimes and an updated version of the Cheng et al. evaporation model was developed (Cheng et al. 2008d). Furthermore, a flow pattern based two-phase pressure drop model was also needed for CO_2 developed by Cheng et al. (2008b).

6.3 A General Gas–Liquid Two-Phase Flow Pattern Map for CO_2 Evaporating Inside Tubes

Flow patterns are very important in understanding the very complex two-phase flow phenomena and heat transfer trends in evaporation heat transfer (Cheng and Xia 2017; Cheng 2016a, b; Cheng et al. 2008c). To predict the local flow patterns in a channel, a flow pattern map is used. In fact, successful flow pattern based evaporation heat transfer and two-phase frictional pressure drop models (Wojtan et al. 2005a, b; Kattan et al. 1998a, b, c; Ould-Didi et al. 2002; Moreno Quibén and Thome 2007a, b; Moreno Quibén et al. 2009a, b) have been proposed in recent years. Over the past decades, many flow pattern maps have been developed to predict two-phase flow patterns in horizontal tubes, such as those by Baker (1954), Taitel and Dukler (1976), Hashizume (1983), Steiner (1993) and so on, just to name a few. Most were developed for adiabatic conditions and then extrapolated by users to diabatic conditions, thereby creating big discrepancies. For this reason, a number of diabatic flow pattern maps related to the corresponding heat transfer mechanisms have been developed (Wojtan et al. 2005a, b; Kattan et al. 1998a, b, c). However, none of these is applicable to CO_2 evaporation in horizontal tubes because the two-phase flow characteristics of CO_2 evaporation are greatly affected by the very high reduced pressures and low surface tensions of CO_2. In addition, the very low viscosity of CO_2 at high reduced pressures may affect the two-phase pressure drop greatly.

Cheng et al. (2008b, d) proposed a new flow pattern map and a new general evaporation heat transfer model, flow pattern map and two-phase frictional pressure drop model for CO_2 in macro- and micro-scale channels to meet the wide range of parameters used in practical applications. The details of the flow pattern map are presented in this section and the evaporation and two-phase pressure drop models are respectively presented in the following sessions. The physical properties of CO_2 have been obtained from REFPROP version 6.01 of NIST (REFPROP 1998). The flow pattern map is intrinsically related to the evaporation heat transfer model in Sect. 6.5 and the two-phase frictional pressure drop model in Sect. 6.6.

For non-circular channels, equivalent diameters rather than hydraulic diameters were used in the flow pattern map (Cheng et al. 2006a, b, 2008b, d; Moreno Quibén et al. 2009a, b) as

$$D_{eq} = \sqrt{\frac{4A}{\pi}} \tag{6.2}$$

Using the equivalent diameter gives the same mass velocity as in the non-circular channel and thus correctly reflects the mean liquid and vapor velocities, something using hydraulic diameter in a two-phase flow does not. In the updated CO_2 flow pattern map, several new features were developed as compared to the Cheng et al. flow pattern map (Cheng et al. 2006a, b):

(1) Combining with the updated flow boiling heat transfer model for CO_2 in Sect. 2.3.2, the annular flow to dryout region (A–D) transition boundary was further modified so as to better fit the sharp changes in flow boiling heat transfer characteristics for higher mass velocities;
(2) Based on experimental heat transfer data, a new criterion for the dryout region to mist flow (D–M) transition was proposed;
(3) Bubbly flow occurs at very high mass velocities and very low vapor qualities and a bubbly flow pattern boundary were integrated into the map to make it more complete.

With these modifications, the updated flow pattern map for CO_2 is now applicable to much higher mass velocities. Complete flow pattern transition criteria of the updated flow pattern map for CO_2 are described below.

As shown in Fig. 6.19, the six dimensionless geometrical parameters used in the flow pattern map are defined as:

$$h_{LD} = \frac{h_L}{D_{eq}} \tag{6.3}$$

$$P_{LD} = \frac{P_L}{D_{eq}} \tag{6.4}$$

$$P_{VD} = \frac{P_V}{D_{eq}} \tag{6.5}$$

$$P_{iD} = \frac{P_i}{D_{eq}} \tag{6.6}$$

$$A_{LD} = \frac{A_L}{D_{eq}^2} \tag{6.7}$$

$$A_{VD} = \frac{A_V}{D_{eq}^2} \tag{6.8}$$

where D_{eq} is the internal tube equivalent diameter (for non-circular channels, equivalent diameter D_{eq} is used. So for circular channels, equivalent diameter D_{eq} equals hydraulic diameter D_h), P_L is the wetted perimeter, P_V is the dry perimeter in contact with vapor, A_L and A_V are the corresponding cross-sectional areas of the liquid and vapor phases, P_i is the length of the phase interface and h_L is the height of the liquid-phase from the bottom of the tube.

As a practical option and for consistency between the flow pattern map and the flow boiling heat transfer model, an easier to implement version of the flow map was proposed by Thome and El Hajal (2002). The void fraction ε which is determined with the Rouhani–Axelsson drift flux model (1970) by Thome and El Hajal is kept the same in the present new flow map for CO_2 as:

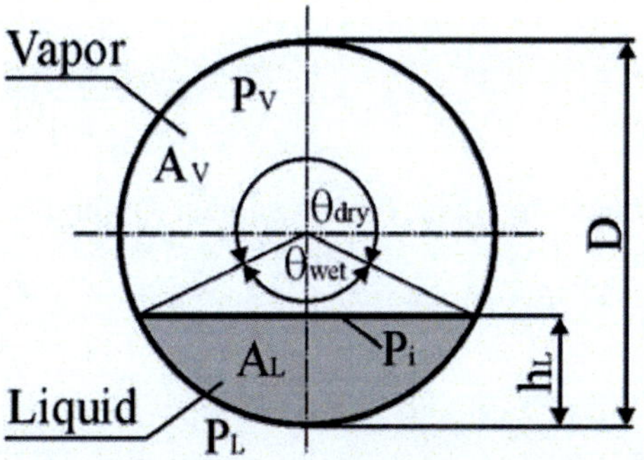

Fig. 6.19 Schematic diagram of stratified two-phase flow in a horizontal channel

$$\varepsilon = \frac{x}{\rho_V}\left[(1+0.12(1-x))\left(\frac{x}{\rho_V}+\frac{1-x}{\rho_L}\right)+\frac{1.18(1-x)[g\sigma(\rho_L-\rho_V)]^{1/4}}{G\rho_L^{1/2}}\right]^{-1} \quad (6.9)$$

Then, the dimensionless parameters are determined as follows:

$$A_{LD} = \frac{A(1-\varepsilon)}{D_{eq}^2} \quad (6.10)$$

$$A_{VD} = \frac{A\varepsilon}{D_{eq}^2} \quad (6.11)$$

$$h_{LD} = 0.5\left(1-\cos\left(\frac{2\pi-\theta_{strat}}{2}\right)\right) \quad (6.12)$$

$$P_{iD} = \sin\left(\frac{2\pi-\theta_{strat}}{2}\right) \quad (6.13)$$

where the stratified angle θ_{strat} (which is the same as θ_{dry} shown in Fig. 6.19) is calculated with the equation proposed by Biberg (1999):

$$\theta_{strat} = 2\pi - 2\left\{\begin{array}{l}\pi(1-\varepsilon)+\left(\frac{3\pi}{2}\right)^{1/3}\left[1-2(1-\varepsilon)+(1-\varepsilon)^{1/3}-\varepsilon^{1/3}\right]\\ -\frac{1}{200}(1-\varepsilon)\varepsilon[1-2(1-\varepsilon)]\left[1+4(1-\varepsilon)^2+\varepsilon^2\right]\end{array}\right\} \quad (6.14)$$

Taking into account the modifications in the annular flow to dryout (A–D), dryout to mist flow (D–M) and intermittent flow to bubbly flow (I-B) transition curves which were newly developed in this study, the implementation procedure of the updated flow pattern map for CO_2 is as follows:

The void fraction ε and dimensionless geometrical parameters A_{LD}, A_{VD}, h_{LD} and P_{iD} are calculated with Eqs. (6.9) to (6.13). The stratified-wavy to intermittent and annular flow (SW-I/A) transition boundary is calculated with the Kattan–Thome–Favrat criterion (Kattan et al. 1998a, b, c):

$$G_{wavy} = \left\{\frac{16A_{VD}^3 g D_{eq}\rho_L\rho_V}{x^2\pi^2\left[1-(2h_{LD}-1)^2\right]^{\frac{1}{2}}}\left[\frac{\pi^2}{25h_{LD}^2}\left(\frac{Fr_L}{We_L}\right)+1\right]\right\}^{\frac{1}{2}} + 50 \quad (6.15)$$

where the liquid Froude number Fr_L and the liquid Weber number We_L are defined as

$$Fr_L = \frac{G^2}{\rho_L^2 g D_{eq}} \quad (6.16)$$

$$We_L = \frac{G^2 D_{eq}}{\rho_L \sigma} \quad (6.17)$$

Then, the stratified-wavy flow region is subdivided into three zones according to the criteria by Wojtan et al. (2005a, b):

- $G > G_{wavy}(x_{IA})$ gives the slug zone;
- $G_{strat} < G < G_{wavy}(x_{IA})$ and $x < x_{IA}$ give the slug/stratified-wavy zone;
- $x \geq x_{IA}$ gives the stratified-wavy zone.

The stratified to stratified-wavy flow (S-SW) transition boundary is calculated with the Kattan–Thome–Favrat criterion (Kattan et al. 1998a, b, c):

$$G_{strat} = \left[\frac{226.3^2 A_{LD} A_{VD}^2 \rho_V (\rho_L - \rho_V) \mu_L g}{x^2 (1-x) \pi^3}\right]^{\frac{1}{3}} \tag{6.18}$$

For the new flow pattern map: $G_{strat} = G_{strat}(x_{IA})$ at $x < x_{IA}$.

The intermittent to annular flow (I–A) transition boundary is calculated with the Cheng et al. criterion (2006a, b):

$$x_{IA} = \left[1.8^{\frac{1}{0.875}} \left(\frac{\rho_V}{\rho_L}\right)^{-\frac{1}{1.75}} \left(\frac{\mu_L}{\mu_V}\right)^{-\frac{1}{7}} + 1\right]^{-1} \tag{6.19}$$

Then, the transition boundary is extended down to its intersection with G_{strat}.

The annular flow to dryout region (A-D) transition boundary is calculated with the new modified criterion of Wojtan et al. (2005a) based on the dryout data of CO_2 in this study:

$$G_{dryout} = \left\{ \begin{array}{c} \frac{1}{0.236}\left[\ln\left(\frac{0.58}{x}\right) + 0.52\right]\left(\frac{D_{eq}}{\rho_V \sigma}\right)^{-0.17}\left[\frac{1}{g D_{eq} \rho_V (\rho_L - \rho_V)}\right]^{-0.17} \\ \times \left(\frac{\rho_V}{\rho_L}\right)^{-0.25}\left(\frac{q}{q_{crit}}\right)^{-0.27} \end{array} \right\}^{1.471} \tag{6.20}$$

which is extracted from the new dryout inception equation in this study:

$$x_{di} = 0.58 e^{\left[0.52 - 0.236 We_V^{0.17} Fr_{V,Mori}^{0.17} \left(\frac{\rho_V}{\rho_L}\right)^{0.25} \left(\frac{q}{q_{crit}}\right)^{0.27}\right]} \tag{6.21}$$

This equation remains the same as in the Wojtan et al. (2005a) flow map for low pressure refrigerants, except that new empirical parameters were obtained based on the CO_2 data since the previous expression did not extrapolate well to reduced pressures far higher than its underlying database. The vapor Weber number We_V and the vapor Froude number $Fr_{V,Mori}$ defined by Mori et al. (2000) are calculated as.

$$We_V = \frac{G^2 D_{eq}}{\rho_V \sigma} \tag{6.22}$$

$$Fr_{V,Mori} = \frac{G^2}{\rho_V (\rho_L - \rho_V) g D_{eq}} \tag{6.23}$$

and the critical heat flux q_{crit} is calculated with the Kutateladze (1948) correlation as

$$q_{crit} = 0.131\rho_V^{0.5}h_{LV}[g\sigma(\rho_L - \rho_V)]^{0.25} \tag{6.24}$$

The dryout region to mist flow (D-M) transition boundary is calculated with the new criterion developed in this study based on the dryout completion data for CO_2:

$$G_M = \left\{ \begin{array}{c} \frac{1}{0.502}\left[\ln\left(\frac{0.61}{x}\right) + 0.57\right]\left(\frac{D_{eq}}{\rho_V\sigma}\right)^{-0.16}\left[\frac{1}{gD_{eq}\rho_V(\rho_L-\rho_V)}\right]^{-0.15} \\ \times\left(\frac{\rho_V}{\rho_L}\right)^{0.09}\left(\frac{q}{q_{crit}}\right)^{-0.72} \end{array} \right\}^{1.613} \tag{6.25}$$

which is extracted from the dryout completion (which means the wall remains completely dry) equation developed in this study by solving for G_M from:

$$x_{de} = 0.61e^{\left[0.57-0.502We_V^{0.16}Fr_{V,Mori}^{0.15}\left(\frac{\rho_V}{\rho_L}\right)^{-0.09}\left(\frac{q}{q_{crit}}\right)^{0.72}\right]} \tag{6.26}$$

Again, this equation and its dimensionless groups remain the same as those used in the previous method (Wojtan et al. 2005a) for conventional low reduced pressure refrigerants and only some empirical values were changed when correlating it to the CO_2 data. The vapor Weber number We_V and the vapor Froude number $Fr_{V,Mori}$ are calculated with Eqs. (6.22) and (6.23).

The intermittent to bubbly flow (I–B) transition boundary is calculated with the criterion which arises at very high mass velocities and low qualities (Kattan et al. 1998a, b, c):

$$G_B = \left[\frac{256A_{VD}A_{LD}^2D_{eq}^{1.25}\rho_L(\rho_L-\rho_V)g}{0.3164(1-x)^{1.75}\pi^2P_{iD}\mu_L^{0.25}}\right]^{1/1.75} \tag{6.27}$$

If $G > G_B$ and $x < x_{IA}$, then the flow is bubbly flow (B).

The following conditions are applied to the transitions in the high vapor quality range:

- If $G_{strat}(x) \geq G_{dryout}(x)$, then $G_{dryout}(x) = G_{strat}(x)$
- If $G_{wavy}(x) \geq G_{dryout}(x)$, then $G_{dryout}(x) = G_{wavy}(x)$
- If $G_{dryout}(x) \geq G_M(x)$, then $G_{dryout}(x) = G_M(x)$

Gashe (2006) recently conducted an experimental study of CO_2 evaporation inside a 0.8 mm diameter rectangular channel for various mass velocities and observed flow patterns by flow visualization as well. The updated CO_2 flow pattern map was compared to his observations. It should be mentioned here that different names for the same flow patterns are used by different authors. Gasche in particular used the definition of plug flow, which is an intermittent flow in our flow pattern map. Just to show one example, Fig. 6.20 shows the observed flow patterns of CO_2 by Gashe for $D_{eq} = 0.833$ mm (equivalent diameter is used here for the rectangular channel). Figure 6.21 shows the observations in Fig. 6.20 compared to the updated flow pattern map (in the flow pattern map, A is annular flow, D is dryout region, I is intermittent flow, M is mist flow, S is stratified flow and SW is stratified-wavy flow.

The stratified to stratified-wavy flow transition is designated as S-SW, the stratified-wavy to intermittent/annular flow transition is designated as SW-I/A, the intermittent to annular flow transition is designated as I-A and so on.). It should be mentioned that the observed slug/annular flow of Gashe is counted as an annular flow in the updated flow pattern map. From the photographs in Fig. 6.20, it seems that the annular flow is the predominant flow in the slug/annular flow defined by Gashe. The observations (3) and (4) are near their correct regimes, especially by the typical flow pattern map standards.

Statistically, 82% of the total 28 flow pattern data of Gashe (2006) are identified correctly by the updated flow map, or more specifically, 75% of the intermittent flows and 88% of the annular (slug/annular flow) flows (Gasche 2006). The updated CO_2 flow pattern map thus predicts the flow patterns observed by Gasche rather well. The lack of other new data in the literature should justify future experimental studies to obtain more. Furthermore, it is commonly understood that flow pattern transitions do not occur abruptly but over a range of conditions to complete the transition from

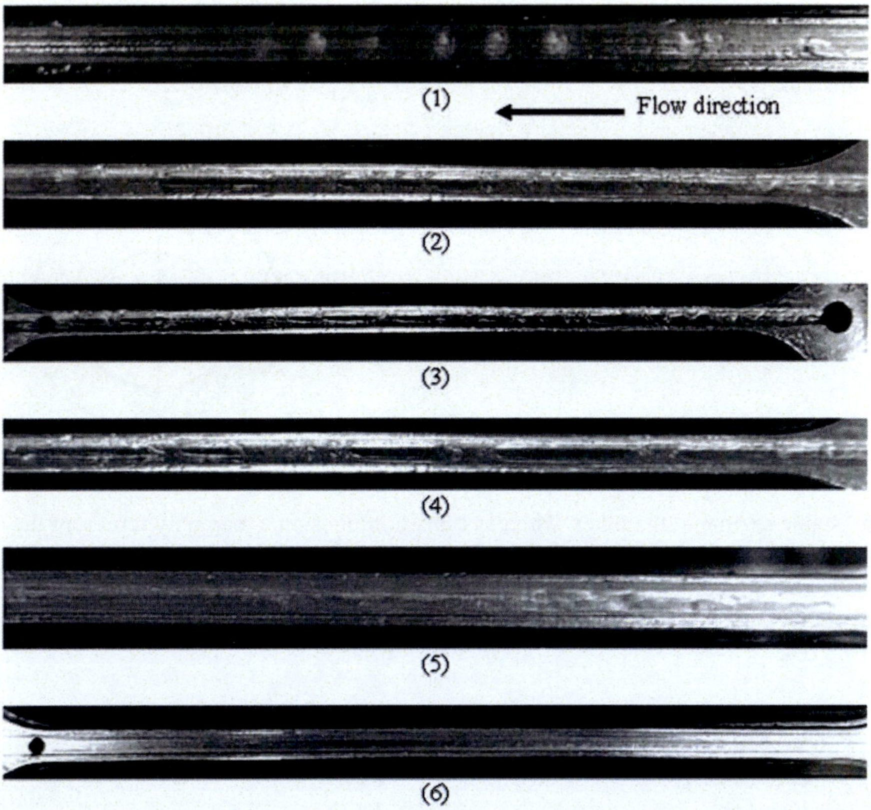

Fig. 6.20 Flow patterns observed by Gasche (Biberg 1999) at the experimental conditions: $G = 149 \text{kg/m}^2 s$, $T_{sat} = 23.3$ °C, $D_{eq} = 0.833$ mm, $q = 1.86$ kW/m^{2}where (1), (2), (3) and (4)—plug flow; (5)—slug/annular flow; (6)—annular flow

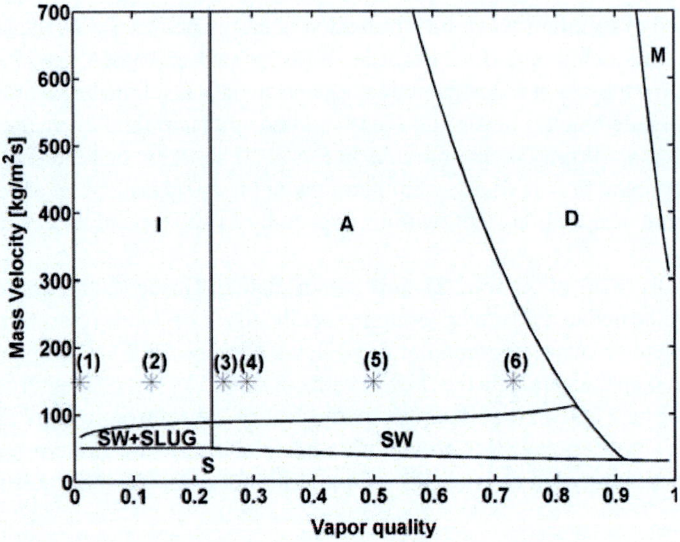

Fig. 6.21 The experimental data of the observed flow patterns by Gashe (2006) in Fig. 2 shown in the updated CO_2 flow pattern map where (1), (2), (3) and (4)—plug flow; (5)—slug/annular flow; (6)—annular flow

Fig. 6.22 Schematic diagram of film thickness

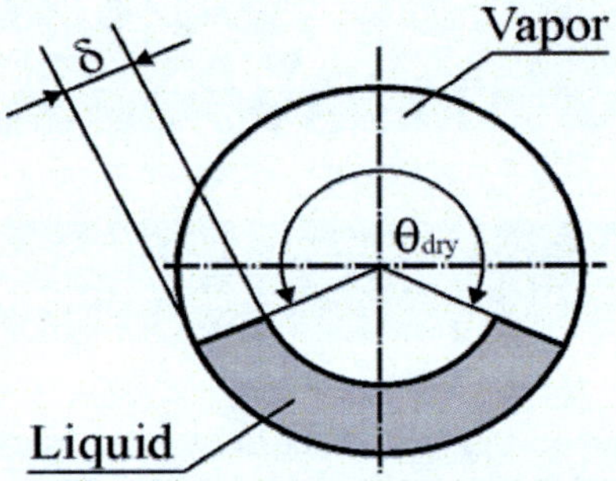

one stable regime to the other, whereas transition lines on a map only represent the probable "centerline" of this transition range. With the limited data available for CO_2 at this point, predicting the "width" of a transition zone around the transition line is not yet feasible, but it should be a good topic for future research.

6.4 A General Flow Pattern Based Evaporation Heat Transfer Model for CO_2

A general updated general evaporation heat transfer model was developed by modifying the Cheng et al. (2006a, b) flow boiling heat transfer model (Cheng et al. 2008b, d). By incorporating the updated new flow pattern map in the above section, the new heat transfer model is physically related to the flow regimes of CO_2 evaporation, and thus correspondingly the new model has been extended to a wider range of conditions and to include new heat transfer methods in mist flow and bubbly flow regimes. The proposed new general flow boiling heat transfer model predicted reasonably well an extensive experimental database derived from the literature.

To develop a general evaporation heat transfer prediction method, it is important that the method is not only numerically accurate but that it also correctly captures the trends in the data to be useful for heat exchanger optimization. Most importantly, the evaporation heat transfer mechanisms should be related to the corresponding flow patterns and be physically explained according to flow pattern transitions. Besides significantly extending the range of the heat transfer database here, several new modifications were implemented in the updated general evaporation heat transfer model and will be presented below. Changes to the flow pattern map also have an effect on the heat transfer model: the new dryout inception vapor quality correlation (Eq. 6.21) and a new dryout completion vapor quality correlation (Eq. 6.26) are used to better segregate the data into these regimes, which have sharply different heat transfer performances. Accordingly, the evaporation heat transfer correlation in the dryout region was updated. In addition, a new mist flow heat transfer correlation for CO_2 was developed based on the CO_2 data and a heat transfer method for bubbly flow was adopted for completeness sake. With these modifications, a new general evaporation heat transfer model for CO_2 was developed to meet a wider range of conditions and to cover all flow regimes (Cheng et al. 2008b, d).

The Kattan–Thome–Favrat (Wojtan et al. 2005a, b; Kattan et al. 1998a, b, c) general equation for the local flow boiling heat transfer coefficients h_{tp} in a horizontal tube is used as the basic flow boiling expression:

$$h_{tp} = \frac{\theta_{dry} h_V + (2\pi - \theta_{dry}) h_{wet}}{2\pi} \tag{6.28}$$

where θ_{dry} is the dry angle defined in Figs. 6.19 and 6.22. The dry angle θ_{dry} defines the flow structures and the ratio of the tube perimeter in contact with liquid and vapor. In stratified flow, θ_{dry} equals the stratified angle θ_{strat} which is calculated with Eq. (6.14). In annular (A), intermittent (I) and bubbly (B) flows, $\theta_{dry} = 0$. For stratified-wavy flow, θ_{dry} varies from zero up to its maximum value θ_{strat}. Stratified-wavy flow has been subdivided into three subzones (slug, slug/stratified-wavy and stratified-wavy) to determine θ_{dry}.

For slug zone (Slug), the high frequency slugs are assumed to maintain a continuous thin liquid layer on the upper tube perimeter. Thus, similar to the intermittent and annular flow regimes, one has:

$$\theta_{dry} = 0 \tag{6.29}$$

For stratified-wavy zone (SW), the following equation is proposed:

$$\theta_{dry} = \theta_{strat}\left(\frac{G_{wavy}-G}{G_{wavy}-G_{strat}}\right)^{0.61} \tag{6.30}$$

For slug-stratified-wavy zone (Slug + SW), the following interpolation between the other two regimes is proposed for x < x_{IA}:

$$\theta_{dry} = \theta_{strat}\frac{x}{x_{IA}}\left(\frac{G_{wavy}-G}{G_{wavy}-G_{strat}}\right)^{0.61} \tag{6.31}$$

The vapor phase heat transfer coefficient on the dry perimeter h_V is calculated with the Dittus–Boelter (1930) correlation assuming tubular flow in the tube:

$$h_V = 0.023Re_V^{0.8}Pr_V^{0.4}\frac{k_V}{D_{eq}} \tag{6.32}$$

where the vapor phase Reynolds number Re_V is defined as follows:

$$Re_V = \frac{GxD_{eq}}{\mu_V\varepsilon} \tag{6.33}$$

The heat transfer coefficient on the wet perimeter h_{wet} is calculated with an asymptotic model that combines the nucleate boiling and convective boiling heat transfer contributions to flow boiling heat transfer by the third power:

$$h_{wet} = \left[(Sh_{nb})^3 + h_{cb}^3\right]^{\frac{1}{3}} \tag{6.34}$$

where h_{nb}, S and h_{cb} are respectively nucleate boiling heat transfer coefficient, nucleate boiling heat transfer suppression factor and convective boiling heat transfer coefficient and are determined in the following equations.

The nucleate boiling heat transfer coefficient h_{nb} is calculated with the Cheng et al. (2006a) nucleate boiling correlation for CO_2 which is a modification of the Cooper (1984) correlation:

$$h_{nb} = 131p_r^{-0.0063}(-\log_{10}p_r)^{-0.55}M^{-0.5}q^{0.58} \tag{6.35}$$

The Cheng et al. (2006a) nucleate boiling heat transfer suppression factor S for CO_2 is applied to reduce the nucleate boiling heat transfer contribution due to the thinning of the annular liquid film:

If $x < x_{IA}$,

$$S = 1 \tag{6.36}$$

If $x \geq x_{IA}$,

$$S = 1 - 1.14\left(\frac{D_{eq}}{0.00753}\right)^2\left(1 - \frac{\delta}{\delta_{IA}}\right)^{2.2} \tag{6.37}$$

Furthermore, if $D_{eq} > 7.53$ mm, then set $D_{eq} = 7.53$ mm. The liquid film thickness δ shown in Fig. 6.22 is calculated with the expression proposed by El Hajal et al. (2003):

$$\delta = \frac{D_{eq}}{2} - \sqrt{\left(\frac{D_{eq}}{2}\right)^2 - \frac{2A_L}{2\pi - \theta_{dry}}} \tag{6.38}$$

where A_L, based on the equivalent diameter, is cross-sectional area occupied by liquid-phase shown in Fig. 6.19. When the liquid occupies more than one-half of the cross section of the tube at low vapor quality, this expression would yield a value of $\delta > D_{eq}/2$, which is not geometrically realistic. Hence, whenever Eq. (6.38) gives $\delta > D_{eq}/2$, δ is set equal to $D_{eq}/2$ (occurs when $\varepsilon < 0.5$). The liquid film δ_{IA} is calculated with Eq. (6.38) at the intermittent (I) to annular flow (A) transition.

The convective boiling heat transfer coefficient hcb is calculated with the following correlation assuming an annular liquid film flow from the original model (Kattan et al. 1998c):

$$h_{cb} = 0.0133 Re_{\delta}^{0.69} Pr_L^{0.4} \frac{k_L}{\delta} \tag{6.39}$$

where the liquid film Reynolds number Re_δ is defined as (Cho and Kim 2007):

$$Re_{\delta} = \frac{4G(1-x)\delta}{\mu_L(1-\varepsilon)} \tag{6.40}$$

The void fraction ε is calculated with Eq. (6.9) and δ is calculated with Eq. (6.38).

The heat transfer coefficient in mist flow is calculated by a new correlation developed in this study, which is a modification of the correlation by Groeneveld (1973), with a new lead constant and a new exponent on Re_H according to CO_2 experimental data:

$$h_M = 2 \times 10^{-8} Re_H^{1.97} Pr_V^{1.06} Y^{-1.83} \frac{k_V}{D_{eq}} \tag{6.41}$$

where the homogeneous Reynolds number Re_H and the correction factor Y are calculated as follows:

$$Re_H = \frac{GD_{eq}}{\mu_V}\left[x + \frac{\rho_V}{\rho_L}(1-x)\right] \tag{6.42}$$

$$Y = 1 - 0.1\left[\left(\frac{\rho_L}{\rho_V} - 1\right)(1-x)\right]^{0.4} \tag{6.43}$$

The heat transfer coefficient in the dryout region is calculated by a linear interpolation proposed by Wojtan et al. (2005b):

$$h_{dryout} = h_{tp}(x_{di}) - \frac{x - x_{di}}{x_{de} - x_{di}}\left[h_{tp}(x_{di}) - h_M(x_{de})\right] \tag{6.44}$$

where $h_{tp}(x_{di})$ is the two-phase heat transfer coefficient calculated with Eq. (6.28) at the dryout inception quality x_{di} and $h_M(x_{de})$ is the mist flow heat transfer coefficient calculated with Eq. (6.41) at the dryout completion quality x_{de}. Dryout inception quality x_{di} and dryout completion quality x_{de} are respectively calculated with Eqs. (6.21) and (6.26). If x_{de} is not defined at the mass velocity being considered, it is assumed that $x_{de} = 0.999$.

A heat transfer model for bubbly flow was added to the model for completeness sake. In the absence of any data, the heat transfer coefficients in bubbly flow regime are calculated by the same method as that in the intermittent flow. Equation (6.28) is used to calculate the local flow boiling heat transfer coefficients. In bubbly (*B*) flow, the dryout angle $\theta_{dry} = 0$.

The updated general flow boiling heat transfer model was compared to an extensive database (Cheng et al. 2008b). Just to show one example here, Fig. 6.23a shows the comparison of the predicted flow boiling heat transfer coefficients to the experimental data of Yun et al. (Cheng et al. 2006a) and Fig. 6.23b shows the corresponding flow map. The updated general flow boiling heat transfer model not only captures the heat transfer trends well but also predicts the experimental heat transfer data well. As it is harder to predict (and harder to accurately measure) heat transfer data in the dryout and mist flow regimes, the updated general heat transfer model does not always predict the experimental data in these two flow regimes satisfactorily. Some examples of such comparisons can be found in Cheng et al. (2008b).

Figure 6.24 shows simulation of the updated flow pattern map and flow boiling model for CO_2 at the indicated conditions, superimposed on the same graphs by Cheng et al. (2007b). The process path for the vapor quality variation from x = 0.01 to x = 0.99 is shown as the horizontal broken line (dash-dot line) while the variation in the heat transfer coefficient as it changes vapor quality and flow pattern is depicted by the dashed line. The flow pattern boundaries are in solid lines. The line (dash line with arrows) indicates the calculated heat transfer coefficient at the indicated mass velocity and vapor quality. Notice the various changes in trends in the heat transfer coefficient as this occurs. For example, when the flow regime passes from annular flow into the dryout regime, there is a sharp inflection in the heat transfer coefficient as the top perimeter of the tube becomes dry.

In further analysis, comparisons have also been made by classes of flows, i.e. the predictions versus all the heat transfer data excluding dryout and mist flow data (essentially the all wet perimeter data), versus all dryout heat transfer data (the partially wet perimeter data) and versus the mist flow data (all dry perimeter data) (Hashitume 1983). Figure 6.25 shows the comparison to the first group, Fig. 6.26 the second and Fig. 6.27 the third. The statistical analysis has shown the following fraction of the database are predicted within ±30%: 71.4% of the entire database (1124 points), 83.2% of all wet wall data points (773 points), 47.6% of the partially wet wall data points (191 points) and 48.2% of the all dry wall data points (160 points).

Overall, the updated general flow boiling heat transfer model predicts the overall database quite well. However, for the dryout and mist flow regimes with partially or all dry perimeters, the heat transfer model is only partially satisfactory. For these last

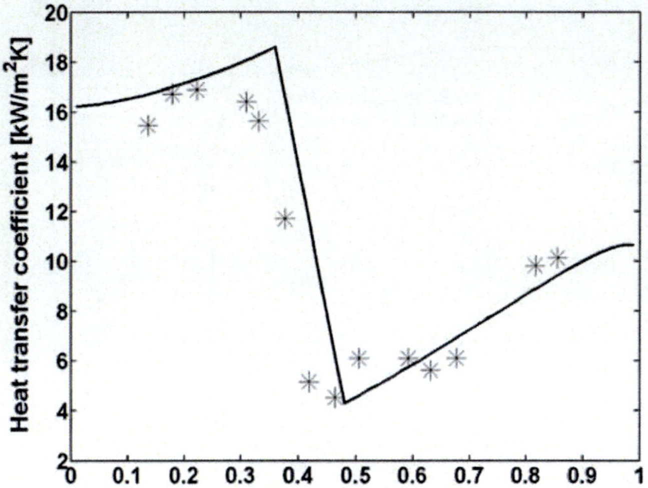

(a)

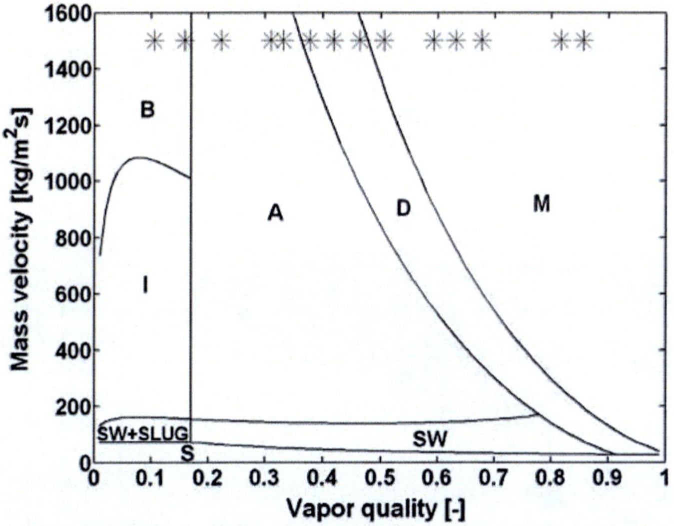

(b)

Fig. 6.23 **a** Comparison of the predicted flow boiling heat transfer coefficients to the experimental data of Yun et al. (Cheng et al. 2006a); **b** the corresponding flow pattern map (at the test conditions: Deq = 2 mm, q = 30 kW/m^2, Tsat = 5 °C and G = 1500 kg/m^{2}s

two regimes, many of the experimental data sets have a level of scatter ranging up 40% themselves. In part, the larger errors are due to the very sharp change in trend in these data with vapor quality, where an error of 2–3% in vapor quality in the energy balance of the experiments or in the prediction of x_{di} and/or x_{de} immediately results in a heat transfer prediction error of 50%. Therefore, more careful experiments are needed in

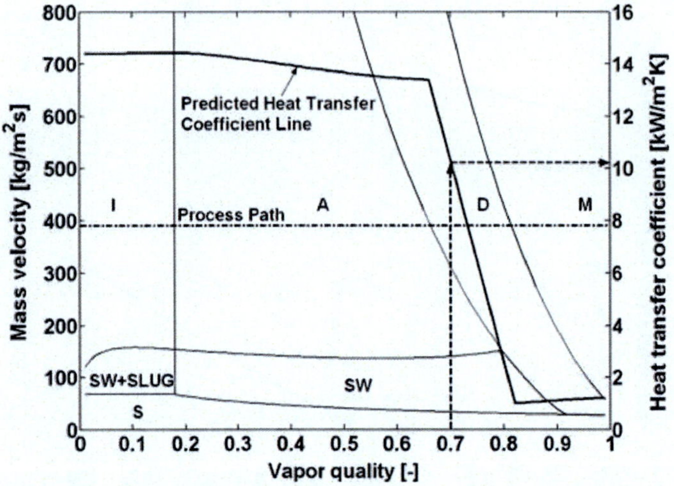

Fig. 6.24 Simulation of flow boiling heat transfer model and flow pattern map for 3 mm channel at the conditions: $q = 20$ kW/m^2, $T_{sat} = 10$ °C and $G = 390$ kg/m^{2}s with indicated value at $x = 0.70$ (Cheng et al. 2007b)

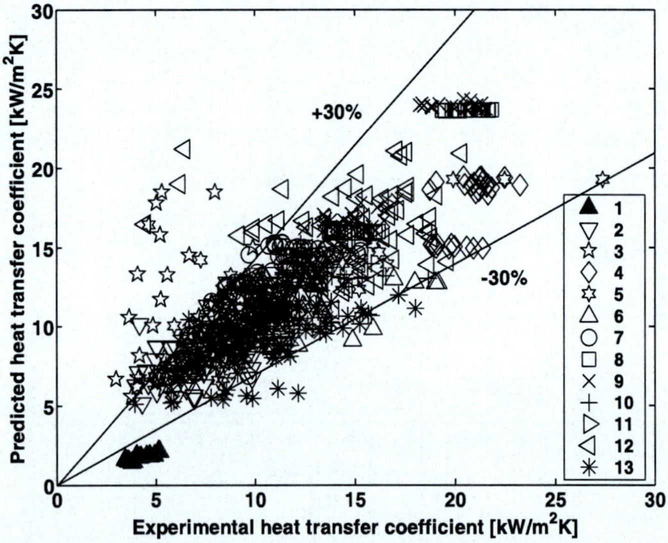

Fig. 6.25 Comparison of the predicted flow boiling heat transfer coefficients to all heat transfer data without the dryout and mist flow data points in the entire database: 1—Knudsen and Jensen (Hihara and Tanaka 2000), 2—Yun et al. (2003), 3—Yoon et al. (2005a), 4—Koyama et al. (2001), 5—Pettersen (2004), 6 — Yun et al. (2005b), 7—Gao and Honda (2005a, 2005b), 8—Tanaka et al. (2001), 9—Hihara and Tanaka (2000), 10—Shinmura et al. (2006), 11—Zhao et al. (2000a, b), 12—Yun et al. (2005a, 2002) and 13—Jeong et al. (2005) (*Note* 1—6 were used in our previous study (Cheng et al. 2006a))

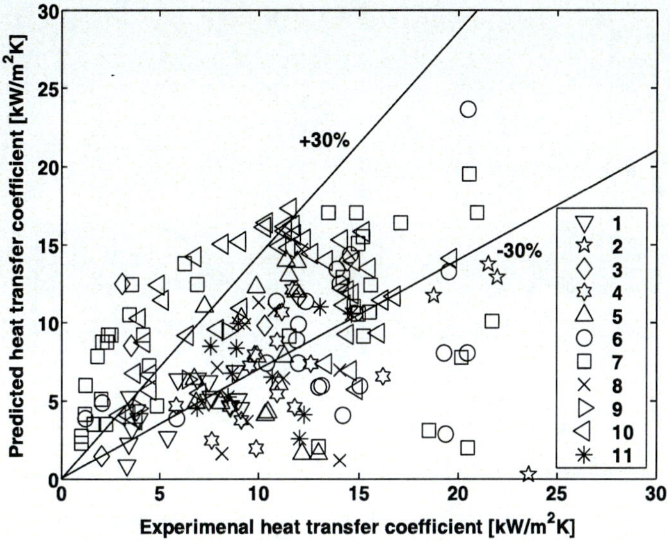

Fig. 6.26 Comparison of the predicted flow boiling heat transfer coefficients to all dryout heat transfer data points in the entire database: 1—Yun et al. (2003), 2—Koyama et al. (2001), 3—Pettersen (2004), 4—Yun et al. (2005b) 5—Gao and Honda (2005a, b), 6—Tanaka et al. (2001), 7—Hihara (Yun et al. 2005a), 8—Shinmura et al. (2006), 9—Zhao et al. (2000a, b), 10—Yun et al. (2005a, 2002) and 11—Jeong et al. (2005) (*Note* 1–4 were used in our previous study (Cheng et al. 2006a))

these two regimes to provide more accurate heat transfer data, with attention to also determine the transitions x_{di} and x_{de}, because they are typical working conditions in the micro-scale channels of extruded multi-port aluminum tubes used for automobile air-conditioners that operate over a wide range of mass velocities up to as high as 1500 kg/m^{2}s.

6.5 A General Flow Pattern Based Two-Phase Frictional Pressure Drop Model for CO_2

The predictions of two-phase flow frictional pressure drops with the leading methods often cause errors of more than 50% (Ould-Didi et al. 2002; Moreno Quibén and Thome 2007a, b), therefore, efforts are increasingly being made to improve the two-phase frictional pressure drop prediction methods and models. Furthermore, the leading two-phase frictional pressure drop prediction methods do not usually contain any flow pattern information, which is intrinsically related to the two-phase frictional pressure drop. Due to the effects of thermal physical and transport properties of CO_2, the leading prediction two-phase frictional pressure drop methods do not work well. The reason is that these methods do not usually cover the much lower liquid-to-vapor

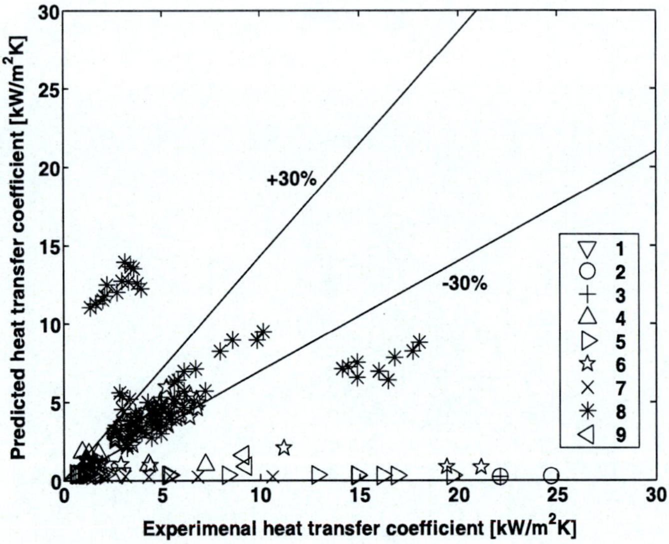

Fig. 6.27 Comparison of the predicted flow boiling heat transfer coefficients to all mist flow heat transfer data points in the entire database: 1—Yun et al. (2003), 2—Koyama et al. (2001), 3— Yun et al. (2005b) 4—Gao and Honda (2005a, 2005b), 5—Tanaka et al. (2001), 6—Hihara (Hihara and Tanaka 2000), 7—Shinmura et al. (2006), 8—Yun et al. (2005a, 2002) and 9—Jeong et al. (2005) (*Note* 1—3 were used in the previous study (Cheng et al. 2006a))

density ratios and very small surface tension characteristics of CO_2 at high pressures. In general, the two-phase frictional pressure drops of CO_2 are much lower than those of other refrigerants (Cheng et al. 2008c, 2008d). Significantly, there is no proven generally applicable two-phase frictional pressure drop prediction method for CO_2, although there are a number of studies on CO_2 two-phase frictional pressure drops in the literature (Yoon et al. 2004; Jeong et al. 2005; Zhao et al. 2000a, b; Pettersen and VestbØstad 2000; Yun and Kim 2003, 2004). Some researchers proposed two-phase frictional pressure drop correlations for CO_2 based on their own experimental data but such methods do not work properly when extrapolated to other conditions. For example, Yoon et al. (2004) proposed a modified Chisholm method (1973) to fit their data in a macro-scale channel but it cannot be applied to other conditions because they tested only one tube diameter. In practical applications, both macro- and micro-scale tubes are used in CO_2 evaporators and heat exchangers.

As opposed to the completely empirical two-phase frictional pressure drop methods, a flow pattern based phenomenological frictional pressure drop model relating the flow patterns to the corresponding two-phase frictional pressure drops is a promising approach in the two-phase pressure drop predictions. Ould Didi et al. (2002) used local flow patterns to analyze two-phase flow pressure drops, which resulted in a significant improvement in accuracy. Based on that, a new flow pattern based phenomenological model of two-phase frictional pressure drops was recently developed by Moreno et al. (2007a, b). The model physically respects the two-phase flow structure of the various flow patterns while maintaining a degree of simplicity as well. The model predicts their experimental data well but not the present CO_2 experimental database.

Cheng et al. (2008d) compiled a large database of CO_2 two-phase frictional pressure drop and compared the database to the leading two-phase frictional pressure drop methods: the empirical two-phase frictional pressure drop methods by Chisholm (1973), Friedel (1979), Grönnerud (1979) and Müller-Steinhagen and Heck (1986), a modified Chisholm correlation by Yoon et al. (2004) and the flow patterned based pressure drop model by Moreno Quibén and Thome (2007a, b). The CO_2 database includes the experimental data of Bredesen et al. (1997), Pettersen (2004), Pettersen and VestbØstad (2000), Zhao et al. (2000a, b) and Yun and Kim (2003, 2004). The test channels include single circular channels and multi-channels with circular, triangular and rectangular cross-sections and electrical and fluid heated test sections. The data were taken from tables where available or by digitizing the pressure drops from graphs in these publications. All together 387 two-phase pressure drop data points were obtained. Figure 6.28 shows the comparison of these leading pressure drop methods to the experimental data of Bredesen et al. (1997) at the indicated test conditions. There are big differences among these methods. Overall, not one of these models is able to predict the CO_2 two-phase frictional pressure drop data well (note that all have been extrapolated beyond their original conditions to make this comparison for CO_2). The Friedel method gave reasonably good predictions, but it failed to predict the pressure drop in smaller channels. Therefore, it is necessary to develop a new general two-phase frictional pressure drop model for CO_2 two-phase flow in macroscale- and micro-scale-channels. Cheng et al. (2008d) developed a flow pattern based two-phase frictional pressure drop model specially for CO_2 using their general flow pattern map for CO_2 presented in Sect. 6.5.

The Cheng et al. two-phase frictional pressure drop model for CO_2 was developed by modifying the model of Moreno Quibén and Thome developed for R-22, R-410a and R134a and incorporating the updated Cheng et al. CO_2 flow pattern map, using the CO_2 pressure drop database by Cheng et al. (2008d). In developing this pressure drop model, two-phase frictional pressure drop data were used. The total pressure drop is the sum of the static pressure drop (gravity pressure drop), the momentum pressure drop (acceleration pressure drop) and the frictional pressure drop:

$$\Delta p_{total} = \Delta p_{static} + \Delta p_m + \Delta p_f \tag{6.45}$$

For horizontal channels, the static pressure drop equals zero. Furthermore, the momentum pressure drop can be calculated as

$$\Delta p_m = G^2 \left\{ \left[\frac{(1-x)^2}{\rho_L(1-\varepsilon)} + \frac{x^2}{\rho_V \varepsilon} \right]_{out} - \left[\frac{(1-x)^2}{\rho_L(1-\varepsilon)} + \frac{x^2}{\rho_V \varepsilon} \right]_{in} \right\} \tag{6.46}$$

Thus, diabatic experimental tests that measure total pressure drops can be reduced using the above expressions to find the two-phase frictional pressure drops.

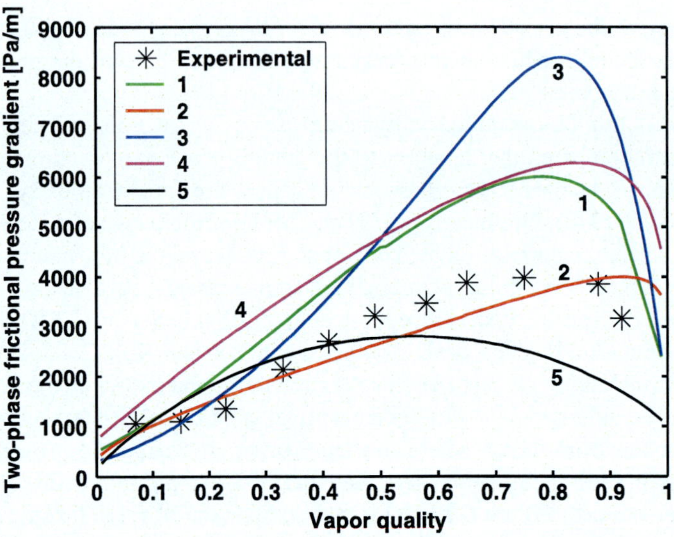

Fig. 6.28 Comparison of the several leading methods to the experimental data of Bredesen et al. (1997) at the experimental conditions: G = 400 kg/m^{2}s, T_{sat} = −10 °C, D_{eq} = 7 mm and q = 6 kW/m^2; 1—The Moreno-Quibén and Thome model (2007a, b); 2—The Friedel method (1979); 3—The Grönnerud method (1979); 4—The Müller-Steighagen-Heck method (1986); 5—The Chisholm method (1973)

The details of the Cheng et al. two-phase flow frictional pressure drop model for CO_2 are as follows:

For non-circular channels, the equivalent diameter D_{eq} is used in the two-phase frictional pressure drop model to remain consistent with that in the flow pattern map. Using the equivalent diameter gives the same mass velocity as in the non-circular channel and thus correctly reflects the mean liquid and vapor velocities, something using hydraulic diameter in a two-phase flow does not. Thus, equivalent diameter D_{eq} is used in the following method.

(1) CO_2 frictional pressure drop model for annular flow (A): The basic equation is the same as that of the Moreno-Quibén and Thome (2007a, b) two-phase frictional pressure drop model:

$$\Delta p_A = 4 f_A \frac{L}{D_{eq}} \frac{\rho_V u_V^2}{2} \tag{6.47}$$

where the two-phase flow friction factor of annular flow f_A was correlated according to CO_2 experimental data here (considering the main parameters which affect the two-phase pressure drops for CO_2) as:

$$f_A = 3.128 Re_V^{-0.454} We_L^{-0.0308} \tag{6.48}$$

This correlation is thus different from that of the Moreno Quibén and Thome (2007a, b) pressure drop model. The mean velocity of the vapor phase u_V is calculated as

$$u_V = \frac{Gx}{\rho_V \varepsilon} \tag{6.49}$$

The void fraction ε is calculated using Eq. (6.9). The vapor phase Reynolds number $Re_V$and the liquid-phase Weber number We_L based on the mean liquid-phase velocity u_L are calculated as

$$Re_V = \frac{GxD_{eq}}{\mu_V \varepsilon} \tag{6.50}$$

$$We_L = \frac{\rho_L u_L^2 D_{eq}}{\sigma} \tag{6.51}$$

$$u_L = \frac{G(1-x)}{\rho_L(1-\varepsilon)} \tag{6.52}$$

(2) CO_2 frictional pressure drop model for slug and intermittent flow (Slug + I): A proration is proposed to avoid any jump in the pressure drops between these two flow patterns, so that the Moreno Quibén and Thome (2007a, b) pressure drop model is updated to become:

$$\Delta p_{SLUG+I} = \Delta p_{LO}\left(1 - \frac{\varepsilon}{\varepsilon_{IA}}\right) + \Delta p_A\left(\frac{\varepsilon}{\varepsilon_{IA}}\right) \tag{6.53}$$

where Δp_A is calculated with Eq. (6.47) and the single-phase frictional pressure drop considering the total vapor–liquid two-phase flow as liquid flow Δp_{LO} is calculated as

$$\Delta p_{LO} = 4 f_{LO} \frac{L}{D_{eq}} \frac{G^2}{2\rho_L} \tag{6.54}$$

The friction factor is calculated with the Blasius equation as

$$f_{LO} = \frac{0.079}{Re_{LO}^{0.25}} \tag{6.55}$$

where Reynolds number Re_{LO} is calculated as

$$Re_{LO} = \frac{GD_{eq}}{\mu_L} \tag{6.56}$$

(3) CO_2 frictional pressure drop model for stratified-wavy flow (SW): The equation is kept the same as that of the Moreno Quibén and Thome (2007a, b) pressure drop model:

$$\Delta p_{SW} = 4 f_{SW} \frac{L}{D_{eq}} \frac{\rho_V u_V^2}{2} \quad (6.57)$$

where the two-phase friction factor of stratified-wavy flow f_{SW} is calculated with the following interpolating expression (a modification of that used in the Moreno Quibén-Thome (Moreno Quibén and Thome 2007a, b) pressure drop model) based on the CO_2 database:

$$f_{SW} = \theta_{dry}^{*}{}^{0.02} f_V + \left(1 - \theta_{dry}^{*}\right)^{0.02} f_A \quad (6.58)$$

and the dimensionless dry angle θ_{dry}^{*} is defined as

$$\theta_{dry}^{*} = \frac{\theta_{dry}}{2\pi} \quad (6.59)$$

where θ_{dry} is the dry angle as shown in Fig. 6.23. The dry angle θ_{dry} defines the flow structure and the ratio of the tube perimeter in contact with vapor. For the stratified-wavy regime (SW), θ_{dry} is calculated with Eq. (6.30).

The single-phase friction factor of the vapor phase f_V is calculated as

$$f_V = \frac{0.079}{Re_V^{0.25}} \quad (6.60)$$

where the vapor Reynolds number Re_v is calculated with Eq. (6.50).

(4) CO_2 frictional pressure drop model for slug-stratified-wavy flow (Slug + SW): It is proposed to avoid any jump in the pressure drops between these two flow patterns and to update the Moreno Quibén and Thome (2007a, b) pressure drop model as:

$$\Delta p_{SLUG+SW} = \Delta p_{LO}\left(1 - \frac{\varepsilon}{\varepsilon_{IA}}\right) + \Delta p_{SW}\left(\frac{\varepsilon}{\varepsilon_{IA}}\right) \quad (6.61)$$

where Δp_{LO} and Δp_{SW} are calculated with Eqs. (6.54) and (6.57), respectively.

(5) CO_2 frictional pressure drop model for mist flow (M): The following expression is kept the same as that in the Moreno-Quibén-Thome (Gungor and Winterton 1986; Kandlikar 1990; Liu and Winterton 1991) pressure drop model:

$$\Delta p_M = 4 f_M \frac{L}{D_{eq}} \frac{G^2}{2\rho_H} \quad (6.62)$$

The homogenous density ρ_H is defined as

$$\rho_H = \rho_L(1 - \varepsilon_H) + \rho_V \varepsilon_H \quad (6.63)$$

where the homogenous void fraction ε_H is calculated as

$$\varepsilon_H = \left(1 + \frac{(1-x)}{x}\frac{\rho_V}{\rho_L}\right)^{-1} \quad (6.64)$$

and the friction factor of mist flow f_M was correlated according to the CO_2 experimental data, which is different from that in the Moreno Quibén and Thome (2007a, b) pressure drop model, as:

$$f_M = \frac{91.2}{Re_M^{0.832}} \quad (6.65)$$

The Reynolds number is defined as:

$$Re_M = \frac{GD_{eq}}{\mu_H} \quad (6.66)$$

where the homogeneous dynamic viscosity is calculated as proposed by Ciccitti et al. (1960):

$$\mu_H = \mu_L(1-x) + \mu_V x \quad (6.67)$$

The constants in Eq. (6.65) are quite different from those in the Blasius equation. The reason is possible because there are limited experimental data in mist flow in the database and also perhaps a lower accuracy of these experimental data. Therefore, more accurate experimental data are needed in mist flow to further verify this correlation or modify it if necessary in the future.

(6) CO_2 frictional pressure drop model for dryout region (D): The linear interpolating expression is kept the same as that in the Moreno Quibén-Thome pressure drop model as:

$$\Delta p_{dryout} = \Delta p_{tp}(x_{di}) - \frac{x - x_{di}}{x_{de} - x_{di}}\left[\Delta p_{tp}(x_{di}) - \Delta p_M(x_{de})\right] \quad (6.68)$$

where $\Delta p_{tp}(x_{di})$ is the frictional pressure drop at the dryout inception quality x_{di} and is calculated with Eq. (6.47) for annular flow or with Eq. (6.57) for stratified-wavy flow, and $\Delta p_M(x_{de})$ is the frictional pressure drop at the completion quality x_{de} and is calculated with Eq. (6.62). x_{di} and x_{de} are respectively calculated with Eqs. (6.21) and (6.26).

(7) CO_2 frictional pressure drop model for stratified flow (S): No data fell into this flow regime but for completeness, the method is kept the same as that in the Moreno Quibén and Thome (2007a, b) pressure drop model as:

For $x \geq x_{IA}$:

$$\Delta p_{strat(x \geq x_{IA})} = 4 f_{strat(x \geq x_{IA})} \frac{L}{D_{eq}} \frac{\rho_V u_V^2}{2} \quad (6.69)$$

where the mean velocity of the vapor phase u_V is calculated with Eq. (6.49) and the two-phase friction factor of stratified flow $f_{strat(x \geq x_{IA})}$ is calculated as

$$f_{strat(x \geq x_{IA})} = \theta^*_{strat} f_V + \left(1 - \theta^*_{strat}\right) f_A \tag{6.70}$$

The single-phase friction factor of the vapor phase f_V and the two-phase friction factor of annular flow f_A are calculated with Eqs. (6.60) and (6.48), respectively, and the dimensionless stratified angle θ^*_{strat} is defined as

$$\theta^*_{strat} = \frac{\theta_{strat}}{2\pi} \tag{6.71}$$

where the stratified angle θ_{strat} is calculated with Eq. (6.14).

For $x < x_{IA}$:

$$\Delta p_{strat(x<x_{IA})} = \Delta p_{LO}\left(1 - \frac{\varepsilon}{\varepsilon_{IA}}\right) + \Delta p_{strat(x \geq x_{IA})}\left(\frac{\varepsilon}{\varepsilon_{IA}}\right) \tag{6.72}$$

where Δp_{LO} and $\Delta p_{strat(x \geq x_{IA})}$ are calculated with Eqs. (6.54) and (6.69), respectively.

(8) CO_2 frictional pressure drop model for bubbly flow (B): No data are available for this regime but keeping consistent with the frictional pressure drops in the neighboring regimes and following the same format as the others without creating a jump at the transition (there is no such a model in the Moreno-Quibén and Thome (2007a, b) pressure drop model), the following expression is used:

$$\Delta p_B = \Delta p_{LO}\left(1 - \frac{\varepsilon}{\varepsilon_{IA}}\right) + \Delta p_A\left(\frac{\varepsilon}{\varepsilon_{IA}}\right) \tag{6.73}$$

where Δp_{LO} and Δp_A are calculated with Eqs. (6.54) and (6.47), respectively. Further experimental data are needed to verify or modify this model.

The Cheng et al. CO_2 two-phase frictional pressure drop model was compared to the CO_2 two-phase pressure drop database (Cheng et al. 2008c). Figure 6.29 shows the comparison of the Cheng et al. CO_2 frictional pressure drop model to the experimental data of Bredesen et al. (1997) at the indicated experimental conditions and the corresponding flow pattern map. The model predicts the data well and also captures the pressure drop trend. Figure 6.34 shows the comparative results of the predictions by the new CO_2 pressure drop model to the entire two-phase pressure drop database. In addition, the statistical results of the predicted and experimental data for individual research according to the percent of data predicted within ± 30% are presented in Fig. 6.30. Furthermore, the detailed breakdown of the statistical analysis for the new pressure drop model is summarized (Cheng et al. 2008c). Most of the experimental data points (75.5%) are in annular flow and 75.7% of experimental data in annular flow are predicted within ±30%. However, the predictions in some regions such as S-Slug and SW are not satisfactory. Generally, the new pressure drop model reasonably predicts the database and importantly captures the trends in the data too. Nonetheless, there are not many experimental data available covering some flow patterns and future experimental work is recommended to address these conditions.

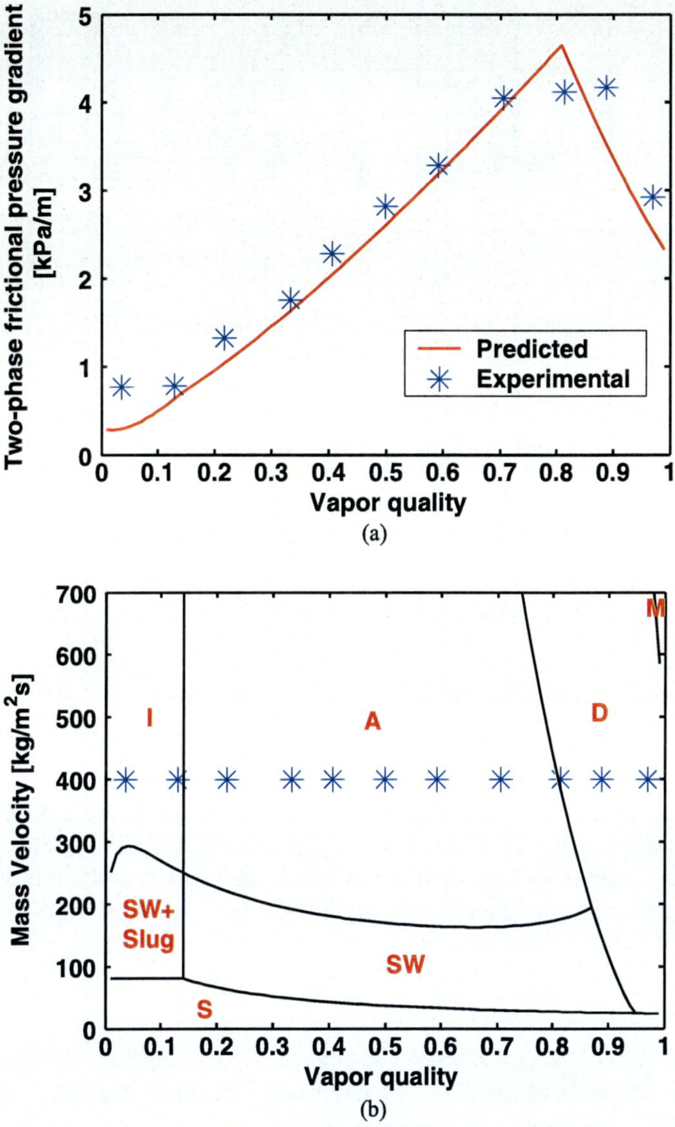

Fig. 6.29 **a** Comparison of the new CO_2 pressure drop model to the experimental data of Bredesen et al. (1997) at the experimental conditions: G = 400 kg/m^{2}s, Tsat = -10 °C, Deq = 7 mm and q = 3 kW/m^2; **b** The corresponding flow pattern map at the same experimental condition as that in (**a**) (I represents intermittent flow, A represents annular flow, D represents dryout region, M represents mist flow, S represents stratified flow and SW represents stratified-wavy flow)

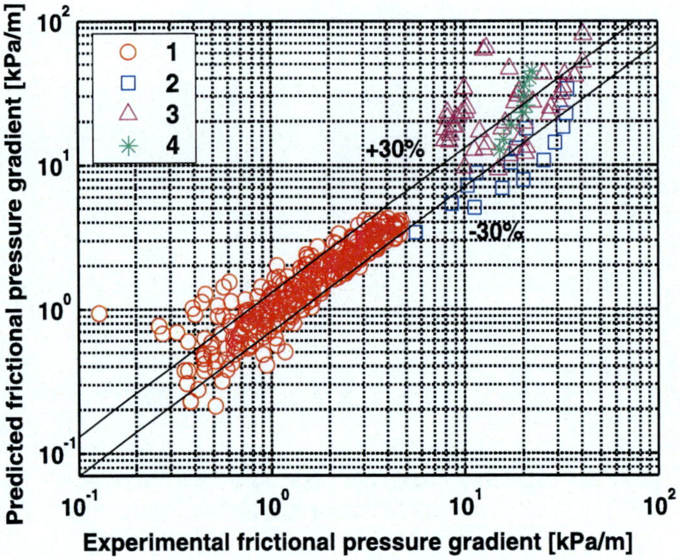

Fig. 6.30 Comparison of the predicted frictional pressure gradients by the new model to the entire database (74.7% of the data are predicted within ± 30%). By individual study, the following percent of data are captured with ±30%: 1—Bredesen et al. (1997) 81.5%, 2—Yun and Kim (2003, 2004) 33.3%, 3—Pettersen (2004) and Pettersen and VestbØstad (2000) 43.2%, and 4—Zhao et al. (2000a, b) 55.6% (Note that 2, 3 and 4 are the data of micro-scale channels)

The Cheng et al. CO_2 two-phase frictional pressure drop model predicts the CO_2 pressure drop database better than the existing methods. Due to the limited and less accurate experimental data in micro-scale channels available in the literature, the CO_2 pressure drop model does not predict these data satisfactorily. It is suggested that additional, more accurate experimental CO_2 pressure drop data be obtained through well designed measurement facilities to further test or improve the model in the future.

6.6 The Oil Effect on CO_2 Two-Phase Pressure Drops and Evaporation Heat Transfer

Evaporation heat transfer and two-phase flow of refrigerant-lubricating oil mixtures are very complex phenomena but very important topics for air-conditioning, heat pumps and refrigeration systems. The presence of lubricating oil may considerably affect the transport properties of a refrigerant and have a significant effect on the evaporation heat transfer, two-phase pressure drop and flow pattern characteristics and thus the design of evaporators in the relevant systems. Due to its large viscosity, surface tension and insulating effects, lubricating oil tends to decrease the evaporation heat transfer coefficient of a refrigerant in most cases while it may increase the evaporation heat transfer coefficient of the refrigerant at low lubricating oil concentrations. Therefore, it is essential to understand how the lubricating oil concentration affects the evaporation heat transfer, two-phase pressure drop and flow regime characteristics and to predict the evaporation heat transfer coefficient and two-phase frictional pressure drop of such mixtures properly.

Refrigerants are classified as completely miscible, partially miscible or immiscible according to their mutual solubility relations with lubricating oils. The miscibility of lubricating oils and refrigerants often has a great effect on the evaporation heat transfer and two-phase frictional pressure drop characteristics. To select the most suitable lubricating oils for CO_2 air-conditioning, refrigeration and heat pump systems, there are several potential issues which should be considered as follows:

(i) **Lubricant transport:** To ensure good oil return to the compressor, the refrigerant oil needs to have either a higher density than liquid CO_2 or good miscibility with it.
(ii) **Wear:** (a) CO_2 is an excellent solvent and this solvency can cause excessive lubricant dilution leading to potential wear and foaming problems; (b) CO_2 requires the use of higher operating pressures (e.g. in automotive air-conditioning system, the working pressure is about one order of magnitude higher compared to R134a). Therefore, high loads increase the stress on bearings which can lead to increased wear.
(iii) **Stability:** CO_2 can react with water to form carbonic acid that can then accelerate potential hydrolysis processes.

A number of lubricants for CO_2 have been studied over the past years. According to the studies of lubricants for CO_2 by Kaneko et al. (2004) and Ikeda et al. (2006), PAGs (Plyalkylene-glycols) are the most suitable lubricants used in CO_2 air-conditioning, refrigeration and heat pump systems so far. Unlike R134a, PAGs are immiscible with CO_2. Therefore, it is essential to understand the evaporation heat transfer and two-phase frictional pressure drop behaviors of CO_2-lubricanting oil mixtures in the evaporation processes so as to design the evaporators properly.

A literature survey of evaporation heat transfer and two-phase frictional pressure drop of CO_2 with lubricating oil is presented in this section. The experimental studies on this topic are very limited. There are several papers on the effect of lubricating oil on CO_2 evaporation heat transfer and two-phase frictional pressure drop. Of these studies, some are related to evaporation heat transfer and two-phase frictional pressure drop in macro-scale channels and others are related to evaporation heat transfer in micro-scale channels. These available studies have been analyzed to obtain useful information for the design of evaporators although the relevant models are limited due to limited experimental data available.

Analysis of these studies on flow boiling and two-phase pressure drop of CO_2-oil mixtures has been carefully conducted to present a state-of-the-art status of the study on these topics. As the lubricating oil (PAG) used in CO_2 air-conditioning, refrigeration and heat pump systems is immiscible with CO_2, emphasis has been placed on evaporation heat transfer and two-phase frictional pressure drop of CO_2 with immiscible oil in micro-scale channels which are of interest to CO_2 air-conditioning and heat pump systems. Unfortunately, there are limited studies of evaporation heat transfer and two-phase frictional pressure drop of CO_2 with immiscible oil in a 3 mm inner diameter tube. With the very limited experimental data available, it is impossible to propose a model for evaporation heat transfer of CO_2 with immiscible oil. In addition, there are also little information of two-phase frictional pressure drop of CO_2 with lubricating oil in micro-scale channels. Accurate and systematic experimental data of evaporation heat transfer and two-phase frictional pressure drop of CO_2 lubricating oil mixtures in both macro-scale and micro-scale tubes are needed. These are the basis for developing relevant prediction methods. Lacking the reliable prediction methods, some empirical methods should be proposed for the design of evaporators as in Sect. 6.8.

Katsuta et al. (2003a) conducted an experimental study of evaporation heat transfer and two-phase frictional pressure drop of CO_2 and lubricating oil (PAG) mixture in a horizontal tube with an inner diameter of 4.59 mm. The oil concentrations were 1% (wt) and 5% (wt), respectively. The oil concentration had little effect on the two-phase pressure drops. However, the two-phase pressure drops increased with increasing the oil concentrations at vapor qualities larger than 0.6. The possible reason of this trend is that the local oil concentration plays an important role in increasing two-phase pressure drops for the low mass velocities. Figure 6.31 shows their pressure drop data at two test conditions. It is interesting to see that at a mass velocity of 800 kg/m^2s, the pressure drops of 5% oil are slightly lower than those of 1%. No explanation for such a result was presented in their paper.

Koyama et al. (2004b, 2005) studied flow boiling of CO_2-PAG mixtures inside a horizontal smooth tube with an inner diameter of 4.42 mm and a micro-fin tube with an inner diameter of 4.76 mm. The authors have concluded that the heat transfer coefficient for the smooth tube decreases abruptly with the addition of a very small amount of oil, while the deterioration rate of heat transfer coefficient for the micro-fin tube is smaller than that of the smooth tube. The two-phase pressure drop in the smooth tube increases with increasing the oil concentration especially at high vapor qualities larger than around 0.6 while that of the micro-fin tube increases slightly with increasing the oil concentration.

Katsuta et al. (2003b, 2004) presented the experimental evaporation heat transfer and two-phase pressure drop data of the lubricating oil concentrations with 0 to 1.1% (wt). For the very low oil concentrations of 0.05% to 0.08% (wt), the flow boiling heat transfer coefficients are decreased by about 40% compared to those without oil. This is quite different from the experimental results of Gao and Honda (2005a, b) for a 3 mm I.D. tube. According to the study of Gao and Honda, the oil concentration does not affect the flow boiling heat transfer coefficients if the oil concentration is less than 0.1% (wt). Furthermore, there is no obvious difference between the flow boiling heat transfer coefficients of the oil concentrations of 0.05% to 0.08% (wt) and those of the oil concentrations with 0.3 to 1.1% (wt). However, when the oil concentration is larger than 1%, heat transfer coefficient decreases greatly.

For all the above studies in macro-scale channels, the oil concentrations were measured by sampling. There is no information on the measurement of the local oil

Fig. 6.31 The effect of the oil on two-phase pressure drop (Katsuta et al. 2003a): **a** G = 400 kg/m^{2}s, q = 10 kW/m^2; **b** G = 800 kg/m^{2}s, q = 10 kW/m^2

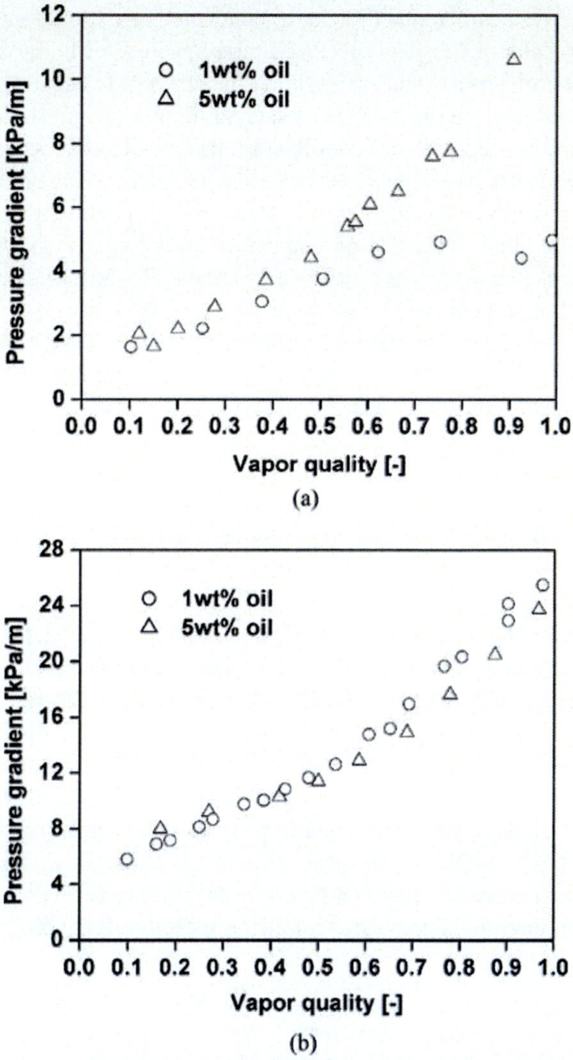

concentration which is of great importance in modeling evaporation heat transfer and two-phase frictional pressure drop.

There are only a few studies of evaporation heat transfer in micro-scale channels and there is no study of two-phase pressure drop in micro-scale channels with oil. Siegismund and Kauffeld (2004) only presented a description of their test facility and the data reduction methods in their paper. No experimental data were presented in their paper although the title of their paper is the influence of lubricant oil on CO_2 evaporation heat transfer in minichannel tubes.

Gao and Honda (2005a, b) conducted an experimental study on evaporation heat transfer of CO_2 and oil (PAG) mixtures inside a horizontal tube with an inner diameter of 3 mm. Figure 6.32 shows the effect of lubricating oil on the local heat transfer coefficient at the indicated test conditions (Gao and Honda 2005a, b). The local evaporation heat transfer coefficient falls considerably when the lubricating oil concentration is larger than 1% (wt). Furthermore, when the lubricating oil concentration increases, the change of heat transfer coefficient with vapor quality becomes small. Therefore, when the oil concentration is larger than 1% (wt), the evaporation heat transfer coefficient is not greatly affected by further increasing the oil concentration. Three oil concentrations were used in their study: less than 0.1% (wt), 1% (wt) and 2% (wt). When the concentration of the lubricating oil is more than 1% (wt), the local heat transfer coefficient was much lower than that without the lubricating oil. According to the authors' analysis, the reason for this decrease is considered due to the nucleate boiling being suppressed by the oil film. Therefore, the evaporation heat transfer is considered to change from the nucleate boiling dominated mechanism to the convective evaporation dominated mechanism. The dryout vapor quality decreases with increasing mass velocity and is a little influenced by the saturation temperature and the lubricating oil concentration. The experimental heat transfer data of the oil concentration less than 0.1% (wt) were already used in the database in our previous study of the CO_2 evaporation heat transfer model (Cheng et al. 2008d). As independent experimental evaporation heat transfer data, the flow pattern based evaporation heat transfer model of Cheng et al. (2008c, d) predicts well the evaporation heat transfer data of the oil concentration less than 0.1% (wt). Therefore, it can be concluded from this study that the oil does not affect the heat transfer when the oil concentration is less than 0.1% (wt) (Cheng et al. 2008d).

Zhao et al. (2002) conducted an experimental study of flow boiling heat transfer of CO_2 with miscible oil in multi micro-channels with an inner diameter of 0.86 mm (an equivalent diameter of 1.15 mm). Figure 6.33 shows the results of the oil effect on heat transfer coefficient at the indicated test conditions. A small concentration

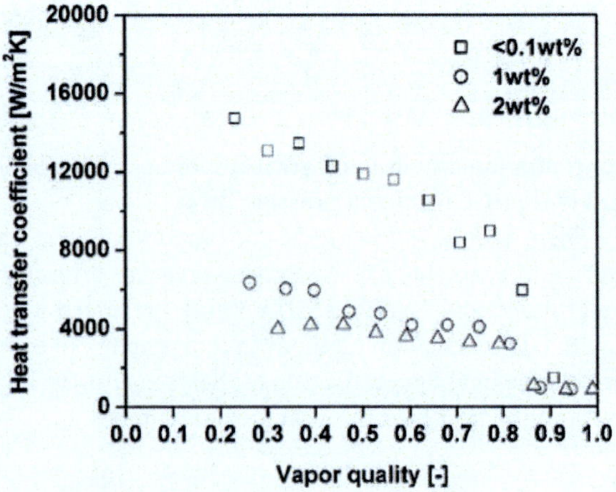

Fig. 6.32 The oil effect on heat transfer coefficient (Gao and Honda 2005a, b): T_{sat} = 10 °C, G = 390 kg/m^{2}s, q = 20 kW/m^2

of the lubricating oil (<3%) increases the flow boiling heat transfer coefficients at low vapor qualities of less than 0.4. The authors attributed this to a foaming effect that increases wall-wetting. In addition, they concluded that increasing mass flux also improved the heat transfer coefficient in the presence of oil since the average vapor quality decreased for constant heat flux and inlet vapor quality. On the other hand, they concluded that increasing the vapor quality and the saturation temperature decreased the heat transfer coefficient in the presence of oil. It must be pointed out here that the authors did not mention what kind of oil was used in their study. It was a miscible oil. Apparently, the effect of the miscible oil on the heat transfer characteristics of CO_2 is quite different from that of an immiscible oil. According to the analysis, for the study of flow boiling heat transfer and two-phase pressure drop of CO_2 with an immiscible oil (PAG) in micro-scale channels, there is no such information in the literature so far.

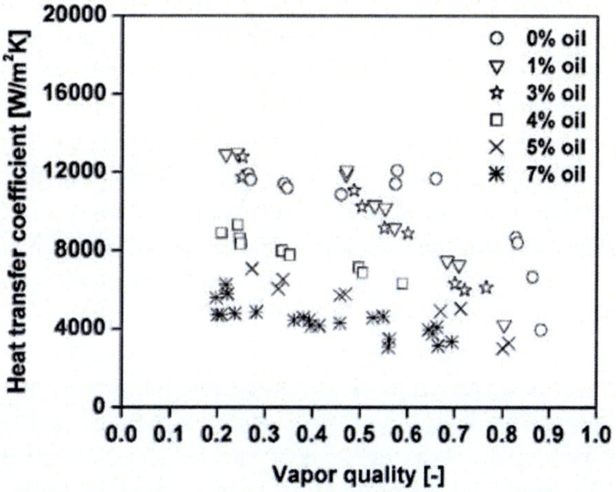

Fig. 6.33 The effect of oil on heat transfer coefficient (Zhao et al. 2002) (T_{sat} = 10 °C, G = 300 kg/m²s, q = 11 kW/m²)

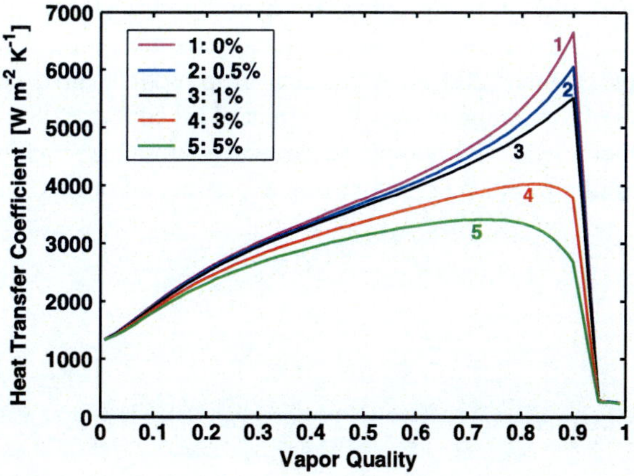

Fig. 6.34 The oil effect (mass concentration from 0 to 5%) on the evaporation heat transfer coefficient of R134a in a horizontal tube (Simulated results by the flow pattern based heat transfer model of Wojtan et al. (Zhao and Bansal 2007; Park and Hrnjak 2005)): G = 300 kg/m²s, T_{sat} = 10 °C, D = 13.84 mm and q = 7.5 kW/m²

In general, very small amounts of the lubricating oil (<0.5% wt) seem to have little effect while larger concentrations (>1% wt) tend to dramatically reduce flow boiling heat transfer coefficients. For two-phase pressure drops, the effect of lubricating oil only occurs at vapor qualities larger than around 0.6.

Based on this analysis, it is recommended that further experimental studies on evaporation and two-phase frictional pressure drop of CO_2-immiscible oil mixtures in both macro-scale and micro-scale channels be conducted over a wide range of test conditions to provide more experimental data on these aspects. In order to understand the evaporation heat transfer mechanisms of CO_2 and immiscible oil mixtures, flow visualization and local measurement of oil concentration are also suggested to be done.

As for the prediction methods for flow boiling heat transfer and two-phase pressure drop with CO_2-immsicible oil mixtures, the flow pattern based flow boiling heat transfer and two-phase pressure drop models of Cheng et al. (Cheng et al. 2008c, 2008d) may be used by introducing the physical properties of CO_2-immiscible lubricating oil mixtures. As the evaporation heat transfer (Cheng et al. 2008d) model predicts well the independent experimental heat transfer data of the lubricating oil concentration less than 0.1% (wt) by Gao and Honda (2005a, b), it is expected that the evaporation heat transfer model may be modified to handle evaporation heat transfer of CO_2-immiscible lubricating oil mixtures. For the prediction of the two-phase pressure drops of CO_2-immiscible lubricating oil mixtures, the flow pattern based two-phase frictional pressure drop model of Cheng et al. (Cheng et al. 2008c) for pure CO_2 may be a good basis to start from when accurate and sufficient experimental data become available.

A review on the oil effects on the evaporation heat transfer, flow regimes and two-phase pressure drops is presented in Bandarra Filho et al. (2009). The evaluation methods for the physical properties are detailed in this paper. Simulations of the oil effects on the evaporation heat transfer and two-phase frictional pressure drops are illustrated here, using the flow pattern based models of Wojtan et al. and Moreno-Quibén-Thome. Figure 6.34 shows the predicted evaporation heat transfer results of R134a-oil using the heat evaporation transfer model of Wojtan et al. (2005a, b) and Fig. 6.35 shows the predicted pressure drop results using the Moreno-Quibén and Thome (2007a, b) pressure drop model with a correct factor (Thome 2004b). These simulations show the effect of oil concentrations on evaporation heat transfer coefficients and two-phase pressure drops. It seems the simulated results are quite similar to those of the existing experimental results.

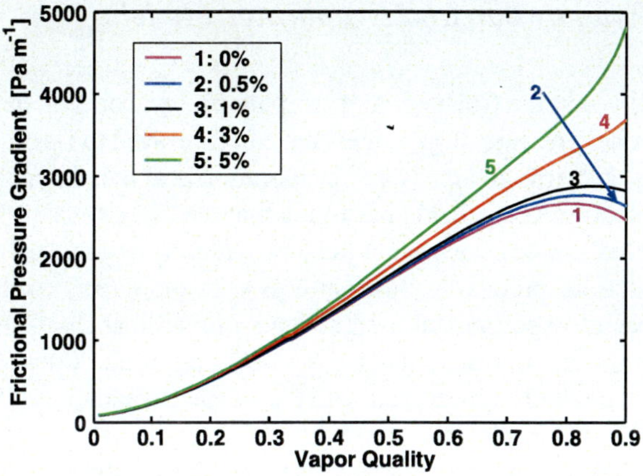

Fig. 6.35 The oil effect (mass concentration from 0 to 5%) on two-phase frictional pressure gradient of R134a in a horizontal tube for flow regimes before dryout region (The two-phase frictional pressure gradients were calculated using the flow pattern based two-phase pressure drop model of Moreno-Quiben and Thome (Gungor and Winterton 1986; Kandlikar 1990) for pure R134a with a correction factor of oil effect on two-phase pressure drop): $G = 300$ kg/m^{2}s, $T_{sat} = 10$ °C, $D = 13.84$ mm and $q = 7.5$ kW/m^2

Due to the immiscibility of lubricant oil and CO_2, whether these methods for miscible oil can be applied to CO_2-lubricant oil or not still need to be further verified. Therefore, extensive experimental data are needed to verify the available methods or to develop new prediction methods for CO_2.

6.7 CO_2 Evaporator Simulation and Design

As a natural working fluid, CO_2 is used in refrigeration, air-conditioning and heat pump systems for better thermal performance, energy efficiency and environmental safety. Most commonly used feed methods for CO_2 are pumped liquid and direct expansion. Various structures of CO_2 evaporators are used in practical applications, e.g. fin tube evaporator, tube in-tube evaporator, plate-fin evaporator etc. CO_2 evaporation occurs inside the tubes and the other side of heat transfer may be air flow, water flow or other waste heat resources. Both macro-scale and micro-scale tubes are used in CO_2 evaporators. The prediction of evaporation heat transfer and two-phase pressure drops inside tubes is critical but difficult due to the effects of various fluid properties and channel sizes (Cheng and Chen 2000; Cheng 2010; Yoshioka et al. 2008). Therefore, proper evaporation heat transfer and two-phase frictional pressure drop models are needed to predict the heat transfer coefficients, two-phase pressure drops and the related flow patterns. Once these are obtained properly. The generally used principles and methods for designing heat exchangers apply for CO_2 evaporators.

6.7.1 CO_2 Evaporation and Evaporator Modeling

Over the past 10 years, the Cheng et al. flow patterns, evaporation heat transfer and two-phase frictional pressure drop models have been proved to favorably agree with the observed flow patterns and measured evaporation heat transfer and pressure drop data. For example, Gao et al. (2011) has found that the Cheng et al. flow pattern map (Cheng et al. 2006a, 2008c) can predict the flow patterns in micro-fin tubes properly. It can also be used for predicting flow patterns with the effect of oil (Wetzel et al. 2014). A number of experimental studies have proved that the Cheng et al. flow pattern based evaporation heat transfer model (Cheng et al. 2006a, 2006b, 2008c, 2008d) is the best method as compared with their experimental data (Gao et al. 2011; Grauso et al. 2013; Mastrullo et al. 2009, 2010; Wu et al. 2011; Patiño et al. 2014). Therefore, the Cheng et al. flow pattern map and flow pattern based evaporation heat transfer model for CO_2 are recommended for identifying the flow patterns and evaluating the heat transfer coefficient of CO_2.

Cheng et al. follow pattern based frictional pressure drop model (Cheng et al. 2008c) and predict the experimental pressure drop data to some extent. In particular, it predicts the micro-scale pressure drop data properly. For CO_2 two-phase pressure drop in macroscale channels, the Friedel correlation (1979) is recommended for calculating the pressure drops up to date. However, the Cheng et al. flow pattern based model should be further improved once extensive accurate experimental data are available.

Marcinichen et al. (2016) developed a sophisticated simulation tool including the latest prediction methods and oil effects and then used this tool to investigate methods to find a much more compact CO_2 evaporator design with a significantly smaller charge of CO_2 for its operation. Compared to an actual CO_2 evaporator design, the new one proposed here reduces the CO_2 charge by 58.7% and also reduces the size (volume) of the CO_2 evaporator by one-half, meaning the refrigeration system becomes much more compact and lighter in weight. Thus, these benefits can be explored to take advantage of a much lower space within the cabinet of the beverage machine for the refrigeration system and also potentially lower manufacturing costs. In their simulation, the leading two-phase flow pattern map and physically-based prediction models for evaporation of CO_2 developed by Cheng et al. (2008c, d) were used. They were also used for modeling oil effects on CO_2 evaporation. The effects of tube size, shape and wettability for significant performance improvements, miscible oil concentration, etc., were addressed while staying within the industry's fabrication and operational limits. The adverse influence of oil was found to reduce the mean CO_2 side heat transfer coefficient and to increase the CO_2 pressure drop by up to 11% and 94%. Figures 6.36 and 6.37 show the considerable negative and escalating impact of these oil concentrations observed on the evaporation heat transfer coefficient and on the two-phase pressure gradient according to the simulation results using the Cheng et al. models. The evaporation heat transfer coefficient decreases significantly at large vapor qualities while the two-phase pressure drops increases significantly in the mean time for higher concentration of oil. Thus, it is suggested that the oil concentration should be kept below 1% to avoid too large of oil penalty.

Marcinichen et al. (2016) presented the simulation results of the design of the evaporators with two different types of fins: plain versus wavy fins. A corrugation angle of 12.5° was assumed for the wavy fin while the fin pitch and thickness remained the same. Also, here the air volumetric flow was fixed to be 250 m^3/h for the simulations. Figure 6.38 shows the better performance of the evaporator with wavy fins in terms of local heat transfer per unit of tube length and outlet air temperature. Also, it is clear to see that only two rows of tubes were more than sufficient to transfer the desired 600 W of heat load. The necessary mass of CO_2 was reduced from 11.47 to 10.43 g when changing from plain fins to wavy fins, which means an additional reduction of about 9.1%.

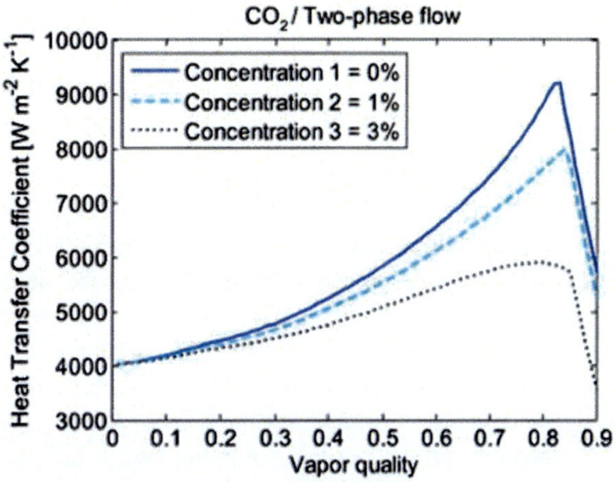

Fig. 6.36 The oil effect (mass concentration from 0 to 3%) on the evaporation heat transfer coefficient of CO_2 in the evaporator tube using the Cheng et al. heat transfer model (Marcinichen et al. 2016)

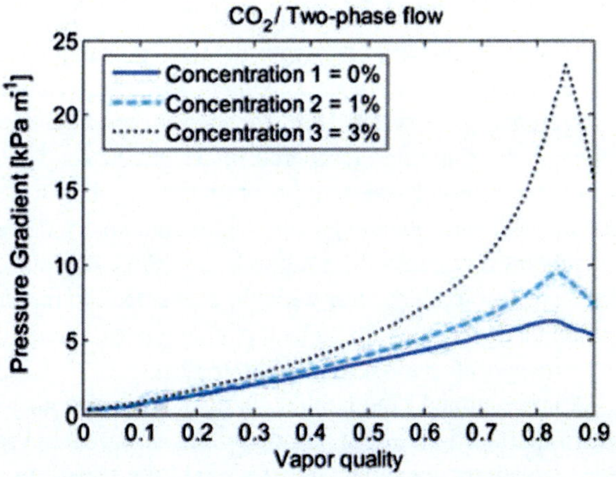

Fig. 6.37 The oil effect (mass concentration from 0 to 3%) on the two-phase frictional pressure gradient of CO_2 in the evaporator tube using the Cheng et al. pressure drop model (Marcinichen et al. 2016)

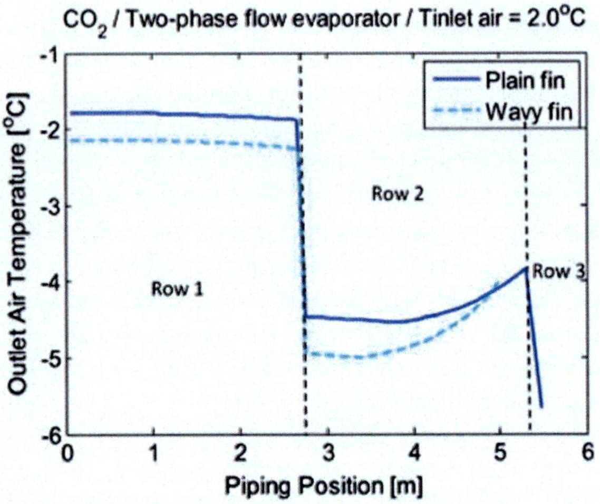

Fig. 6.38 Simulation results of the outlet air temperature per row for the plain fin and the wavy fin evaporators (Marcinichen et al. 2016)

6.7.2 Simulations of CO_2 Thermal Systems Using the Cheng et al Models

The Cheng et al. flow pattern based evaporation heat transfer and two-phase frictional pressure drop models have been widely used for simulations of CO_2 refrigeration, air-conditioning and heat pump systems by a number of researchers (Yamaguchi et al. 2011; Wang et al. 2013; Faria et al. 2016). For example, in order to improve the system performance of the CO_2 heat pump, it is necessary to develop an optimum design and a control method for the CO_2 heat pump water heater.

Yamaguchi et al. (2011) have developed a high-precision and general-purpose system simulation model for the CO_2 heat pump water heater and investigated the validity of the model with detailed experiments. Figure 6.39 shows the system flow diagram of the CO_2 heat pump water heater in their study This system consists of a gas cooler, an evaporator, an internal heat exchanger, a compressor, and an expansion valve. In the internal heat exchanger, the refrigerant from the evaporator cools the refrigerant that flows in from the gas cooler, and then flows into the compressor, where its pressure and temperature are increased. The heated refrigerant then flows into the gas cooler, where it heats up the supplied water. Subsequently, it flows into the internal heat exchanger and is expanded by the expansion valve. Finally, the refrigerant flows into the evaporator. In the evaporator, the refrigerant absorbs heat from the ambient air, after which it flows back to the internal heat exchanger. In this system, a cross-finned tube heat exchanger with smooth plate fins is adapted for use in the evaporator as shown in Fig. 6.39.

A cross-finned tube heat exchanger with smooth plate fins is adapted for use in the evaporator in the heat pump system, as shown in Fig. 6.39. In the development of their model, the local heat transfer coefficient and two-phase frictional pressure drop inside the tubes are calculated using the Cheng et al. flow pattern based evaporation

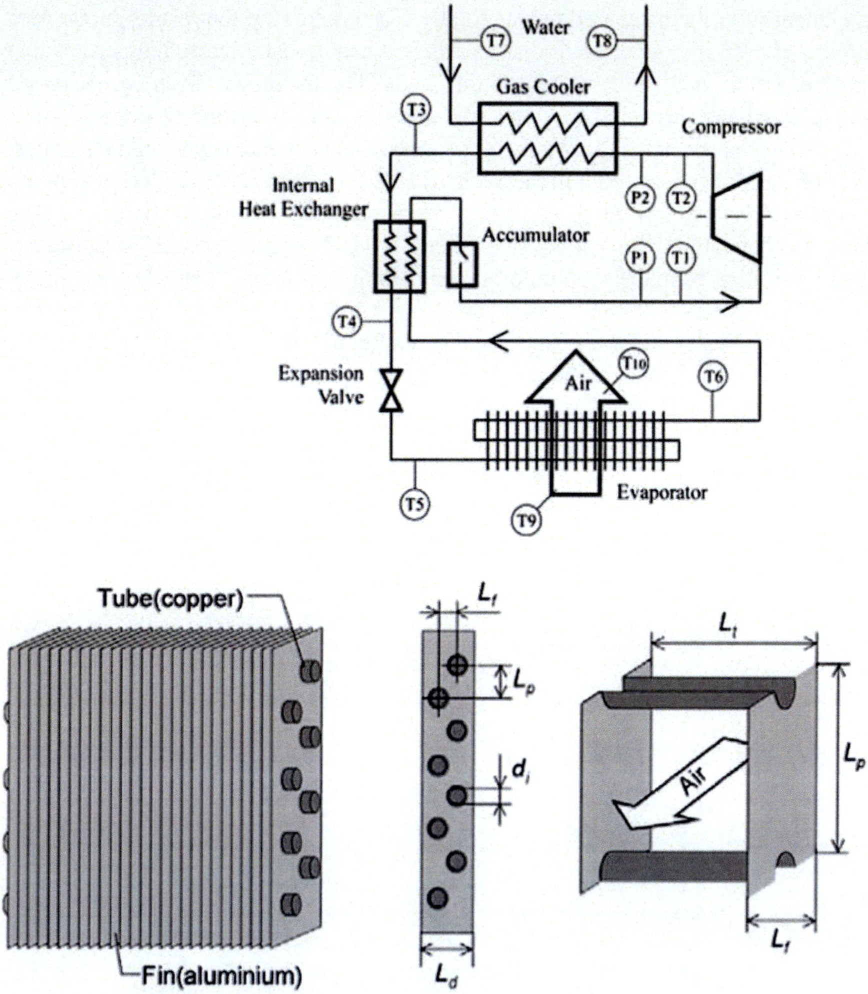

Fig. 6.39 System flow diagram of CO_2 heat pump and measuring points in the study of Yamaguchi et al. (2011)

heat transfer model (Cheng et al. 2008c, 2008d). These models were developed specifically for CO_2 to predict the evaporation heat transfer coefficient and the two-phase frictional pressure drop on the basis of the flow pattern map for CO_2. They also considered the oil effect on the performance. According to study of the PAG oil effect on heat transfer by Yoshioka et al. (Tong et al. 2016), the heat transfer coefficient of the evaporator with 5.0 wt.% PAG oil was almost 25% lower than that of the evaporator without oil. This value was used in their study. Figure 6.40 shows the effect of the inlet water temperature on the COP, heat exchange rate in the gas cooler and evaporator, and power consumption of the compressor. In comparison of predicted and measured values, it is confirmed that the predicted values are in good agreement with the measured values totally. For evaporator, the results show good agreement between the measured and predicted heat transfer rates. This means that the Cheng et al. models work well in this study. The maximum difference between the predicted and measured COP is 5.4%, and the average difference is 0.9%.

Considering the actual heat transfer processes of heat exchangers and characteristics of components adopted in the cycle of air-source trans-critical CO_2 heat pump water heater system, Wang et al. (2013) conducted a simulation for the heat pump cycle. In their simulation model, the Cheng et al. flow pattern based evaporation model was adopted in the evaporator model in their simulation. It shows good agreement with the experimental results. Faria et al. (2016) investigated the behavior of the solar evaporator and expansion valve assembly of a transcritical CO_2 heat pump in transient and steady conditions. The dynamic behavior of systems using CO_2 as a refrigerant is significantly influenced by the dynamics of the heat transfer mechanisms. The solar evaporator model is based on the equations of conservation of mass, momentum, and energy. The model validation is realized by comparing simulation results with the Cheng et al. models and the experimental data. A model is a useful

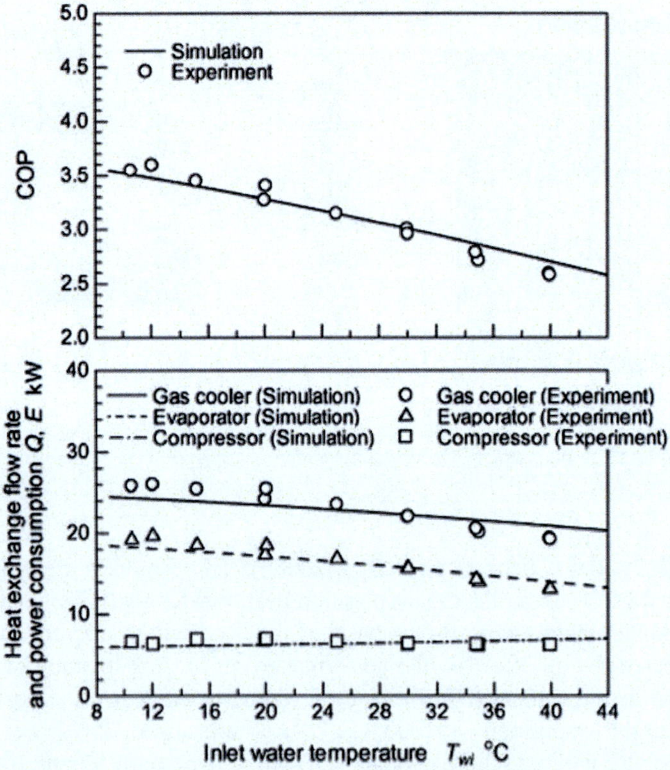

Fig. 6.40 The effect of the water inlet temperature the COP and heat flow rate (Yamaguchi et al. 2011)

tool for analyzing the behavior in transient and steady conditions simulating various operating conditions of the heat pump including solar radiation, ambient temperature, wind speed and atmospheric conditions. Yang et al. (2016) investigated experimentally a prototype of a combined R134a and transcritical CO_2 heat pump. The experimental results demonstrated that the combined system could operate reliably and supply stable temperature hot water over a wide range of ambient temperatures and feed water temperatures. The experimental results also showed that ambient temperature had a large effect on the system performance. Furthermore, they investigated the improvement of the combined system in comparison to the standard transcritical CO_2 heat pump, a simple mathematical model was developed and validated by the experimental data. The comparison results showed that the combined system could offer higher system COP at experimental ambient and feed water temperatures. In their evaporator model, the Cheng et al. models were used in the simulations. In general, the Cheng et al. flow pattern based evaporation heat transfer and two-phase pressure drop models (Cheng et al. 2006a, 2006b, 2008c, 2008d) show good agreement with the experimental data in the existing studies when used in modeling the evaporators in these systems.

As an excellent coolant, CO_2 may be used in electronic cooling (Cheng and Thome 2009), two-phase thermosyphon loop (Tong et al. 2016; Xia et al. 2017) and evaporative CO_2 cooling system for the upgrade of the CMS pixel detector (Daguin et al. 2012) etc. Furthermore, new applications with CO_2 in various thermal systems with micro-scale channel evaporators should be further explored (Cheng 2010; Cheng and Liu 2011) to improve the energy efficiency and environmental safety. According to the existing studies, it has been confirmed that the Cheng et al. models favorably predict the evaporation heat transfer, two-phase pressure drop and flow patterns in heat exchanger tubes. This also means that the mechanistic models relating the flow patterns to the corresponding heat transfer coefficients and two-phase pressure drops are promising methods. Therefore, more accurate flow pattern observations and heat transfer and pressure drop experimental data under a wide range of test conditions are still needed. Furthermore, understanding the oil effects on the two-phase flow and heat transfer of CO_2 and relevant practical models are also needed. New thermal and energy systems should also be targeted to extend the applications of CO_2 in industry.

6.7.3 CO_2 Evaporator Design and Selection

Using the proper heat transfer models is the key to the thermal design of CO_2 evaporators. The oil effect on the evaporation heat transfer and two-phase pressure drops should be considered as the fouling resistance if the oil presents. The Cheng et al. evaporation heat transfer and pressure drop models (Cheng et al. 2006a, b, 2008b, d) should be used to evaluate the heat transfer coefficients and two-phase pressure drops for the working conditions. The heat transfer coefficients on the other side of the evaporators should be evaluated with the proper correlations (Bergman et al. 2017; Thulukkanam 2013). Once the tube material, diameter and wall thickness are selected, the overall heat transfer coefficient may be determined. The conventional heat exchanger design methods may be applied for the CO_2 evaporator. For a more accurate evaluation of the heat transfer rate, local heat transfer may be evaluated using the relevant model.

Besides the thermal design of a CO_2 evaporator, other aspects such as material selection, operation pressure, optimum minimum mass flow rate and defrost etc. should be considered in the selection/design of the evaporator. The process of designing/selecting air cooling evaporators to operate in CO_2 refrigeration, air-conditioning and heat pump systems is very similar to designing/selecting evaporators for ammonia and other refrigerants according to the ASHRAE standard (2013). Typically, the following inputs are required for properly selecting either CO_2 evaporators:

(a) Elevation, (b) Return air ('Air on') temperature, (c) Return air relative humidity, (d) Evaporating temperature, (e) Type of feed, (f) Overfeed rate (if pumped feed), (g) Liquid pressure and temperature at the expansion valve (if direct expansion), (h) Required cooling duty, (i). Type of defrost, (j) Supply voltage, (k) Materials of construction, (l) Required MAWP (Maximum Allowable Working Pressure).

Other inputs may include: (m) Maximum allowable air velocity, (n) Minimum air flow rate, (o) Maximum allowable fan speed, (p) Maximum allowable sound pressure (usually in dB(A)), (q) Minimum air throw distance, (r) Minimum number of fans, (s) Dimensional constraints (maximum height or length limitations).

The Output data typically includes: (t) Actual cooling duty, (u) Air flow rate and air velocity, (v) Leaving air temperature, (w) Leaving air relative humidity, (x) Sound pressure level, (y) Air throw distance, (z) Dimensional characteristics and (aa) Electrical characteristics.

There are many similarities in evaporator rating methods and construction for ammonia and other refrigerants, however, the very different thermodynamic and chemical characteristics of CO_2 compared to ammonia and other refrigerants require special attention with regard to the following aspects:

- **Material Compatibility:** CO_2 can be used safely with copper and copper-bearing alloys. Actually, dry CO_2 is quite inert and can be used with all commonly used base metals; copper, carbon steel, stainless steel, and aluminum. Care must be taken to select materials with sufficient strength to withstand the higher MAWP required for CO_2. This normally rules out the use of aluminum with CO_2.
- **Pressure:** CO_2 evaporators will operate at significantly higher pressures than ammonia and other refrigerants for a given temperature.
- **Heat Transfer:** Thermodynamic and transport properties are very different for CO_2 compared to ammonia and other refrigerants and result in very different evaporator circuiting arrangements to achieve equivalent cooling capacity. The Cheng et al. models (Cheng et al. 2006a, b, 2008b, d) are recommended for calculation of evaporation heat transfer coefficient.

Both copper and stainless steel tubing and pipe are recommended for use in CO_2 evaporators provided the diameters and wall thicknesses meet the required design pressures. When using copper, a non-phosphorous bearing brazing alloy is recommended. This is needed to limit the risk of leaks caused by acidic conditions resulting from the presence of carbonic acid. Carbon steel is not recommended for use in CO_2 evaporators due to (a) susceptibility to corrosion in the presence of carbonic acid, and (b) embrittlement at low temperatures. Aluminum is not recommended for use in CO_2 evaporators due to its lower yield and tensile strength characteristics. ASHRAE Standard 15 (2013) establishes the design pressure requirements for CO_2 systems:

- ASHRAE Standard 15 requires the design pressure for CO_2 evaporators to be "at least 20% higher than the saturation pressure corresponding to the warmest location in the circuit." The "warmest location in the circuit" is interpreted as the warmest anticipated room temperature in which the evaporator(s) will operate. Although minimum tube wall thicknesses are recommended (ASHRAE 2013), the evaporator manufacturer must insure that all pressure bearing components in the coil, including headers and pipe connections, are designed correctly.
- The temperature used to establish design pressure must be carefully selected to account for conditions which include (but are not necessarily limited to) those shown here: (a) Startup conditions, (b) Peak load operation, (c) Abnormal loads (process temperature excursions) and (d) Standby conditions that occur frequently (i) Power outages limited in time duration but which may happen with some frequency and (ii) Shutdown during cleanup.

Reducing the overfeed rate in pumped refrigerant systems is desirable because pumping power will be reduced by the cube of the ratio of the reduction in flowrate. As the liquid overfeed rate is reduced, however, the risk of operating evaporators with the refrigerant in separated flow patterns (stratified/wavy) increases. Cooling capacity of the evaporator falls off dramatically when this occurs. Therefore, it is essential to maintain the minimum mass flow rate in the evaporators. For instance, with CO_2 in an evaporator having 5/8" tubes, a minimum mass flux of 200 kg/m^2s is required to avoid stratified/wavy flow. Optimum Overfeed Rates are recommended as follows:

- Pumped CO_2 systems can be successfully operated with lower overfeed rates compared to ammonia and other refrigerants.
- Recommended overfeed rates for pumped CO_2 evaporators are 1.5:1 for coolers and 2:1 for freezers.

CO_2 evaporators can be operated with direct expansion feed. Care must be taken by the evaporator manufacturer to circuit the coil in such a way that the refrigerant mass flux is kept above 200 kg/m^2s in order to avoid stratified/wavy flow. This becomes challenging with larger diameter tubes (greater than 5/8"). At very low temperatures, enhanced tubes (micro-fin copper) are recommended as a way to mitigate separated flow patterns and improve performance.

Defrost is another important aspect in CO_2 evaporator design and selection. CO_2 evaporators are commonly defrosted using the following methods: (a) Air, (b) Water and (c) Electric Resistance. Control valve groups for these methods of defrost are very simple and low cost. Hot gas defrost with CO_2 evaporators is not commonly used. In a cascade system, the intermediate CO_2 temperature/pressure is normally too low to allow the CO_2 from that circuit to be used for defrost. This then requires a separate high pressure (capable of 50 bar) compressor with sufficient capacity to be installed expressly for the purposes of providing hot gas for defrost. Other means of generating hot CO_2 gas for defrost include the use of a heat-driven boiler vessel, typically heated by discharge gas from the high side of the cascade system. The complexity and added expense of hot gas defrost with CO_2 have limited its application.

6.8 Concluding Remarks

A general CO_2 two-phase flow patterns and models of CO_2 evaporation heat transfer and two phase frictional pressure drops in macroscale- and micro-scale- channels and their applications in modeling and designing CO_2 evaporators in refrigeration, air-conditioning and heat pump systems are presented and discussed in this chapter. The available experimental studies on the relevant topics are analyzed and summarized at first. Then, the Cheng et al. models of two-phase flow pattern map for CO_2 evaporation, flow pattern based flow boiling heat transfer and pressure drop models are presented. The oil effects on CO_2 evaporation heat transfer and two-phase frictional pressure drops are discussed as well. Furthermore, modeling of CO_2 evaporation and simulation/design of CO_2 evaporators are reviewed and analyzed. The research needs in the future are identified. Finally, CO_2 evaporator design and selection are presented for practical application. The main conclusions are summarized as follows:

(1) According to comprehensive literature review, the available experimental results have shown different evaporation heat transfer and two-phase flow characteristics of CO_2 at high and low reduced pressures. The evaporation heat transfer and two-phase flow of CO_2 at saturation temperatures ranging from 0 to 25 °C show different characteristics from those of conventional refrigerants due to the significant differences in thermal physical and transport properties. CO_2 has much higher evaporation heat transfer and much lower two-phase frictional pressure drops than other low pressure refrigerants such as R134a and ammonia etc. The evaporation heat transfer mechanisms are the dominance of the nucleate boiling at low/moderate vapor qualities prior to dryout and the occurrence of dryout in CO_2 at relatively lower vapor qualities than conventional refrigerants. Furthermore, the effect of the saturation temperature on the evaporation heat transfer coefficient is more noticeable. At high saturation temperatures, the nucleate boiling mechanism is more pronounced at low vapor qualities.

(2) The experimental data from the different independent studies show somewhat different evaporation heat transfer trends at similar test conditions. It is difficult to explain the differences at similar conditions or the parameter effects on the evaporation heat transfer behaviors according to the physical mechanisms. This might be caused by large measurement uncertainties. Therefore, more

accurate experimental data are needed for CO_2 evaporation in both macro- and micro-channels through well designed test facilities and careful experiments. In particular, a wide range of test conditions should be designed to meet the practical needs in industry.

(3) Overall, the Cheng et al. general flow pattern map and flow pattern based evaporation heat transfer model for CO_2 reasonably predicts the observed flow patterns and experimental heat transfer coefficients in the literature and also capture properly the parametric trends. However, for the dryout and mist flow regimes with partially or all dry perimeters, the heat transfer model is only partially satisfactory. Therefore, more careful experiments are needed in these two regimes to provide more accurate heat transfer data, with attention to also determine the transitions x_{di} and x_{de}, because they are typically working conditions in the micro-scale channels of extruded multi-port tubes used for refrigerant, automobile air-conditioning and refrigeration systems that operate over a wide range of mass velocities up to as high as 1500 kg/m^{2}s.

(4) The Cheng et al. flow pattern based CO_2 two-phase flow pressure drop model predicts the CO_2 pressure drop database better than the existing methods. Due to the very few and less accurate experimental data in micro-scale channels currently available, it is suggested that additional, more accurate experimental CO_2 pressure drop data be obtained to further test or improve the model in the future.

(5) Oil has significant effects on CO_2 evaporation heat transfer and two-phase pressure drops. In general, very small amounts of the lubricating oil (<0.5% wt) seem to have little effect while larger concentrations (>1% wt) tend to dramatically reduce flow boiling heat transfer coefficients. For two-phase pressure drops, the effect of lubricating oil only occurs at vapor qualities larger than around 0.6. Further experimental studies on evaporation heat transfer and two-phase frictional pressure drops of CO_2-immiscible oil mixtures in both macro-scale and micro-scale channels be conducted over a wide range of test conditions to provide more experimental data on these aspects. In order to understand the heat transfer mechanisms of evaporation with CO_2 and immiscible oil, flow visualization and local measurement of oil concentration are also suggested to be done. Due to the immiscibility of lubricant oil and CO_2, whether the available methods for miscible oil can be applied to CO_2-lubricant oil or not still need to be further verified. Therefore, extensive experimental data are needed to verify the available methods or to develop new methods for CO_2.

(6) The Cheng et al. flow pattern based CO_2 evaporation heat transfer and two-phase flow pressure drop models have been adopted in the simulations of evaporators in various thermal systems by a number of researchers. It shows that the experimental results favorably agree with the simulation results. Therefore, it is recommended that Cheng et al. methods be used in the design and simulation of CO_2 evaporators.

(7) As an excellent working fluid, CO_2 has also been investigated in other practical applications such as electronic chip cooling, evaporative CO_2 cooling system for the upgrade of the CMS pixel detector and potential for renewable energy and

power generation. Therefore, effort should be made to explore new applications of CO_2 evaporation in the relevant fields.

(8) CO_2 evaporator design and selection issues such as material selection, operation pressure, optimum minimum mass flow rate and defrost etc. have been presented for the practical design in industry.

References

ASHRAE, ANSI/ASHRAE Standard 15-2013 (2013) Safety Standard for Refrigeration Systems. Atlanta, GA, American Society of Heating Refrigerating and Air-Conditioning Engineers

Baker O (1954) Design of pipe lines for simultaneous flow of oil and gas. Oil Gas J 53:185–190

Bandarra Filho EP, Cheng L, Thome JR (2009) Flow boiling characteristics and flow pattern visualization of refrigerant/lubricant oil mixtures. Int J Refrig 32:185–2002

Bergman TL, Lavine A, Incropera FP, DeWitt DP (2017) Fundamentals of heat and mass transfer, 8th edn. John Wiley

Biberg D (1999) An explicit approximation for the wetted angle in two-phase stratified pipe flow. Can J Chem Eng 77:1221–1224

Bredesen A, Hafner A, Pettersen J, Neksa P, Aflekt K (1997) Heat transfer and pressure drop for in-tube evaporation of CO_2. In: Proceedings of the international conference on heat transfer issues in natural refrigerants. University of Maryland, USA, pp 1–15

Chen JC (1966) Correlation for boiling heat transfer to saturated fluids in convective flow. Ind Chem Eng Des Dev 5:322–339

Cheng L (2010) Microscale nanoscale thermal and fluid transport phenomena: rapidly developing research fields. Int J Microscale Nanoscale Therm Trans 1(1):3–6

Cheng L (2013) Fundamental issues of critical heat flux phenomena during flow boiling in microscale-channels and nucleate pool boiling in confined spaces. Heat Transfer Eng 34(13):1011–1043

Cheng L (2014) Evaluation of correlations for supercritical CO_2 cooling convective heat transfer and pressure drop in macro- and micro-scale tubes. Int J Microscale Nanoscale Thermal Fluid Transport Phenomena 5(2):113–126

Cheng L (2016a) Microscale flow patterns and bubble growth in microchannels. In: Saha SK (ed) Microchannel phase change heat transfer. Elsevier Publisher, pp 91–140

Cheng L (2016b) Flow boiling heat transfer with models in microchannels. In: Saha SK (ed) Microchannel phase change heat transfer. Elsevier Publisher, pp 141–191

Cheng L, Chen T (2000) Comparison of six typical correlations for upward flow boiling heat transfer with kerosene in a vertical smooth tube. Heat Transfer Eng 21(5):27–34

Cheng L, Mewes D (2006) Review of two-phase flow and flow boiling of mixtures in small and mini channels. Int J Multiphase Flow 32:183–207

Cheng L, Thome JR (2009) Cooling of microprocessors using flow boiling of CO_2 in micro-evaporators: Preliminary analysis and performance comparison. Appl Therm Eng 29:2426–2432

Cheng L, Liu L (2011) Analysis and evaluation of gas liquid two phase frictional pressure drop prediction methods for microscale channels. Int J Microscale Nanoscale Therm Fluid Trans Phen 2(4):257–280

Cheng L, Xia G (2017) Fundamental issues, mechanisms and models of flow boiling heat transfer in microscale channels. Int J Heat Mass Transfer 108 (Part A):97–127

Cheng L, Mewes D, Luke A (2007a) Boiling phenomena with surfactants and polymeric additives: a state-of-the-art review. Int J Heat Mass Transfer 50:2744–2771

Cheng L, Ribatski G, Thome JR (2007b) On the prediction of flow boiling heat transfer of CO_2. In: The 22nd IIR international congress of refrigeration, Beijing, P.R. China, 21–26 Aug 2007

Cheng L, Ribatski G, Thome JR (2008a) Analysis of supercritical CO_2 cooling in macro- and micro Channels. Int J Refrig 31:1301–1316

Cheng L, Ribatski G, Thome JR (2008b) New prediction methods for CO_2 evaporation inside tubes: part II — a general flow boiling heat transfer model based on flow patterns. Int J Heat Mass Transfer 51:125–135

Cheng L, Ribatski G, Thome JR (2008c) Gas-liquid two-phase flow patterns and flow pattern maps: fundamentals and applications. ASME Appl Mech Rev 61, 050802-1–050802-28

Cheng L, Ribatski G, Moreno J, Thome JR (2008d) New prediction methods for CO_2 evaporation inside tubes: part I - A general two-phase flow pattern map and development of a phenomenological model of two-phase flow frictional pressure drop. Int J Heat Mass Transfer 51:111–124

Cheng L, Ribatski G, Wojtan L, Thome JR (2006a) New flow boiling heat transfer model and flow pattern map for carbon dioxide evaporating inside horizontal tubes. Int J Heat Mass Transfer 49:4082–4094

Cheng L, Ribatski G, Wojtan L, Thome JR (2006b) Erratum to: new flow boiling heat transfer model and flow pattern map for carbon dioxide evaporating inside tubes [Heat Mass Transfer 49(21–22) (2006b)4082–4094]. Int J Heat Mass Transfer 50:391

Chisholm D (1973) Pressure gradients due to friction during the flow of evaporating two-phase mixtures in smooth tubes and channels. Int J Heat Mass Transfer 16:347–358

Cho JM, Kim MS (2007) Experimental studies on the evaporative heat transfer and pressure drop of CO_2 in smooth and micro-fin tubes of the diameters of 5 and 9.52 mm. Int J Refrig 30:986–994

Ciccitti A, Lombardi C, Silvestri M, Soldaini G, Zavattarelli R (1960) Two-phase cooling experiments—pressure drop, heat transfer and burnout measurements. Energ Nucleare 7(6):407–425

Cooper MG (1984) Saturation nucleate pool boiling: a simple correlation. In: 1st U.K. national conference on heat transfer, vol 2, pp 785–793

Daguin J, Arndt K, Bertl W, Noite J, Petagna P, Postema H, Tropea P, Verlaat B (2012) Evaporative CO_2 cooling system for the upgrade of the CMS pixel detector at CERN. In: 13th InterSociety conference on thermal and thermomechanical phenomena in electronic systems, 30 May–1 June 2012. https://ieeexplore.ieee.org/abstract/document/6231499

Dittus FW, Boelter LMK (1930) Heat transfer in automobile radiator of the tubular type. Univ Calif Publ Eng 2:443–461

EES (2005) Engineering equation solver. F-Chart Software Inc., Wisconsin (USA)

El Hajal J, Thome JR, Cavallini A (2003) Condensation in horizontal tubes, part 2: new heat transfer model based on flow regimes. Int J Heat Mass Transfer 46:3365–3387

Faria RN, Nunes RO, Koury RNN, Machado L (2016) Dynamic modeling study for a solar evaporator with expansion valve assembly of a transcritical CO_2 heat pump. Int J Refrig 64:203–213

Friedel L (1979) Improved friction drop correlations for horizontal and vertical two-phase pipe flow. In: European two-phase flow group meeting, Ispra, Italy, paper E2

Gao L, Honda T (2005a) Effects of lubricant oil on boiling heat transfer of CO_2 inside a horizontal smooth tube. In: 42nd national heat transfer symposium of Japan, pp 261–270

Gao L, Honda T (2005b) An experimental study on flow boiling heat transfer of carbon dioxide and oil mixtures inside a horizontal smooth tube. In: IIR Vicenza conference-thermophysical properties and transfer processes of refrigerants, Vicenza, Italy, pp 237–243

Gao L, Watanabe Y, Honda T (2011) CO_2 flow boiling in small diameter smooth and micro-fin tubes. J JSME 11:SS66–SS72

Gasche JL (2006) Carbon dioxide evaporation in a single micro-channel. J Braz Soc Mech Sci Eng 28:69–83

Grauso S, Mastrullo R, Mauro AW, Vanoli GP (2013) Flow boiling of R410A and CO_2 from low to medium reduced pressures in macro channels: experiments and assessment of prediction methods. Int J Heat Mass Transfer 56(Issues 1–2):107–118

Groeneveld DC (1973) Post dry-out heat transfer at reactor operating conditions. In: ANS topical meeting on water reactor safety, Salt Lake City

Grönnerud R (1979) Investigation of liquid hold-up, flow-resistance and heat transfer in circulation type of evaporators, part IV: two-phase flow resistance in boiling refrigerants. In: Annexe 1972–1, Bull. De l'Inst. Du Froid

Gungor KE, Winterton RHS (1986) A general correlation for flow boiling in tubes and annuli. Int J Heat Mass Transfer 29:351–358

Hashitume K (1983) Flow pattern and void fraction of refrigerant two-phase flow in a horizontal pipe. Bull JSME 26:1597–1602

Hihara E (2006) Fundamental technology for carbon dioxide operated heat pumps. In: JSAE automotive air-conditioning symposium, pp 243–262

Hihara E, Tanaka S (2000) Boiling heat transfer of carbon dioxide in horizontal tubes. In: Preliminary proceedings of IIR Gustav Lorentzen conference on natural working fluids, Purdue, pp 279–284

Ikeda H, Kaneko M, Tokiai T, Yoshii A, Suto H (2006) Evaluation of lubricants for carbon dioxide automobile A/C system. In: SAE 7th alternate refrigerant system symposium

Jeong S, Cho E, Kim H (2005) Evaporative heat transfer and pressure drop in a microchannel tube. In: Proceedings of the 3rd international conference on microchannels and minichannels, Toronto, Ontario, Canada, Part B, pp 103–108

Kandlikar SG (1990) A general correlation for saturated two-phase flow boiling heat transfer inside horizontal and vertical tubes. J Heat Transfer 112:219–228

Kandlikar SG (2002) Fundamental issues related to flow boiling in minichannels and microchannels. Exp Therm Fluid Sci 26:389–407

Kaneko M, Ikeda H, Kawaguchi Y, Suto H, Schweim E (2004) The development of capped-PAG refrigeration lubricants for automotive A/C with CO_2 and HFC152a. In: SAE 5th alternate refrigerant systems symposium

Katsuta M, Kinpara H, Yagi S, Mukaiyama H (2004) The effect of oil contamination on evaporator heat transfer characteristics of CO_2 refrigeration cycle. Trans Jpn Soc Refrig Air Cond Eng (JSRAE) 21(3):275–283

Katsuta M, Takeo N, Kamimura I, Mukaiyama H (2003a) A study on the evaporator of refrigerant cycle: characteristics of heat transfer coefficient and pressure drop on mixing CO_2 and oil (PAG). Trans Jpn Soc Refrig Air Cond Eng (JSRAE) 20(3):397–405

Katsuta M, Kinpara H, Aoyagi T, Yagi S, Kamimura I, Mukaiyama H (2003b) The effect of oil contamination on evaporator heat transfer characteristics of CO_2 refrigeration cycle. In: 40th National heat transfer symposium of Japan

Kattan N, Thome JR, Favrat D (1998a) Flow boiling in horizontal tubes. Part 1: Development of a diabatic two-phase flow pattern map. J Heat Transfer 120:140–147

Kattan N, Thome JR, Favrat D (1998b) Flow boiling in horizontal tubes: part 2-new heat transfer data for five refrigerants. J Heat Transfer 120:148–155

Kattan N, Thome JR, Favrat D (1998c) Flow boiling in horizontal tubes: part-3: development of a new heat transfer model based on flow patterns. J Heat Transfer 120:156–165

Kew PA, Cornwell K (1997) Correlations for the prediction of boiling heat transfer in small-diameter channels. Appl Therm Eng 17:705–715

Kim MH, Pettersen J, Bullard CW (2000) Fundamental process and system design issues in CO_2 vapor compression systems. Prog Energy Combust Sci 30:119–174

Knudsen HJ, Jensen RH (1997) Heat transfer coefficient for boiling carbon dioxide. In: Workshop proceedings—CO_2 technologies in refrigeration, heat pumps and air conditioning systems, Trondheim, Norway, pp 319–328

Koyama S, Kuwahara K, Shinmura E, Ikeda S (2001) Experimental study on flow boiling of carbon dioxide in a horizontal small diameter tube. In: IIR commission B1 meeting, Paderborn, Germany, pp 526–533

Koyama S, Lee S, Ito D, Kuwahara K, Ogawa H (2004a) Experimental study on flow boiling of pure CO_2 and CO_2-oil mixtures inside horizontal smooth and micro-fin copper tubes. In: Proceedings of 6th IIT-Gustav Lorentzen conference, Glasgow, UK

Koyama S, Lee SM, Ito D, Kuwahara K, Ogawa H (2004b) Experimental study on flow boiling of pure CO_2 and CO_2–oil mixture inside horizontal smooth and micro-fin tubes. In: 6th IIR-Gustav Lorentzen conference, Glasgow, UK, Aug 29 –Sept 1st 2004b

Koyama S, Ito D, Lee SM, Kuwahara K, Saeki C, Ogawa H (2005) Experimental investigation for flow boiling of pure CO_2 and CO_2–oil mixture inside horizontal copper tubes. In: IIR 2005 Vicenza conference-thermophysical properties and transfer processes of refrigerants, Vicenza, Italy

Kutateladze SS (1948) On the transition to film boiling under natural convection. Kotloturbostroenie 3:10–12

Liu Z, Winterton RHS (1991) A general correlation for saturated and subcooled flow boiling in tubes and annuli based on a nucleate pool boiling equation. Int J Heat Mass Transfer 34:2759–2766

Lorentzen G (1994) Revival of carbon dioxide as al refrigerant. Int J Refrig 17:292–330

Lorentzen G (1995) The use of natural refrigerants: a complete solution to the CFC/HCFC predicament. Int J Refrig 18:190–197

Marcinichen JB, Thome JR, Pereira RH (2016) Working fluid charge reduction, part 1: CO_2 fin tube evaporator designed for light commercial appliances utilizing flow pattern based phenomenological prediction models. Int J Refrig 65:258–272

Mastrullo R, Mauro AW, Rosato A, Vanoli GP (2009) Carbon dioxide local heat transfer coefficients during flow boiling in a horizontal circular smooth tube. Int J Heat Mass Transfer 52(19–20):4189–4194

Mastrullo R, Mauro AW, Rosato A, Vanoli GP (2010) Carbon dioxide heat transfer coefficients and pressure drops during flow boiling: assessment of predictive methods. Int J Heat Mass Transfer 6:1068–1085

Mehendale SS, Jacobi AM, Ahah RK (2000) Fluid flow and heat transfer at micro- and meso-scales with application to heat exchanger design. ASME Appl Mech Rev 53:175–193

Moreno Quibén J, Thome JR (2007a) Flow pattern based two-phase frictional pressure drop model for horizontal tubes, part i: diabatic and adiabatic experimental study. Int J Heat Fluid Flow 28:1049–1059

Moreno Quibén J, Thome JR (2007b) Flow pattern based two-phase frictional pressure drop model for horizontal tubes, part ii: new phenomenological model. Int J Heat Fluid Flow 28:1060–1072

Moreno Quibén J, Cheng L, da Silva Lima RJ, Thome JR (2009a) Flow boiling in horizontal flattened tubes: part i — two-phase frictional pressure drop results and model. Int J Heat Mass Transfer 52:3634–3644

Moreno Quibén J, Cheng L, da Silva Lima RJ, Thome J (2009b) R Flow boiling in horizontal flattened tubes: part ii — flow boiling heat transfer results and model. Int J Heat Mass Transfer 52:3645–3653

Mori H, Yoshida S, Ohishi K, Kokimoto Y (2000) Dryout quality and post dryout heat transfer coefficient in horizontal evaporator tubes. In: Proceeding of 3rd European thermal sciences conference, pp 839–844

Müller-Steinhagen H, Heck K (1986) A simple friction pressure drop correlation for two-phase flow in pipes. Chem Eng Process 20:297–308

Ould-Didi MB, Kattan N, Thome JR (2002) Prediction of two-phase pressure gradients of refrigerants in horizontal tubes. Int J Refrig 25:935–947

Park CY, Hrnjak PS (2005) Flow boiling heat transfer of CO_2 at low temperatures in a horizontal smooth tube. J Heat Transfer 127:1305–1312

Patiño J, Llopis R, Sánchez D, Sanz-Kock C, Cabello R, Torrella E (2014) A comparative analysis of a CO_2 evaporator model using experimental heat transfer correlations and a flow pattern map. Int J Heat Mass Transfer 71:361–375

Pettersen J (2004) Flow vaporization of CO_2 in microchannel tubes. Exp Therm Fluid Sci 28:111–121

Pettersen J, VestbØstad K (2000) Heat transfer and pressure drop for flow of supercritical and subcritical in microchannel tubes. In: Workshop proceedings-selected issue on CO_2 as working fluid in compression system. Trondheim, Norway, pp 101–114

REFPROP (1998) NIST Refrigerant properties database 23. Gaithersburg, MD, Version 6.01

REFPROP (2002) NIST Refrigerant properties database 23. Gaithersburg, MD, Version 7.0

Rouhani Z, Axelsson E (1970) Calculation of volume void fraction in a subcooled and quality region. Int J Heat Mass Transfer 17:383–393

Sawalha S (2005) Using CO_2 in supermarket refrigeration. ASHRAE J 47(8):26–30

Shah MM (1982) Chart correlation for saturated boiling heat transfer: equations and further study. ASHRAE Trans 88:185–196

Shinmura E, Take K, Koyama S (2006) Development of high-performance CO_2 evaporator. In: JSAE automotive air-conditioning symposium, pp 217–227

Siegismund V, Kauffeld M (2004) Influence of lubricant oil o CO_2 heat transfer in minichannel tubes. In: 6th IIR-Gustav Lorentzen Conference, Glasgow, UK, Aug 29–Sept 1 2004

Steiner D (1993) VDI-Wärmeatlas, Verein Deutscher Ingenieure VDI-Gessellschaft Verfahrenstechnik und Chemieingenieurwessen (GCV), Düsseldorf, Ch. Hbb

Taitel Y, Dukler AE (1976) A model for predicting flow regime transitions in horizontal and near horizontal gas-liquid flow. AIChE J 22:47–55

Tanak S, Daiguji H, Takemura F, Hihara E (2001) Boiling heat transfer of carbon dioxide in horizontal tubes. In: 38th National heat transfer symposium of Japan, pp 899–900

Thome JR (1996) Boiling of new refrigerants: a state-of-the-art review. Int J Refrig 19:435–457

Thome JR (2004a) Boiling in microchannels: a review of experiment and theory. Int J Heat Fluid Flow 25:128–139

Thome JR (2004b) Engineering data book III. Wolverine Tube Inc. http://www.wlv.com/products/databook/db3/DataBookIII.pdf

Thome JR (2006) State-of-the art overview of boiling and two-phase flows in microchannels. Heat Transfer Eng 27(9):4–19

Thome JR, El Hajal J (2002) Two-phase flow pattern map for evaporation in horizontal tubes: latest version. In: 1st International conference on heat transfer, fluid mechanics and thermodynamics, Kruger Park, South Africa, 8–10 Apr 2002, pp 182–188

Thome JR, Ribatski G (2005) State-of-the-art of two-phase flow and flow boiling heat transfer and pressure drop of CO_2 in macro- and micro-channels. Int J Refrig 28:1149–1168

Thome JR, Cheng L, Ribatski G, Vales LF (2008) Flow boiling of ammonia and hydrocarbons: a state-of-the-art review. Int J Refrig 31:603–620

Thulukkanam K (2013) Heat exchanger design handbook, 2nd edn CRC Press

Tong Z, Liu XH, Li Z, Jiang Y (2016) Experimental study on the effect of fill ratio on an R744 two-phase thermosyphon loop. Appl Therm Eng 99:302–312

Wang S, Tuo H, Cao F, Xing Z (2013) Experimental investigation on air-source transcritical CO_2 heat pump water heater system at a fixed water inlet temperature. Int J Refrig 36:701–716

Wetzel M, Dietrich B, Wetzel T (2014) Influence of oil on heat transfer and pressure drop during flow boiling of CO_2 at low temperatures. Exp Therm Fluid Sci 59:202–212

Wojtan L, Ursenbacher T, Thome JR (2005a) Investigation of flow boiling in horizontal tubes: part i—A new diabatic two-phase flow pattern map. Int J Heat Mass Transfer 48:2955–2969

Wojtan L, Ursenbacher T, Thome JR (2005b) Investigation of flow boiling in horizontal tubes: part ii—development of a new heat transfer model for stratified-wawy, dryout and mist flow regimes. Int J Heat Mass Transfer 48:2970–2985

Wu J, Koettig T, Ch, Franke, Helmer D, Eisel T, Haug F, Bremer J (2011) Investigation of heat transfer and pressure drop of CO_2 two-phase flow in a horizontal minichannel. Int J Heat Mass Transfer 54(9–10):154–2162

Xia G, Wang W, Cheng L, Ma D (2017) Visualization study on the instabilities of phase-change heat transfer in a flat two-phase closed thermosyphon. Appl Therm Eng 116:392–405

Yamaguchi S, Kato D, Saito K, Kawai S (2011) Development and validation of static simulation model for CO_2 heat pump. Int J Heat Mass Transfer 54:1986–1906

Yang D, Song Y, Cao F, Jin L, Wang X (2016) Theoretical and experimental investigation of a combined R134a and transcritical CO_2 heat pump for space heating. Int J Refrig 72:156–170

Yoon SH, Cho ES, Hwang YW, Kim MS, Min K, Kim Y (2004) Characteristics of evaporative heat transfer and pressure drop of carbon dioxide and correlation development. Int J Refrig 27:111–119
Yoshioka S, Kim HY, Kasai K (2008) Heat transfer performance and oil behaviour for R744 with PAG oil in air-cooled heat exchanger, In: 8th IIR Gustav lorentzen conference on natural working fluids, pp 389–396
Yun R, Kim Y (2003) Two-phase pressure drop of CO_2 in mini tubes and microchannels. In: First international conference on microchannels and minichannels, Rochester, NY, pp 507–511
Yun R, Kim Y (2004) Two-phase pressure drop of CO_2 in mini tubes and microchannels: Microscale Thermophysical Eng 8:259–270
Yun R, Choi C, Kim Y (2002) Convective boiling heat transfer of carbondioxide in horizontal small diameter tubes. In: IIR/IIF-Commission B1, B2, E1 and E2-Guangzhou, China, pp 293–303
Yun R, Kim Y, Kim MS (2005a) Flow boiling heat transfer of carbon dioxide in horizontal mini tubes. Int J Heat Fluid Flow 26:801–809
Yun R, Kim Y, Kim MS (2005b) Convective boiling heat transfer characteristics of CO_2 in microchannels. Int J Heat Mass Transfer 48:235–242
Yun R, Kim Y, Kim MS, Choi Y (2003) Boiling heat transfer and dryout phenomenon of CO_2 in a horizontal smooth tube. Int J Heat Mass Transfer 46:2353–2361
Zhao X, Bansal PK (2007) Flow boiling heat transfer characteristics of CO_2 at low temperatures. Int J Refrig 30:937–945
Zhao Y, Molki M, Ohadi MM, Dessiatoun SV (2000a) Flow boiling of CO_2 in microchannels. ASHRAE Trans 106(Part I):437–445
Zhao Y, Molki M, Ohadi MM (2000b) Heat transfer and pressure drop of CO_2 flow boiling in microchannels. In: Proceedings of the ASME heat transfer division, vol 2, pp 243–249
Zhao Y, Molki M, Ohadi MM, Fanca FHR, Radermacher R, Cho K (2002) Flow boiling of CO_2 with miscible oil in microchannels. ASHRAE Trans 108(Part I):135–144

Chapter 7
CO_2 Commercial Refrigeration Cycle and Systems

Trygve Magne Eikevik

The retail sector is one of the most important applications for CO_2 refrigeration cycles. In this chapter, the systems used for this application are detailed, presenting the potential uses of this refrigerant and delving into different configurations. The developments associated with an improved performance at locations with warm and hot climates are discussed. Potential integration of HVAC needs with the refrigeration system is also explored. This chapter ends with a review of future trends and research challenges that will be addressed in the near future.

7.1 Main Features of Commercial Refrigeration and Systems Used

Nowadays supermarket sector is fundamental in most of Europe and other developed countries, and it is growing and spreading fast in other developing countries, such as India or China. The factors that can explain this trend are urbanization, globalization and an emerging middle class in developing countries (Karampour et al. 2016). Supermarkets are buildings with high energy consumption per unit of area, and they use, on average, 3–4% of the electricity generated in industrialized countries. The heating, ventilation, air conditioning and refrigeration (HVAC&R) loads of a supermarket may account for more than 50% of the total energy consumption of a supermarket.

In addition to the indirect CO_2 emissions associated with electricity generation for supermarkets, there are direct CO_2 emissions caused by leakages of the working fluids of HVAC&R units. Still, many systems use hydrofluorocarbons (HFCs) with very

T. M. Eikevik (✉)
Department of Energy and Process Engineering, Norwegian University of Science and Technology (NTNU), Trondheim, Norway
e-mail: trygve.m.eikevik@ntnu.no

X.-R. Zhang and T. M. Eikevik (eds.), *CO_2 Refrigeration Cycle and Systems*, Lecture Notes in Energy 96, https://doi.org/10.1007/978-3-031-22512-3_7

Table 7.1 Temperature levels for product and for evaporation usually found in commercial refrigeration

Product	Product temperature [°C]	Conventional evaporation temperature [°C]	Objective evaporation temperature [°C]
Cooling	1–14	−10 to −5	−2
Freezing	−12 to −18	−35 to −30	−25

high global warming potential (GWP) and important leakage rates. An increasing number of countries worldwide are committed to follow the Kigali amendment of the Montreal Protocol, which aims at reducing the use of HFCs by 85% by 2050 to limit the global temperature rise to 2 °C (https://www.unenvironment.org). Regulations have been enforced to eliminate or at least reduce the use (and refrigerant charge) of HFCs in favor of natural refrigerants, such as ammonia (NH_3), CO_2 or hydrocarbons (HC), or the new generation of low-GWP HFCs, which are also called hydrofluoroolefins (HFOs). An example in Europe is the F-Gas Regulation (https://eur-lex.europa.eu/legal-content/EN/TXT/?uri=uriserv:OJ.L_.2014.150.01.0195.01.ENG), combined with special taxes for HFCs in some countries.

Refrigeration in the commercial sector is used to preserve food in the display cabinets or the cold chambers in the supermarket. Products can be stored at temperatures above 0 °C (cooling), usually regarded as medium temperature (MT) refrigeration, or below 0 °C (freezing), which is low temperature (LT) refrigeration. Table 7.1 describes the typical preservation temperatures for both cooled and frozen products (Kauko et al. 2016). In addition, the table shows the conventional evaporation temperatures required to maintain the temperatures of the air inside the cabinets and cold chambers. Measures such as an appropriate layout of the supermarket, using efficient cabinets with doors or lids, flooded evaporators, air humidity control in the sales area, etc. help to increase the evaporation temperatures and reduce the energy consumption of the supermarket.

Commercial refrigeration systems, which are the main focus of this chapter, can be classified into three different types: multiplex systems, distributed compressor systems and secondary loop systems (Lazzarin et al. 2018). A short description of each type of commercial refrigeration system is included below. CO_2 can be found as refrigerant in any of these, but it is now preferred as a refrigerant for multiplex (centralized) units.

Multiplex systems consist of at least a compressor rack (depending on the refrigerant choice), which is usually located in the machine room, outside the sales area. The liquid line with liquid refrigerant is supplied to different air-to-refrigerant evaporator coils located in retail cabinets or store coolers. On the other hand, the condenser(s) of the compressor pack is often located on the roof or at a side wall of the supermarket building. Typically, R404A has been used, but the recent increase in prices of HFC refrigerants, particularly in Europe, have made that CO_2 is increasing its share of the market.

Multiplex systems are relatively easy to build, and the control of their refrigeration capacity is simple, which makes them very attractive. In addition, since the compressors are located outside the sales area, their noise does not disturb customers. Moreover, heat recovery from their condenser/gas cooler can be implemented, which reduces the payback time of the whole unit and eliminates the need for auxiliary boilers or heat pumps.

The main disadvantages of Multiplex refrigeration systems are associated with the long lines that come from the machine room to different display cabinets and evaporators and return. First, large refrigerant charge quantities are required, usually in the order of magnitude of thousands of kilograms. In case of leakage, total or partial, refrigerant needs to be filled to guarantee the right operation of the unit. This is particularly critical in refrigeration systems using R404A, due to its high environmental impact and cost. The impact of CO_2 systems is much lower due to the relatively low cost and low GWP of the refrigerant. The scenario of total loss of the charge of the refrigeration unit is unprocurable, and even though the tightness of systems is improving, the annual refrigerant leakage rate is reported to be as high as 15%, with some leaking as much as 30% of the charge per year (Devin et al. 2015). In addition, long refrigerant lines can cause high pressure drops, which deteriorate the COP depending on the refrigerant more or less(pressure drop in CO_2 refrigeration units has relative effect on the efficiency). The recovery of oil can also be complicated due to the long piping used in Multiplex systems.

Distributed compressor systems' compressor racks (from 3 to 5) are distributed throughout the supermarket area, nearby the group of display cabinets where they supply. Even though there are different possibilities in terms of heat rejection at the condenser of each compressor rack, a common solution is to add a secondary loop that brings the heat to a rooftop unit. This reduces even further than the refrigerant charge that the units need. The main advantage of such a configuration is that the refrigerant lines can be much shorter than that of Multiplex systems, which reduces the charge of refrigerant, pressure drops and leakage rates largely. However, their main inconvenience is that the noise in the sales area is increased.

Plug-in cabinets, with at least one self-contained vapor compression loop, are distributed compressor systems brought to the extreme. Each cabinet is equipped with a finned-coil evaporator, a compressor, an expansion valve and a condenser. The condenser can be air-cooled, with the air from the shopping area as heat sink, or can have a water loop to bring the heat from the supermarket to an air-cooler located outdoors or to a chilling unit. Water-to-water chillers are expected for the LT display cabinets and refrigeration systems, which use heat sinks with lower temperature to keep a reasonable pressure ratio. The power consumption of the water pumps and water chillers cannot be neglected when water-cooled configurations are evaluated.

Plug-ins are particularly interesting for small size retail, where a multiplex/centralized solution is not feasible, or for specific display cabinets distributed in different locations of the supermarket which are not easily accessible for the centralized unit. In addition, they allow independent selection of the evaporation temperature in each cabinet as a function of the product preserved, which is a significant advantage.

R290 (propane) is currently proposed as a very efficient refrigerant for plug-ins due to its outstanding performance and the low-charges required, which can be below the 150 g limit per loop indicated in the IEC 60,335–2-89 (https://webstore.iec.ch/publication/22446). There was an initial attempt of introducing CO_2 for this application (beverage refrigerators), but it was soon dropped.

Secondary loop systems are the units that incorporate an additional working fluid between the refrigerant and the air in the display cabinets or cooling/freezing rooms. This additional fluid, normally called secondary fluid, is cooled down by the refrigerant at the evaporator of the indirect refrigeration system and pumped to the different display cabinets and chambers, in which it absorbs heat from the air, and then returns to the evaporator.

The main purpose of systems with a secondary loop is to reduce the refrigerant charge and leakage rate, which could improve safety and reduce environmental impact and cost. It has been common practice to use as secondary fluids aqueous solutions of salts or alcohols. There also have been some attempts of using ice slurries as secondary fluids (Hägg 2005). A third option is to utilize the phase change of CO_2, as described more in detail in the next section, condensing it in the evaporator of the main refrigeration system, pumping liquid CO_2 to the cabinets, and evaporating it in the evaporator.

The main disadvantage of secondary loop systems is that an additional heat exchange exists, which reduces the overall performance of the unit. In addition, the brines normally used are very viscous at working temperatures, increasing the pumping power. CO_2 solves these brine-related issues, but it is also more delicate to operate.

7.2 Evolution from Secondary Refrigerant to All-CO_2 Commercial Refrigeration Systems

As mentioned above, CO_2 can be found as refrigerant for commercial applications in secondary loops, in the lower-stage of cascade refrigeration systems or only as a refrigerant. This subsection explores the main characteristics of these systems using carbon dioxide.

7.2.1 CO_2 as Secondary Fluid

Indirect refrigeration systems use a secondary loop to connect the refrigeration system and the display cabinets and chambers where the products are preserved. CO_2 has been suggested as an appropriate secondary fluid operating in the liquid–vapor phase change region, due to the high heat transfer coefficients of CO_2 condensing in the evaporator of the main refrigeration system and evaporating at the cabinets (Fig. 7.1).

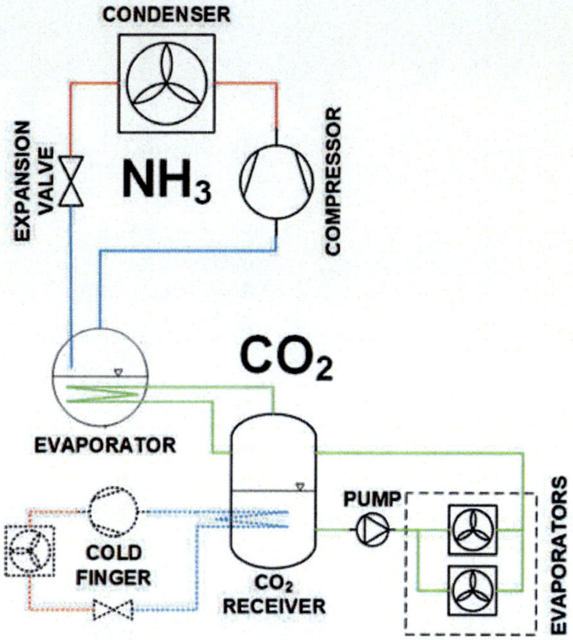

Fig. 7.1 Indirect ammonia (NH_3) refrigeration system using CO_2 as secondary fluid (in dotted lines the cold finger that could be installed to keep the pressure of the CO_2 loop)

Different from conventional secondary fluids (brines) that use sensible heat, Operating with phase change involves the use of latent heat of CO_2, which means the mass flow can be reduced importantly, even if the evaporators are overfed. This, combined with the much lower viscosity and higher density of CO_2 compared with other brines, reduces the pumping power significantly. Table 7.2 includes the values of kinematic viscosity of several of these brines at different temperature levels. The higher the viscosity is, the more pumping power is required for the secondary loop.

The combination of heat exchange processes involving phase change and low viscosity of CO_2 is positive for the performance of indirect systems with CO_2 secondary loops, competing with other solutions for direct heat exchange with air in terms of energy performance. Besides, CO_2 under this configuration operates oil-free, meaning that there are no challenges of oil-return and the performance of heat exchangers will not be affected by it.

Another positive characteristic of CO_2 secondary systems is that they can be manufactured with components of smaller sizes than those that use conventional brines. Using smaller valves and piping allows the costs of the secondary loop to be reduced. Moreover, smaller components withstand the operating pressures of CO_2 better.

The standstill pressure of CO_2 at ambient temperature cannot be disregarded. If the main refrigeration system is going to be paused for a certain period, a Cooling Finger is recommended (indicated in Fig. 7.1 with dotted lines). This could restrict the maximum CO_2 pressure, preventing it from blow-off. It is a solution also used in cascade systems and transcritical systems. The other option, particularly for smaller

Table 7.2 Kinematic viscosity as a function of temperature of liquid CO_2 (saturated) and other solutions that can be used as secondary fluids (brines). EG stands for ethylene glycol, PG for propylene glycol. A dash in a cell means that the fluid is not suitable for the temperature indicated

	Kinematic viscosity [mm^2/s]								
Temperature [°C]	CO_2	EG [40 vol%]	EG [50 vol%]	PG [40 vol%]	PG [50 vol%]	$CaCl_2$ [20 mass%]	$CaCl_2$ [30 mass%]	Ethanol [40 mass%]	Ethanol [50 mass%]
−10	0.12	8.47	11.75	22.18	32.9	4.09	9.04	13.40	11.99
−20	0.14	14.69	20.30	–	68.79	–	14.27	25.33	21.11
−30	0.15	–	40.39	–	–	–	–	–	42.48

systems, is to design the CO_2 loop in a way that could withstand the pressure as a consequence of the warmest day possible. This can be done with the current technology, since there are components such as receivers, pumps, and evaporators, rated for pressures of 80 bar or higher. In addition, the copper alloy K65 ($CuFe_2P$) allows pressures as high as 120 bar to be reached in piping and fittings.

Electric defrost is the preferred solution for evaporators with CO_2 secondary loops due to its simplicity. However, there are alternatives such as auxiliary loops with warm glycol or hot gas defrost.

7.2.2 *Cascade Systems with CO_2*

Compared with other refrigerants, the relatively low critical point of CO_2 (31.1 °C and 73.9 bar) means that transcritical operation is very likely to happen if a commercial refrigeration system has an air-cooled gas cooler, i.e. it is rejecting heat to the ambient. This is the solution for transcritical CO_2 systems (all-CO_2 systems), which are described in detail further down in this section.

An alternative to avoid the transcritical operation of CO_2 systems and keep them in the region where CO_2 is most efficient is to install a cascade as shown in Fig. 7.2. Since the focus of this book chapter is commercial refrigeration, all the figures from here and on will include evaporators at two temperature levels that usually exist in commercial refrigeration systems: MT and LT evaporators, for cooling and freezing of products, respectively.

The cascade consists of two separate refrigerant loops, the top stage and the low stage, connected by a heat exchanger, the cascade heat exchanger. The cascade heat exchanger operates as an evaporator for the high stage, absorbing heat from the low stage, in which refrigerant is condensing.

This solution allows the selection of refrigerants with favorable properties for cooling and freezing at the top and low stages, respectively. CO_2 is, indisputably, the most common solution for the low stage. In the top stage, there are systems with ammonia, different HFCs or propane, among others. The combination NH_3/CO_2 is regularly used in the industrial sector. Using the secondary loop for the MT evaporation would allow the reduction of charge of the cascade top stage, which is beneficial for the different reasons already mentioned. The secondary loop selected for the cooling level could be either a brine (as represented in Fig. 7.2) or a system with CO_2 (as represented in Fig. 7.3).

Even though in all the cascade solutions presented so far, the heat transfer between the top and low stage is direct (in the cascade heat exchanger), in the past it was usual that there was an intermediate fluid between the two stages to increase the safety and simplify regulation. The cascade heat exchanger was the weakest point and this was a solution to solve potential leakages from the CO_2 to ammonia loop. If this happens, a chemical reaction occurs and creates a solid product, ammonium carbamate that clogs filters and stops the system.

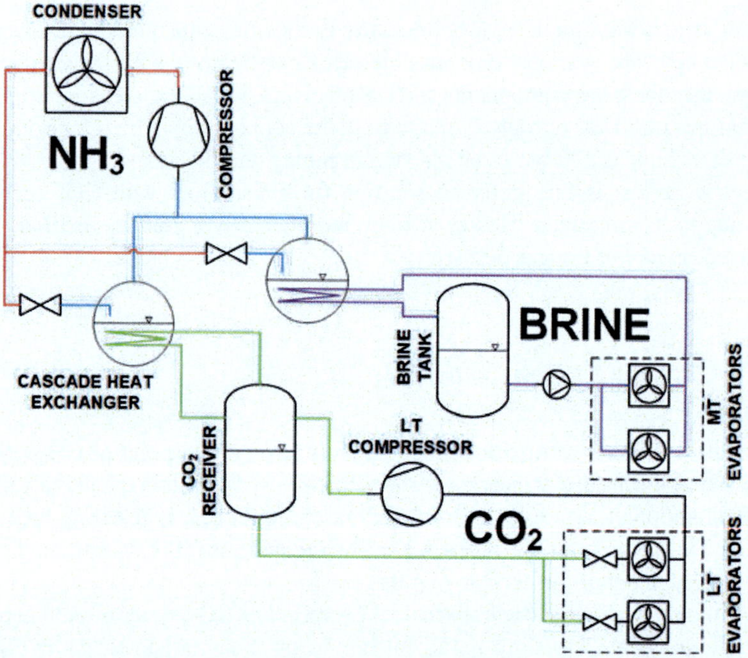

Fig. 7.2 Cascade system with ammonia (NH_3) and CO_2. A brine is used in this case to exchange heat from the MT evaporators to the ammonia. CO_2 is directly distributed to the LT evaporators

The system with the intermediate fluid would reduce the performance of a cascade system (additional temperature differences are needed) and increase the cost. Another option that is more common nowadays is to use double-wall tubes in the cascade heat exchanger, with the connection of the space between tubes to the environment. It is also possible to have an additional cascade heat exchanger in parallel to the regular one, which is used when the regular one needs to be maintained.

7.2.3 *All-CO_2 Systems*

The simplest solution for transcritical all-CO_2 systems is the concept developed by Prof. Gustav Lorentzen and his co-workers around 1988, which was designed particularly for mobile air conditioning (MAC). This idea is described in detail in Chap. 9 and it will just be mentioned here.

The main concept is to use a low pressure receiver and one expansion valve to expand CO_2 from the high pressure to the evaporator pressure level. This expansion valve regulates the high pressure, in contrast with conventional systems where the flow through the evaporator is adjusted to achieve certain conditions at the evaporator

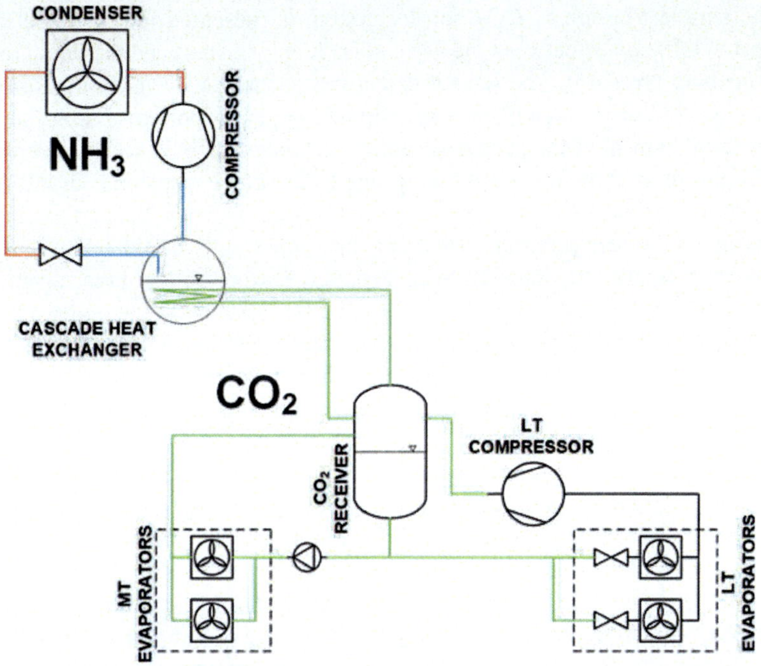

Fig. 7.3 Cascade system with ammonia (NH_3) and CO_2 both as secondary loop for cooling purpose (MT evaporators) and as direct refrigerant for freezing (LT evaporators)

outlet. The right balance of charge allows overfed evaporation, which improves the performance of the solution.

It was proposed as a solution for small units (plug-ins) with one evaporator, but most companies are heading towards propane (low-charge self-contained loops) for this application. Concerning about MAC, several manufacturers in Europe are using or will use this configuration for CO_2, since R134a is not an option anymore (Directive 2006/40/EC of the European Parliament and of the Council of 17 May 2006 relating to emissions from air conditioning systems in motor vehicles and amending Council Directive 70/156/EEC (Text with EEA relevance)) and there are discussions about the safety of R1234yf, an HFO (hydrofluoroolefin) classified as A2L (Standard 15–2016 (packaged w, Standard 34–2016)—Safety Standard for Refrigeration Systems and Designation and Classification of Refrigerants) (mildly flammable).

For commercial refrigeration systems, where there are at least two evaporation levels (cooling and freezing or MT and LT evaporation), the basic configuration of the all-CO_2 unit is denoted Booster. It is a logical evolution of the concept of cascade system with two CO_2 loops, one at the top stage and another at the bottom stage, but eliminating the cascade heat exchanger. It involves expansion in two stages and a liquid receiver between these two expansions. Booster systems are normally multiplex, with the compressor pack located at the machine room, and distribution of liquid is to the different cabinets and cooling/freezing chambers at the supermarket.

The simplest option of CO_2 booster system is presented in Fig. 7.4. The first expansion takes place between the gas cooler (high pressure) and the liquid receiver (intermediate pressure). The second expansion occurs between the liquid receiver and the evaporators (typically one expansion valve per evaporator). Thus, there is an independent control of the conditions at the evaporator and the high pressure.

In a booster system, the valve taking care of the first expansion is usually called high pressure valve (HPV). As detailed previously in this book, the performance of a transcritical CO_2 refrigeration system depends importantly on the relation between gas cooler pressure and temperature at the outlet. The controller of the valve follows an algorithm that connects both variables. The temperature at the outlet of the gas cooler is generally associated with the temperature outdoors and the size of the gas cooler. In case there is heat recovery from the high pressure side, the setpoint for the pressure might be independent of the ambient conditions. This case will be discussed in a subsequent section.

In order to meet the requirements of cooling and freezing, there are usually two groups of evaporators in a booster system: MT evaporators and LT evaporators, respectively. The liquid coming from the liquid receiver is expanded by the expansion valves of each evaporator. The mass flow circulating through each valve is adjusted as a function of the load at the evaporator. The basic strategy is to operate the evaporator as direct expansion (DX), i.e. achieving a certain superheat at the outlet of the evaporator. The idea is to guarantee the safe operation of the corresponding compressors, without liquid droplets at their suction, and it is the usual solution for conventional HFC systems. However, this solution is particularly inefficient in CO_2, since the use of the evaporator for superheating involves a reduction of the evaporation temperature. Every 1 K reduction accounts for almost 1 bar in pressure reduction, in contrast with other fluids, which increases compressors' power consumption. The energy-efficient alternative is to overfeed evaporators, and this issue is detailed in Sect. 7.3.3.

Each group of compressors is in charge of the control of the corresponding evaporation temperature (typical setpoints indicated in Table 7.1). The controller adjusts the capacity of the compressor group as a function of the mass flows of refrigerant circulating at each pressure level. The regulation of capacity can be obtained by setting more than one compressor of different sizes and/or an inverter-driven compressor, i.e. using a variable speed drive (VSD). Actually, the most common solution is to have the smallest compressor of the group with VSD, combined with larger compressors running at fixed speed. A smooth control of the capacity will be linked to a selection where the capacity gap between consecutive combinations of compressors is minimum. In this line, it is also important not to oversize the compressor group.

LT compressors suck the refrigerant from the LT evaporators and compress it, usually, to the suction of the MT compressors. A desuperheater (presented in Fig. 7.4 with dotted line) is used in some systems to reduce the temperature of the refrigerant discharged by the LT compressors. It is important when the share of LT demand to MT demand of the unit is relatively high since the superheating degree at the suction of the MT compressors could be high, reducing the performance of the compressors and even putting the integrity of the oil at stake. The heat at the desuperheater could

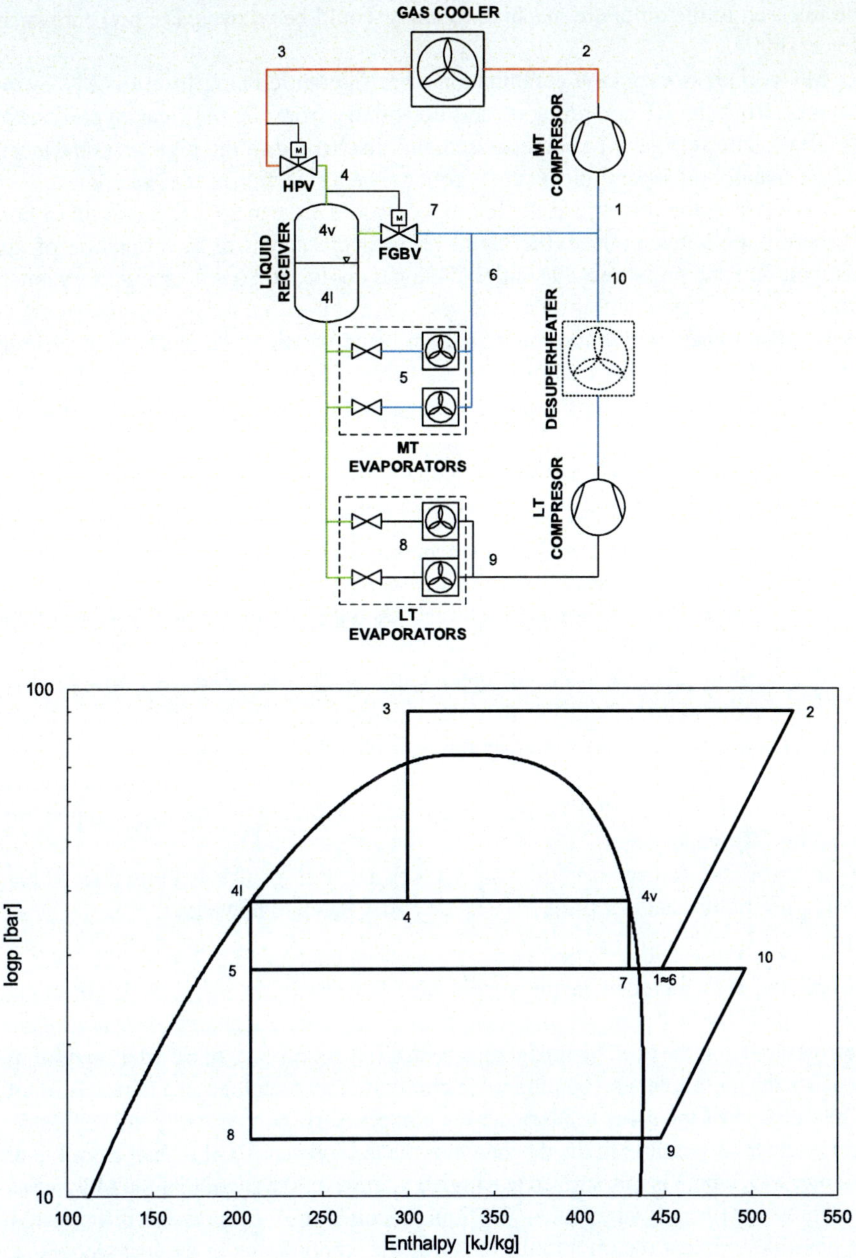

Fig. 7.4 Booster system with active control of the liquid receiver pressure: left, simplified sketch; right, logp-h diagram

be released to the air at the machine room, or could be recovered to preheat water, for example.

MT compressors suck a combination of refrigerant coming from the MT evaporators, from the LT compressors, and potentially from the flash gas bypass valve (FGBV). The refrigerant is compressed and discharged to the high pressure level, where usually oil separation occurs before the heat rejection at the gas cooler.

Concerning the pressure at the liquid receiver, there can be active control or not. If there is no active control, the liquid receiver pressure will be a function of the temperature and pressure at the outlet of the gas cooler, and it will change very much during a year or even within a day. However, in commercial refrigeration systems, it is common to have active control of the liquid receiver pressure. In a booster system, this is performed by the flash gas bypass valve (FGBV), as presented in Fig. 7.4. The flash gas at the top of the liquid receiver is throttled to the suction of the MT compressors.

The strategy of active control of the liquid receiver pressure entails certain considerations, which are listed below:

- A relatively constant liquid receiver pressure simplifies the control and stable operation of the expansion valves upstream of the evaporators.
- The receiver pressure setpoint has a very limited effect on the performance of the refrigeration system.
- The receiver pressure has to be high enough so that the expansion valves at MT evaporators operate with the minimum requested pressure difference (a lower limit could be 4 bar (GEA Refrigeration Technologies 2016)).
- The receiver pressure should not be too high, since it could exceed the maximum operation pressure (MOP) suggested by the manufacturer of the expansion valves at the LT evaporators.
- It is possible to use standard copper tubes and fittings if it is limited to 40 bar. Higher liquid receiver pressures can be easily handled nowadays.

Finally, the oil distribution system, not presented in Fig. 7.4 for the sake of simplicity, was included in the partial diagram shown in Fig. 7.5. Right at the discharge of the MT compressors, there is an oil separator, in charge of separating the refrigerant from the oil. The refrigerant, with a low percentage of oil after separation, reaches the gas cooler and continues the circuit; the oil circulates in a parallel circuit. The purpose of the oil is to lubricate the compressors, consequently, oil separators are as close as possible to the discharge of the compressors. Oil is then stored in an oil receiver, which is pressurized at a higher value than the suction of the MT and LT compressors (usually higher than the liquid receiver too). By pressure difference, it is possible to inject the oil into the carter of the compressors in the just amount, so that they are not flooded with oil. Level switches adjust the injection, and they can disable a compressor in case there is not enough oil to run it safely.

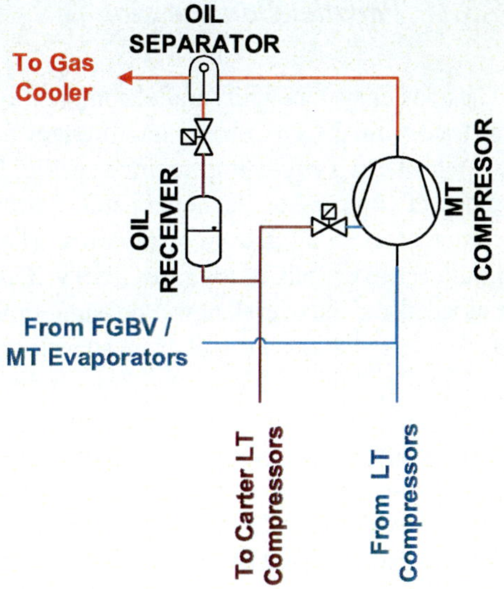

Fig. 7.5 Sketch showing the oil loop for an all-CO_2 booster system

7.3 System Adaptation to Warm Climates: Main Approaches

Transcritical all-CO_2 refrigeration systems have shown their reliability and performance at locations with cold and mild climates. However, a Booster system, as described above, cannot compete with other refrigerants and configurations in warm and hot climates unless it is conveniently adapted.

The reason why CO_2 is not the optimal choice when rejecting heat at high temperatures and pressures can be discussed as followed:

- The significant rise of the high pressure, associated with heat rejection at high ambient temperature, leads to huge expansion losses at the HPV.
- The expansion of the high pressure and high temperature CO_2 through the HPV originates from a large amount of flash gas at the liquid receiver. This flash gas fraction needs to be removed from the liquid receiver to keep the pressure. In a booster system, the flash gas is expanded to the suction of the MT compressors, and recompressed, which is a very inefficient solution when there is an important fraction of flash gas.
- The relation between temperature at the outlet of the gas cooler and its pressure is critical for systems with CO_2 as refrigerant. In addition, the optimization algorithm is system-dependent to a certain extent.

In the following subsections, the reader will find different technologies that have been proved to improve the performance of CO_2 commercial refrigeration systems in warm and hot climates.

7.3.1 Parallel Compression

The booster system with parallel compression is considered as the basic configuration of transcritical CO_2 refrigeration system for warm and hot climates. A simplified sketch of the system is presented in Fig. 7.6. The main development is the new group of compressors, denoted parallel compressors, intermediate temperature (IT) compressors or auxiliary compressors. Their task is to regulate the pressure at the liquid receiver, substituting the FGBV if possible. Thus, the MT compressors are unloaded and the overall power consumption decreases. There is a general agreement about the remarkable effect of this development.

The implementation of parallel compressors does not allow the FGBV to be removed. The reason is that the amount of flash gas produced at low ambient temperatures might not be enough to keep the smallest parallel compressor in operation, i.e. the FGBV regulates the liquid receiver pressure when there is insufficient flash gas. This also explains why parallel compression is not implemented in cold climates: it is not possible to find an efficient compressor of sufficiently low capacity to run with the little flash gas available during most of the year and the existing compressors would operate a reduced number of hours, which is not cost-effective.

7.3.2 Ejectors

Ejectors are devices that utilize the pressure and temperature of the primary stream (motive flow) to bring a low pressure stream to an intermediate pressure level. In the case of CO_2 commercial refrigeration systems, ejectors have been regarded as a solution to the large throttling losses of the HPV. The concept is that ejectors can utilize part of the expansion work available from the high pressure, at the outlet of the gas cooler, to lift the pressure of the CO_2 at lower pressure levels at the system.

Figure 7.7 presents an example of the use of vapor ejector in a CO_2 booster system with parallel compression for commercial refrigeration. In such a system, the main goal of the ejector is to regulate the high pressure. While doing so and with the flow rate coming at high pressure and relatively high temperature, it is also able to suck CO_2 from the liquid separator. The design of the geometry inside the ejector allows the pressure of the mixed streams to be lifted up to the liquid receiver, unloading partially the MT compressors and increasing the requirement for parallel compression. This "free-compression" should have a positive effect on the performance of the refrigeration system.

As indicated above, the main goal of the ejector is to control the high pressure at every condition of load and temperature at the outlet of the gas cooler. Typically, ejectors have a fixed geometry, meaning that with a single ejector, it is impossible to adjust the pressure by changing mass flow rates through the motive nozzle (high pressure stream of the ejector). Manufacturers have realized this issue and solved it in two different ways:

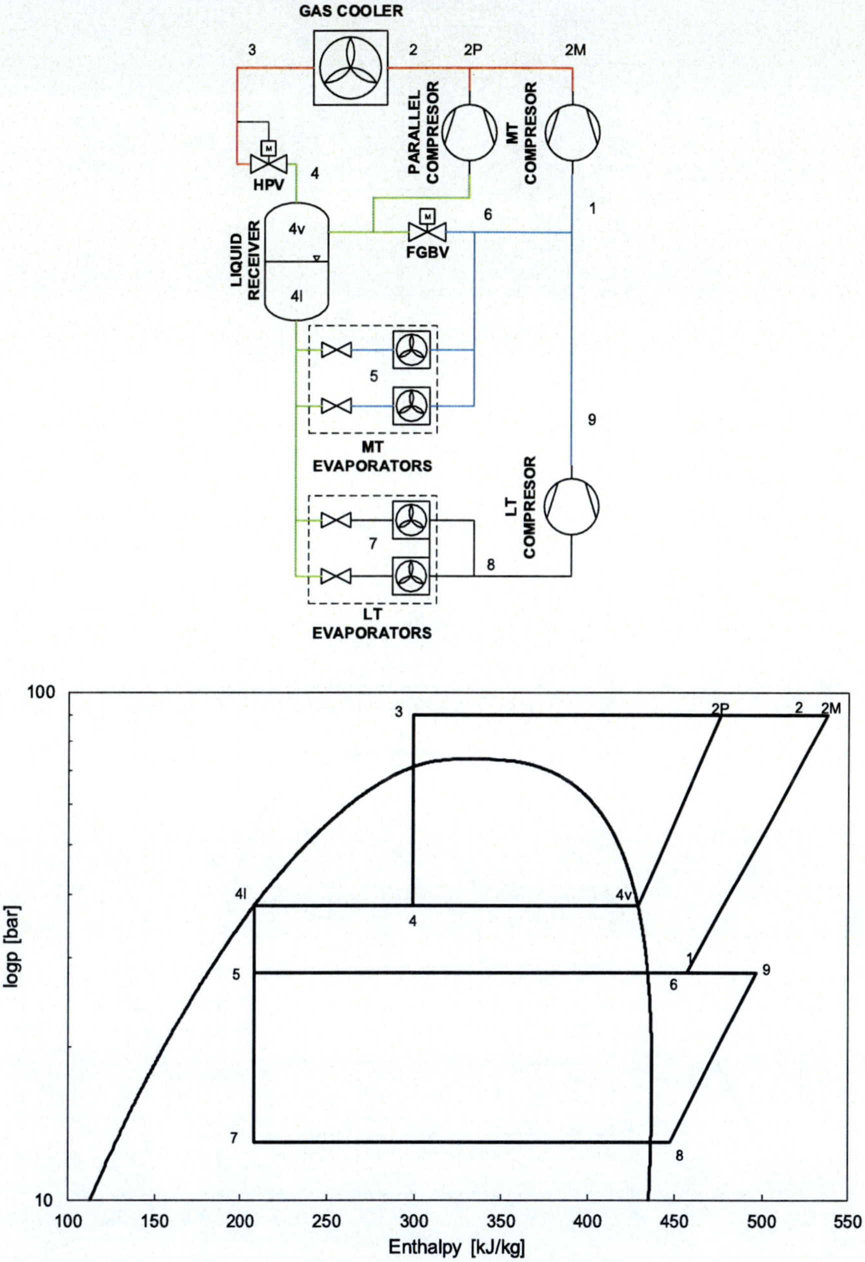

Fig. 7.6 Booster system with parallel compression: left, simplified sketch; right, logp-h diagram

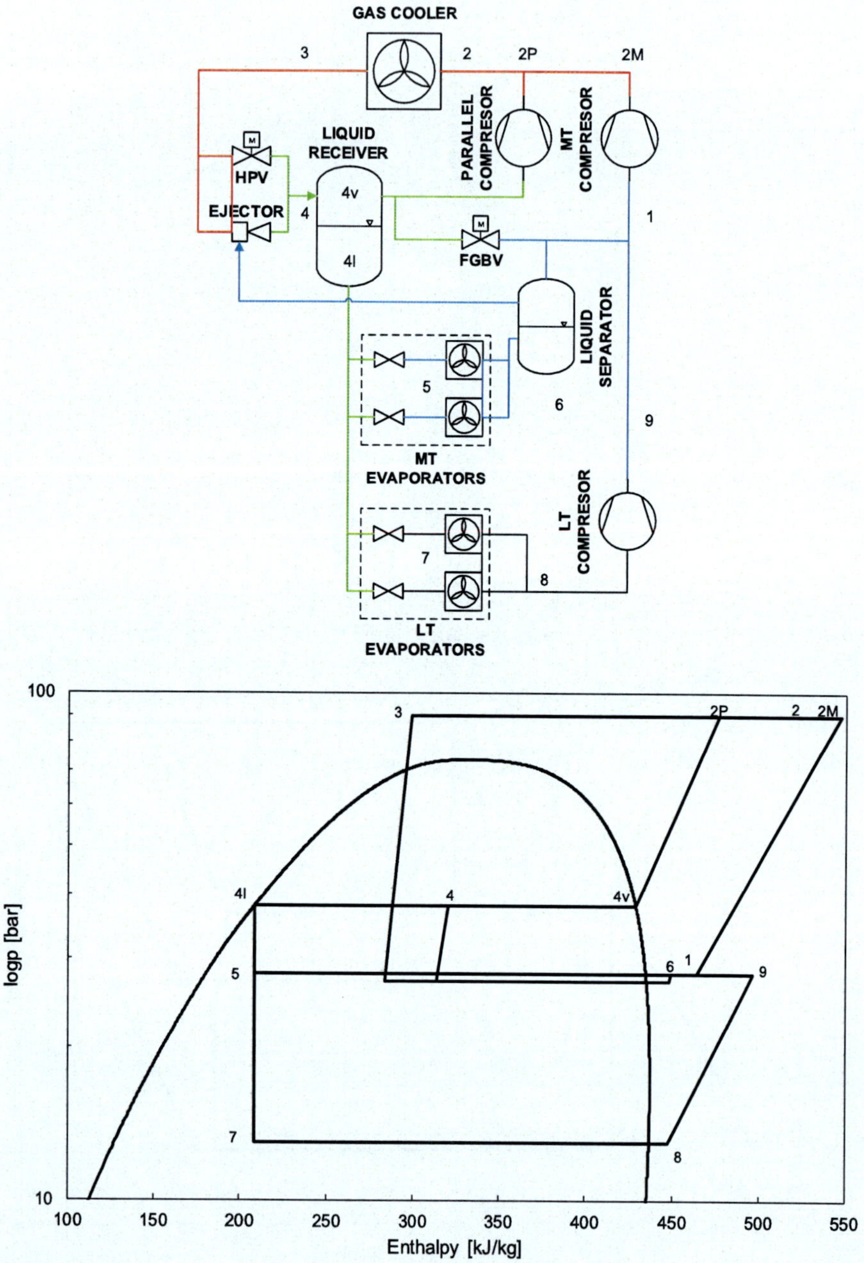

Fig. 7.7 Booster system with parallel compression and high pressure lift multiejector block: left, simplified sketch; right, logp-h diagram

- Fixed-geometry ejectors in parallel. The solution is based on locating several fixed geometry ejectors of different sizes in parallel. This entails a stepwise control of the capacity and high pressure by activation and combination of the ejectors. Danfoss performs this with the "Multi Ejector" block (https://www.danfoss.com/en/products/valves/dcs/electric-expansion-valves,multi-ejector-solution,#tab-overview), with all the ejector cartridges inserted in an aluminum block where the common connection ports (and a filter) are assembled. Each ejector cartridge is switched on with the corresponding solenoid valve. Another approach is by Enex with the "Enjectors" (https://www.enex.it/en/enjector). Individual ejectors are placed in parallel, with a solenoid valve each at upstream of the motive nozzle of the ejector.
- Electronic modulating ejectors. These ejectors have a fixed geometry but are able to regulate the mass flow through the motive nozzle continuously with a needle that adjusts the cross section. The manufacturer Carel supplies such a solution, with the commercial name of EmJ (https://www.carel.com/emj-electronic-modulating-ejector). It is also possible to set several of these modulating ejectors in parallel, with two main goals. On the one hand, reaching higher capacities. On the other hand, limiting the modulation range of the ejector in order to operate the ejectors with opening degrees that maximize their efficiency.

Ejectors should substitute the HPV fully, which could be removed from CO_2 refrigeration system. However, it is usual that manufacturers keep it in parallel to the ejector (as indicated in Fig. 7.7) for safety reasons or to extend the capacity range of the refrigeration system if the ejector is undersized.

Ejectors have been seen to perform best at relatively high ambient temperatures, i.e. high temperatures at the outlet of the gas cooler. At lower ambient temperatures and pressures, the ejector could potentially be unable to entrain any refrigerant through the suction port, throttling the high pressure fluid as in the case of the HPV.

The ejector design described so far is usually denoted high pressure lift (low entrainment ratio) ejector. It operates with a relatively high pressure lift from the suction port to the discharge port, in the range of 8–12 bar. Consequently, the ratio of mass flow sucked to the motive mass flow, a.k.a. entrainment ratio, is not higher than 0.2 or 0.3, depending on the operating conditions. In contrast with this design, there are also low pressure lift (high entrainment ratio) ejectors, which have reduced pressure lift (3–5 bar, approximately) but have entrainment ratios higher than 0.7.

7.3.3 Flooded Evaporation

Overfed/flooded evaporation is a successful enhancement for CO_2 refrigeration systems worldwide and it appears as an alternative to the commonly used direct expansion (DX) evaporators. In a system with DX evaporators, the mass flow rate of refrigerant through each evaporator is adjusted by the opening degree of the expansion valve located upstream. The strategy is to set a certain degree of superheat, from

8 to 12 K, at the outlet of the evaporators and guarantee the safe operation of the corresponding compressors (for example in Figs. 7.4 or 7.6). The degree of superheat is usually calculated by subtracting the evaporation temperature, evaluated with the evaporation pressure measured by a pressure transducer, from the temperature measurement obtained by a probe at the outlet of each evaporator.

In DX evaporators, an important fraction of the total heat transfer surface, normally between 10 and 30%, is operating with superheated refrigerant. The major consequence of this is that the evaporation temperature needs to be reduced to maintain the refrigeration capacity and keep the temperature of the product inside the cabinets within the correct range. As represented in Fig. 7.8 left, if the evaporator (considered as counter flow) is designed with a pinch point equal to 4 K and the degree of superheat is 8 K, the evaporation temperature should be -8 °C. This reduces the efficiency of the refrigeration system and increases the formation of frost, which leads to more frequent defrosting periods.

With flooded (overfed) evaporators, the strategy is to avoid superheat operation. Thus, there is a more efficient use of the heat transfer surface, which entails a higher evaporation temperature than using DX evaporators. In the case simulated in Fig. 7.8 right, if the pinch point is also equal to 4 K and with the same air temperature profile, the evaporation temperature could be increased up to −2 °C. The higher evaporation temperature reduces the power consumption of the compressors and the need for defrosting, which also has a positive impact on the stability and efficiency of the system. It is estimated that 2% power consumption reduction can be obtained for each 1 K increase in the evaporation temperature.

The use of flooded evaporators involves that the quality of the refrigerant at the outlet of the evaporators is smaller than 1, i.e. there is a fraction of liquid in the vapor refrigerant. Thus, the refrigerant cannot be sucked directly by the compressors, since their integrity would be compromised. A liquid separator is typically implemented downstream of the MT evaporators to prevent unsafe operation. The challenge is to control the liquid level in this reservoir. There are three main trends to do it:

- Liquid ejectors. Liquid ejectors pump the liquid at the bottom of the liquid separator to the liquid receiver using the expansion of the motive flow (Fig. 7.9 left). This solution is compatible with vapor ejectors, which are actually capable of pumping liquid refrigerant when the motive flow has a relatively high temperature and pressure. However, dedicated and specially designed liquid ejectors are recommended to operate flooded evaporators when the motive stream is at less favorable conditions.
- Pump. A pump can be used to return the liquid refrigerant from the liquid separator to the liquid receiver. This solution entails an extra power consumption, compared with the liquid ejectors, but it should be relatively low due to the properties of CO_2.
- Liquid line valves to supply LT evaporators. This solution is the patented system FTE (Full Transcritical Efficiency) by the Italian company EPTA (https://www.eptarefrigeration.com/en/node/739). Instead of returning the liquid from the separator to the receiver, the idea is to utilize the liquid at the LT evaporators. To do

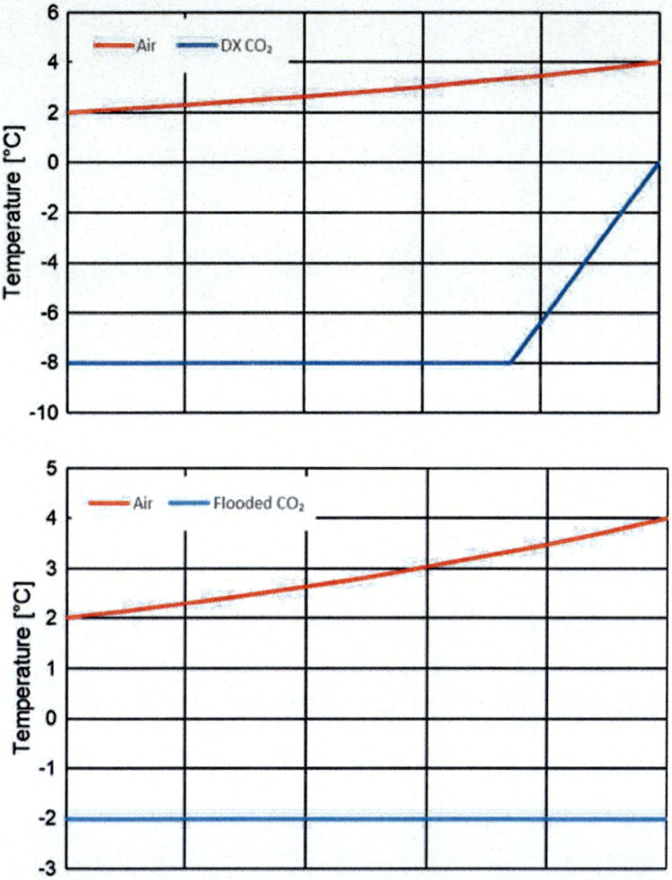

Fig. 7.8 Simplified temperature profiles of air and refrigerant in a DX evaporator (left) and in an overfed evaporator (right). The degree of superheat in the DX evaporator was set at 8 K. The pinch point at both cases was set to 4 K

so, two valves are needed, so that liquid can be supplied from the liquid separator, if there is any available, or in the conventional pattern from the liquid receiver. Its main advantage is simplicity, since it avoids the use of auxiliary devices such as liquid ejectors or pumps, keeping the efficient operation of flooded evaporators. However, the stability of the flow through the LT evaporators could be compromised when the supply of liquid switches from one line to the other (different pressure levels upstream of the expansion valves of the LT evaporators).

So far, flooded evaporation has been presented for MT evaporators. However, analogous benefits are achievable for LT evaporators. The challenge is associated again with handling the liquid that could potentially reach the suction of the LT compressors. To the best of the authors' knowledge, the only technical solution available is to install at least one liquid separator downstream of the LT evaporators,

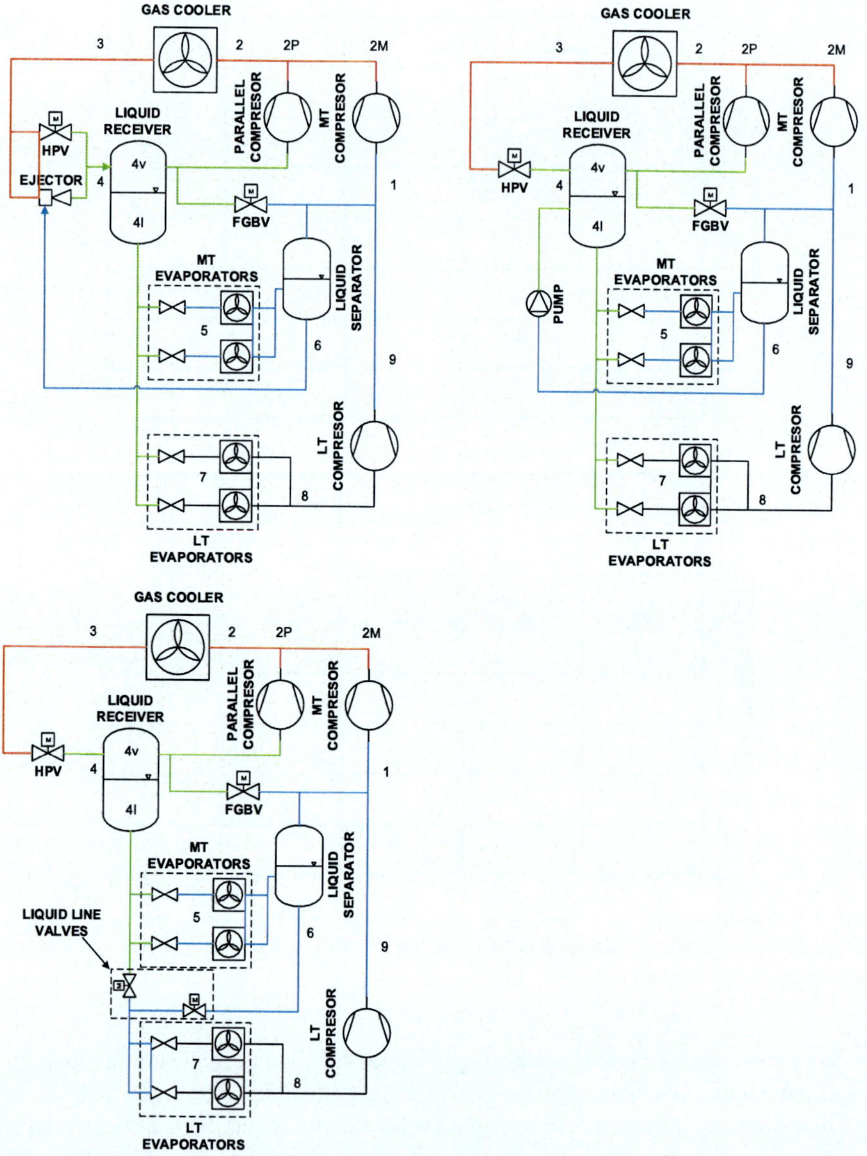

Fig. 7.9 Strategies for the control of the liquid level in liquid separators of CO_2 commercial refrigeration systems: left, liquid ejectors; middle, pump; right, liquid line valves to supply LT evaporators

and combine it with an internal heat exchanger that superheats the stream coming from the LT evaporators by subcooling the liquid line supply.

7.3.4 Mechanical Subcooling

Mechanical subcooling is another approach to improve the performance of CO_2 booster systems for commercial refrigeration, and it is presented in Fig. 7.10. An additional heat exchanger is implemented in the CO_2 loop, downstream of the gas cooler, rejecting heat to an auxiliary chiller. Thus, it is possible to reduce the temperature of the CO_2 below the ambient temperature and decrease the high pressure. Immediately, the power consumption of the compressors at the CO_2 loop is diminished.

The main drawback of mechanical subcooling is that there exists the additional power consumption of the auxiliary chiller, and particularly of its compressor. This extra energy should be taken into account in the evaluation of such CO_2 units. Even with the power consumption of the auxiliary unit, there is agreement that mechanical subcooling is a beneficial solution compared with the booster system alone.

The auxiliary system is usually as simple as that presented in Fig. 7.10: evaporator/subcooler (SUBC), compressor, condenser and expansion valve. Different refrigerants have been considered for this unit, including HFCs (R-134a, R-152a or R-1234yf) and natural working fluid(R-290).

7.3.5 Other Solutions

Booster-cascade configuration

Recently, a variation of the conventional cascade system has been proposed, combining the all-CO_2 booster system and the cascade one (Purohit et al. 2018). Figure 7.11 presents the relevant part of this configuration, i.e. focus is on the high pressure side. The concept is to have a booster system which is forced to operate at subcritical conditions, with a top stage using another refrigerant (NH_3, HCs or HFCs). The booster unit has two condensers in parallel and a three-way valve placed upstream of both to divert the flow to either one or the other, depending on the ambient conditions. The first rejects heat to the ambient, and will be active whenever the air temperature is low enough to keep a relatively low condensing temperature or not. The second is a cascade heat exchanger (condenser for the CO_2, evaporator for the top stage). The temperature to switch from one system to the other should be optimized in order to minimize the power consumption of the whole unit (CO_2 booster system + auxiliary unit).

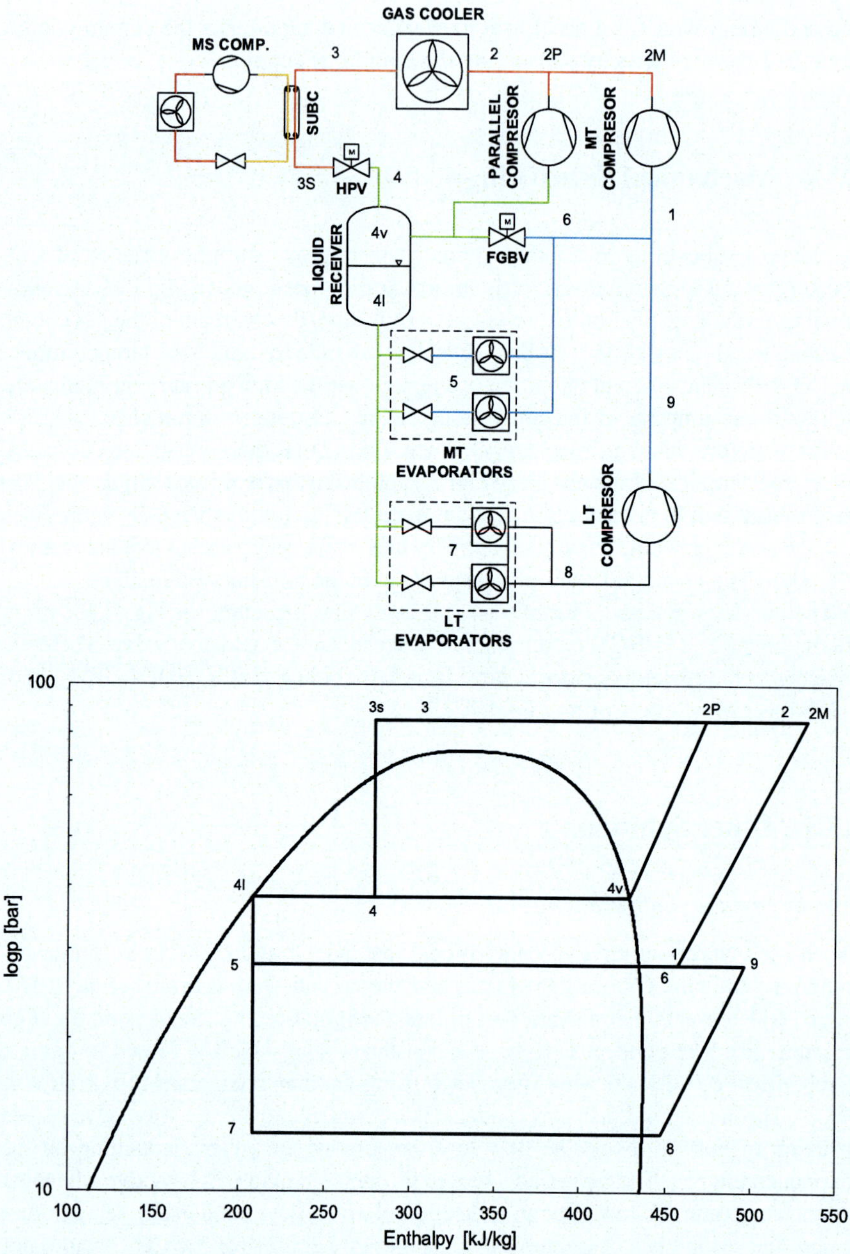

Fig. 7.10 Booster system with parallel compression and mechanical subcooling: left, simplified sketch; right, logp-h diagram

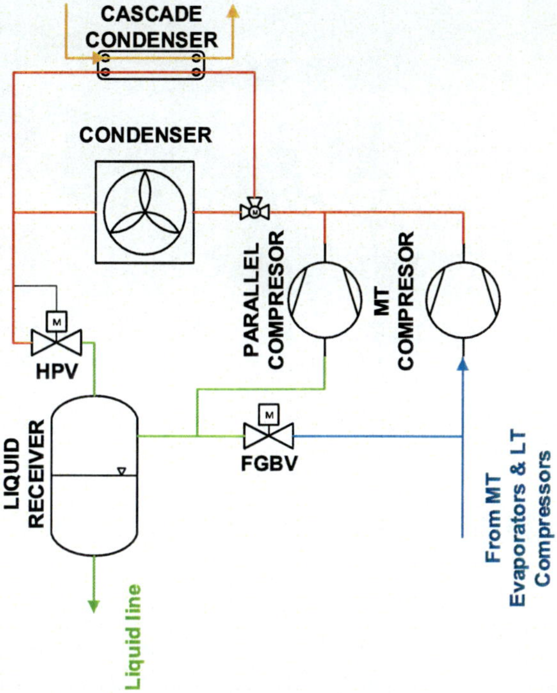

Fig. 7.11 CO_2 booster system with cascade condenser

Adiabatic gas cooler

With the same aim of reducing the gas cooler outlet temperature, the manufacturers of air-cooled gas coolers have developed adiabatic gas coolers, which allows the outlet temperature below the ambient temperature (dry-bulb temperature). An example of an adiabatic gas cooler is shown in Fig. 7.12 (https://www.guntnersolutions.us/products/condensers/compact/v-shape-compact-hydroblu-con). It possesses a pre-cooling pad on the air side, upstream of the conventional air-cooled gas cooler. Water is distributed and wets the surface of the pad, so that air is pre-cooled by evaporative cooling.

The pre-cooling of the air involves a reduction of the gas cooler outlet temperature, which is beneficial to increase the number of operating hours in the subcritical region and reduce the peak temperatures in transcritical state. In addition, the pre-cooling pad eliminates the requirements of water pre-treatment, in contrast with other methods that distribute water in the tubes of the gas cooler and could potentially lead to scaling on the surface. Evaporative pre-cooling of the air is limited by the conditions of air relative humidity. It is possible to reach air temperatures close to the wet-bulb temperature, but in very humid climates, where the relative humidity is close to 100%, there is no difference between the dry-bulb and wet-bulb temperatures.

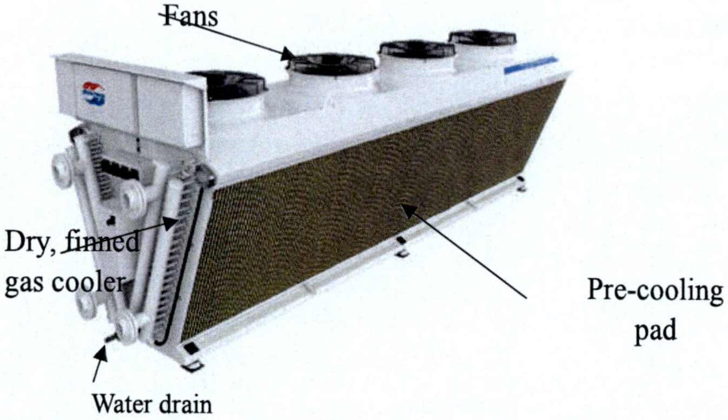

Fig. 7.12 Adiabatic gas cooler for CO_2 transcritical system with pre-cooling pad for the air (modified from Güntner (Purohit et al. 2018))

7.4 Integration of HVAC Demands into the CO_2 Commercial Refrigeration System

As seen so far, there have been many different developments to improve the efficiency of CO_2 commercial refrigeration systems at locations with warm and hot climates. However, there is another path to improve the overall performance of a CO_2 unit, and it is to utilize all the potential available and integrate the heating, ventilation and air conditioning demands of the supermarket and the surrounding building into the commercial refrigeration system. This eliminates the need for auxiliary HVAC systems, which could be boilers or other vapor compression systems using refrigerants that will not be an option in the future, such as R410A or R134a (HFCs).

Heat recovery with CO_2

Recovering heat from the CO_2 booster system is a straightforward solution to improve its overall performance. Supermarkets need heating (and tap water in some cases) and this heat is available in the gas cooler (high pressure) in large amounts and has relatively high temperatures compared to other working fluids.

Figure 7.13 shows the simplified logp-h diagram of a CO_2 booster system with heat recovery at different levels of high pressure and considering a temperature at the outlet of the gas cooler equal to 10 °C. The simple form of heat reclaims with CO_2 used by several manufacturers is to increase the high pressure setpoint up to transcritical conditions, e.g. 90 bar (red line), in order to guarantee the availability of heat at high enough temperatures. However, such a solution entails an increase in power consumption of the MT compressors that could be unnecessary for most heat reclaim loads.

A more efficient strategy would involve adjusting the high pressure setpoint to the actual need of heat reclaim (using the HPV) and rejecting the rest of the heat to the ambient. In Fig. 7.13, it is considered that heat is only useful at CO_2 temperatures

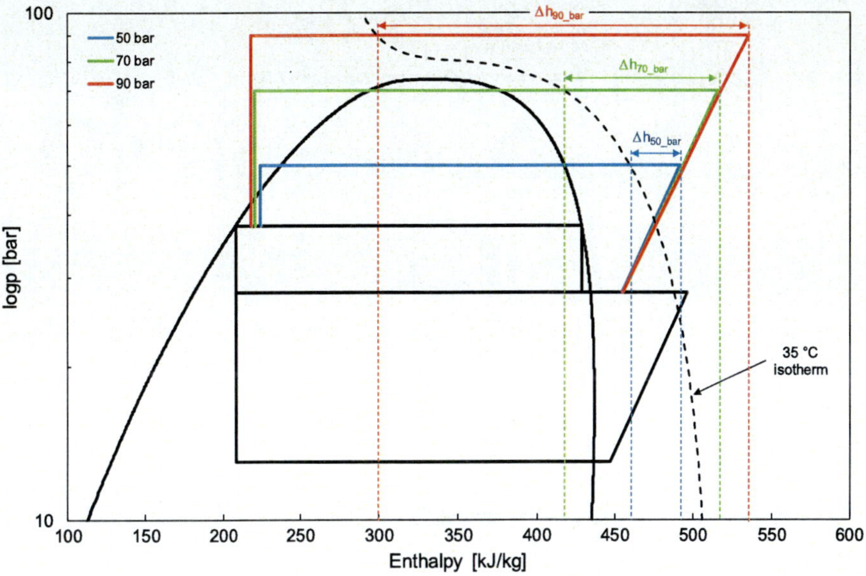

Fig. 7.13 logp-h diagram of a booster system with heat recovery at different levels of high pressure and with temperature at the outlet of the gas cooler equal to 10 °C.

equal to or higher than 35 °C, which could be used for radiant floor heating solution or ventilation. For low heat recovery load, the system should be able to operate at 50 bar of high pressure, where there is around 10% of specific heating capacity above 35 °C and with a maximum temperature close to 60 °C. If more heat is needed, the setpoint should be increased, for example up to 70 bar (still subcritical). At this pressure, there is already 30% of specific heating capacity at more than 35 °C (maximum temperature of 91 °C), i.e. 3 times as much as that of 50 bar and with a specific power consumption increase close to 60%. For extreme heat reclaim demand, transcritical operation could be achieved (90 bar in the figure). 75% of the specific heating capacity is above 35 °C now, with a specific power consumption of compressors 30% higher than that of 70 bar.

A layout of the booster system (potentially with parallel compression) to recover effectively CO_2 is shown in Fig. 7.14. A first gas cooler would be devoted to the production of tap water, which in principle is heated up to temperatures higher than that needed for space heating or ventilation. A second gas cooler would be in charge of the production of space heating/ventilation. Finally, there would be the air-cooled gas cooler, rejecting all the unused heat to the ambient. Three-way valves are recommended to bypass the heat recovery gas coolers if they are not in operation. The control of the secondary loops (pumps and valves in water loops, fans and flaps in air loops) in the heat recovery gas coolers will depend on the particular installation.

Thermal storage needs to be considered for the best use of the heat reclaim integrated in the CO_2 refrigeration system, particularly for tap water. The aim is to decouple heat recovery and use, so that it is possible to produce tap water with the

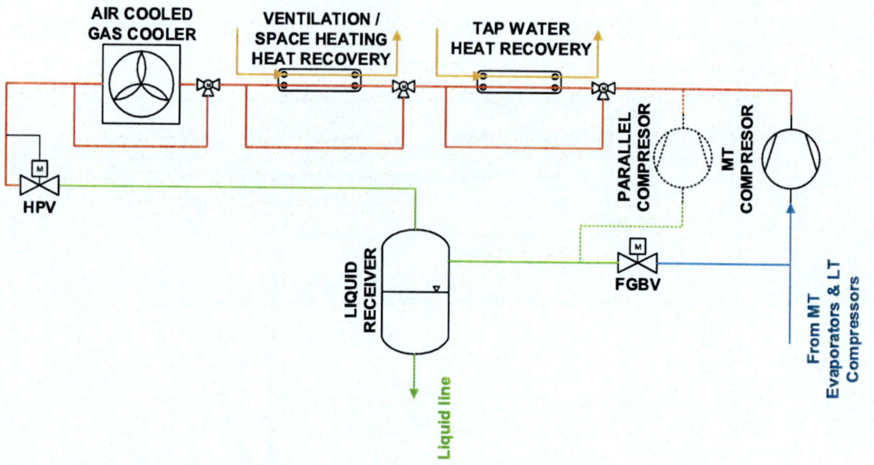

Fig. 7.14 Example of heat recovery with a CO_2 booster system. Heat is recovered for tap water production and for space heating and/or ventilation

heat available under normal operation, and minimize the number of hours when the high pressure is forced upwards.

More complex solutions for heat recovery, involving more gas coolers, could be installed. Further heat rejection/recovery is possible to the ground through boreholes, to melt the snow in front of the supermarkets (in cold climates), or to preheat the tap water. Preheating of tap water could also be performed with the heat available at the desuperheater downstream of the LT compressors (see Fig. 7.4).

A particularity of supermarkets in cold climates, with very high heating loads, is the need for an additional evaporator for heat recovery. The heat absorbed at the evaporators located in the display cabinets and cold chambers could be insufficient to meet the needs of heat reclaim. An evaporator located between the HPV and the liquid receiver could bring the extra capacity needed. The source of heat for the evaporator could be from either the ground (boreholes) or the ambient air (integrating the coils in the air-cooled gas cooler).

Air conditioning integration with CO_2

Air conditioning production may be integrated in an all-CO_2 refrigeration system in different ways, as indicated in Fig. 7.15. The first option (left) is based on the implementation of the AC evaporator between the HPV and the liquid receiver, with a three-valve to allow the bypass of the heat exchanger in case air conditioning is not demanded. In the second solution, the AC evaporator is located in parallel to the liquid receiver, and it is usually called gravity driven. The evaporator is fed from the bottom of the liquid receiver by gravity. The secondary stream at the AC evaporator will exchange heat with the CO_2 when air conditioning is requested, producing the vaporization of CO_2. Due to the density differences, vapor CO_2 returns to the top of the liquid receiver, meanwhile, new liquid coming from the same reservoir fills the evaporator (https://www.enex.it/en/refrigeration/yukon). Independently of the method used, parallel compression becomes necessary with AC integration due to the amount of flash gas reaching the liquid receiver.

The aforementioned configurations entail AC evaporation at the pressure level of the liquid receiver. However, it is also possible to decouple, at least partially, the

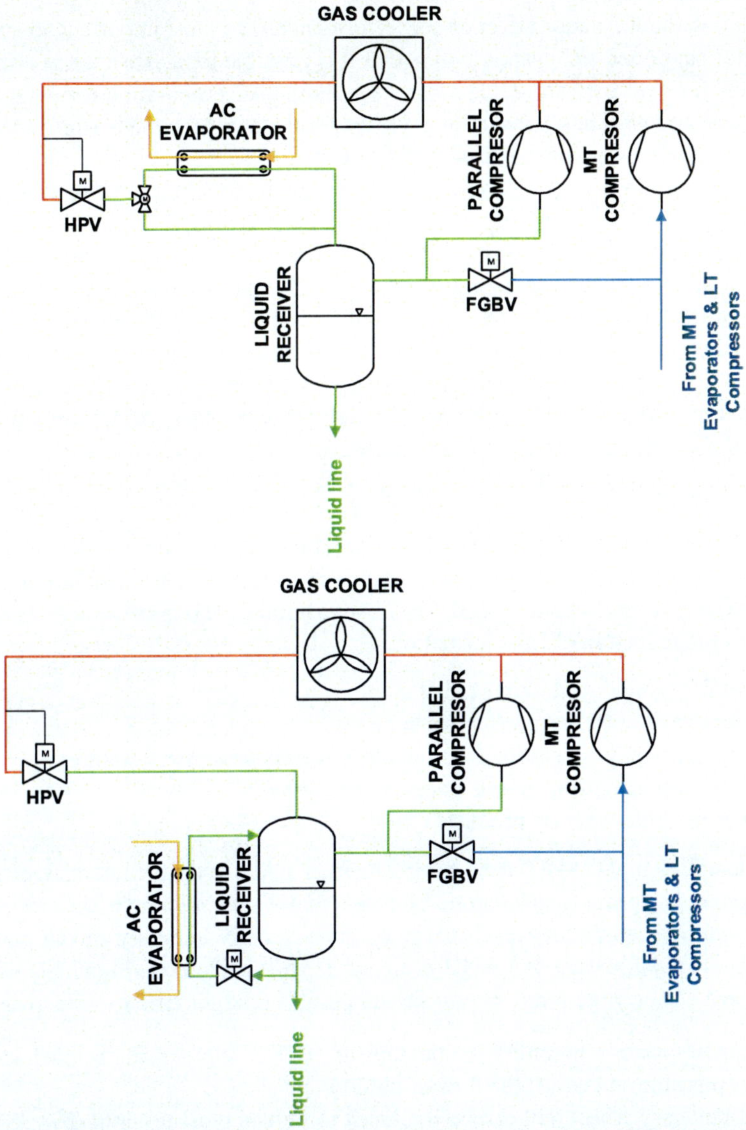

Fig. 7.15 Integration of air conditioning production (AC evaporator) in a CO_2 booster system with parallel compression. Left, AC evaporator between the HPV and the liquid receiver; right, gravity driven AC evaporator

evaporation pressure of the AC evaporator from that of the liquid receiver by using an ejector (Fig. 7.16 left). The AC evaporator is located downstream of the reservoir, with an expansion valve to adjust the flow of liquid CO_2. The vapor (or vapor–liquid mixture) at the outlet of the evaporator is entrained by the ejector, as explained in Sect. 7.3.2. Typically, a low pressure lift (high entrainment ratio) ejector will be designed for such application.

This method also allows overfed evaporation, since the ejector is capable of wet operation without a major effect on its performance. The most important advantage of ejector supported AC evaporation is that it is possible to increase the pressure at the liquid receiver between 2 and 5 bar, keeping the AC evaporation temperature at the required range. Thus, the suction pressure for the parallel compressors increases and the overall power consumption diminishes.

Another possibility of integration of AC production, still under development, is based on the integration of any of the options indicated in Fig. 7.15, with ejector supported AC evaporation (Fig. 7.16 right) (Hafner 2018). The evaporators are arranged in parallel from the point of view of CO_2 and in series for the secondary stream (water/glycol solution in this case). The warm secondary stream flows first through the gravity driven evaporator, which operates at the liquid receiver pressure, and then through the ejector supported evaporator, running at a lower evaporation temperature. The two levels of evaporation temperature make this a more efficient solution, if compared with the previous ones.

Air handling units and direct heating/cooling

The control of air humidity is particularly important due to the implications it has on product quality and the refrigeration loads. Even though very dry conditions could have an effect on the weight of food, low relative humidity brings several advantages linked to the reduction of frost formation:

- Fewer defrosting cycles are needed.
- Higher evaporation temperatures can be set.
- More quality in preservation of products, more stable air temperatures.
- Less condensation/frost on the doors of the cabinets.
- Less formation of ice on products.

Air handling units (AHU) can be implemented linked to the commercial refrigeration system and provide an integrated solution for the conditioning of the indoor air, i.e. maintaining satisfactory levels of temperature, relative humidity and air composition in the supermarket. An AHU designed for a supermarket (Fig. 7.17) with a centralized CO_2 refrigeration system should include (Hafner 2013):

- Air quality control (humidity, temperature and CO_2 levels) to reuse as much indoor air as possible and adjust the flow of outdoor air.
- Heat recovery wheel that can be bypassed to reduce pressure drop (bypass also applicable to other heat exchangers).
- Implementation of a heat exchanger to use the heat available from the refrigeration system.

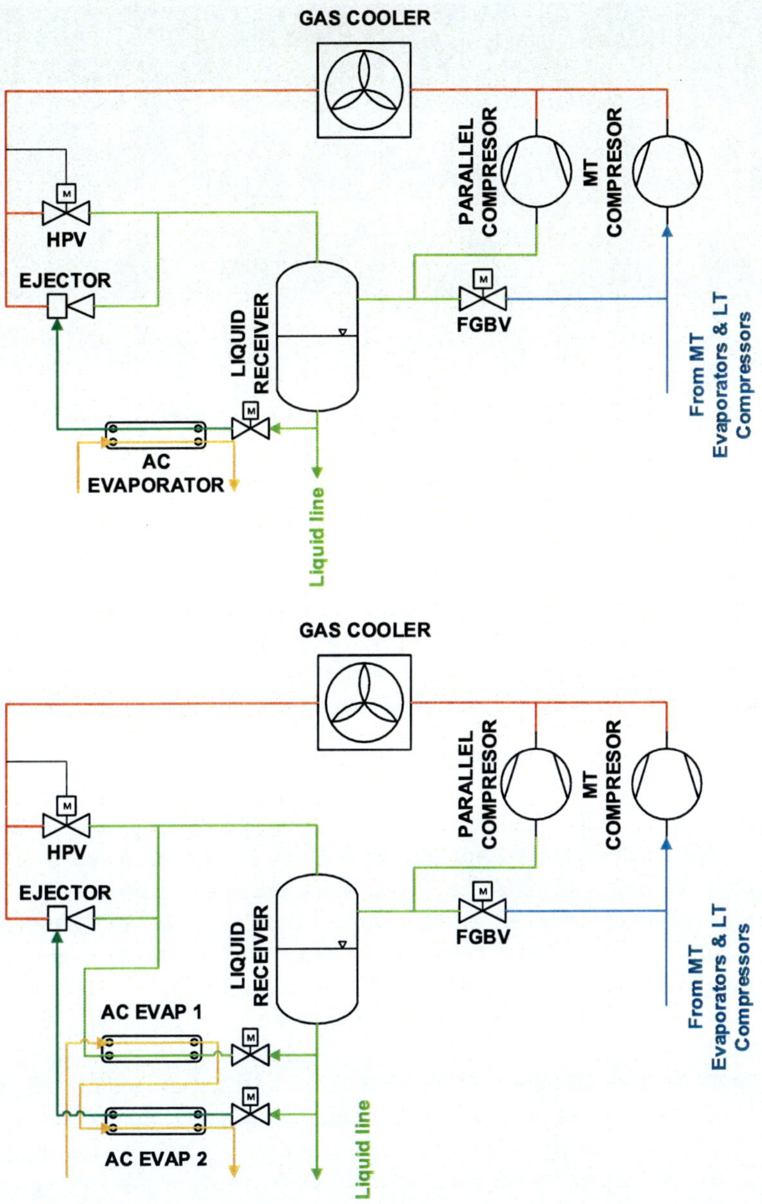

Fig. 7.16 Implementation of ejector for the integration of air conditioning production in a CO_2 booster system with parallel compression. Left, ejector supported AC evaporator; right, combination of gravity driven and ejector supported AC evaporators

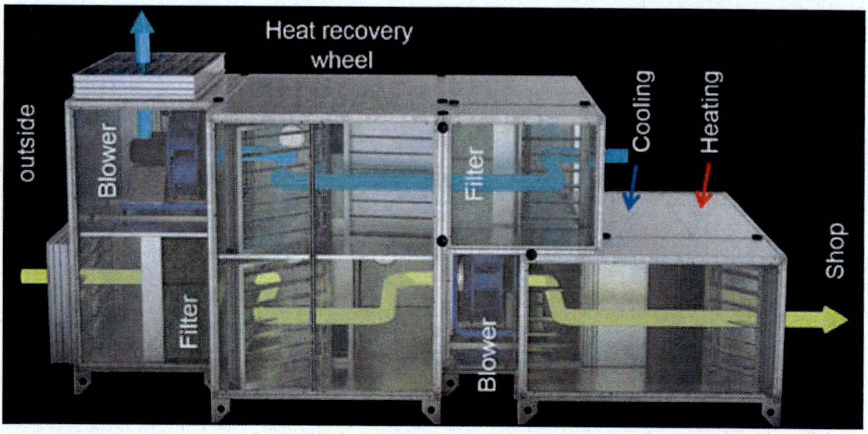

Fig. 7.17 Air handling unit (AHU) for supermarket air conditioning and heating with a heat recovery wheel (as seen in Hafner 2013 (https://www.enex.it/en/refrigeration/yukon))

- Dehumidification with a heat exchanger integrated in the AHU, using cooling from the CO_2 unit. Alternatively, a desiccant wheel could be introduced, using a water-absorbing material (e.g. silica gel) (Munters 2011).
- Cooling from the CO_2 unit in summer operation.
- Fans with permanent magnet electrical motor (high efficiency, low vibration and noise).

AHUs can be combined with other heating/cooling systems to achieve an even distribution of the indoor temperature. Floor heating has been proposed, particularly for cold climates, but it could also be done by air heaters/coolers distributed throughout the shop. The efficiency of such solution can be maximized with direct heat exchange between the air in the shop and the CO_2 of the refrigeration unit (Girotto 2016). For that purpose, the Italian manufacturer Enex has developed specific units that can be used for both heating and cooling (Fig. 7.18). Air coolers/heaters can also operate on their own in small convenience stores, where an AHU would not be cost-efficient.

There are several advantages of eliminating the intermediate loop (water, water glycol) and having direct heat transfer CO_2–air. On the one hand, the temperature lift or drop between refrigerant and air can be reduced, which lessens the power consumption of the compressor pack. In addition, the need for water pumps, tanks and plumbing disappears, which has a positive impact on the cost and power use of the overall system, and entails that only one contractor is needed for the whole installation.

A similar strategy of direct heat exchange CO_2–air could be applied within the coils in the AHU, being the advantages similar to those mentioned above.

Fig. 7.18 Fan coil unit for heating/cooling of the sales area with direct heat transfer CO_2–air (Hafner 2018)

7.5 Future Developments and Lines of Research

Future developments and research address the three main concerns of supermarket owners regarding all-CO_2 commercial refrigeration systems: cost, efficiency and complexity of the refrigeration loop. Even though CO_2 components and systems are still more costly than that of other refrigerants, the difference is decreasing as more systems are out in the market and more manufacturers compete. CO_2 systems are currently competitive in terms of efficiency, but this better performance has been achieved with complicated layouts. The tendency should be to reduce this complexity without compromising the efficiency, which in turn will have a positive effect on the cost of units.

Parallel operation of ejectors devoted to different applications

It has been discussed in previous subsections that ejectors can be used in CO_2 booster systems to lift the pressure of liquid or vapor from the liquid separator to the liquid receiver (high pressure lift ejectors), and entrain the refrigerant from the AC evaporator in ejector supported configurations (low pressure lift ejectors). If both solutions are installed in parallel in the refrigeration system, the so called AC (low pressure lift) ejector is in charge of adjusting the conditions of the refrigerant at the AC evaporator as a function of the load. Simultaneously, the high pressure lift ejector regulates the high pressure, following the setpoint indicated by the controller and substituting the HPV.

Both ejectors share the same high pressure stream (motive flow). The control of the high pressure could be challenging due to the common use of this mass flow and this situation has been observed in system simulations under the following circumstances:

- when the share of AC load to the total refrigeration load of the installation was relatively high,
- and when the AC ejector performs inefficiently.

A control strategy to deal with such situation has been proposed (Pardiñas et al. 2018) after numerical simulations, but should be field-tested. It includes adjusting the AC evaporation pressure and pressure lift of the AC ejector to the ambient conditions, since its efficiency is very much dependent on pressure and temperature at the motive nozzle. If the high pressure control is still impossible, the liquid receiver setpoint should be reduced to lower the pressure lift at the AC ejector (fixed AC evaporation temperature). This would increase the entrainment ratio and reduce the flow requested at the motive nozzle, bringing the control of the high pressure back to the high pressure lift ejector.

Control of liquid level at liquid separator to supply LT evaporators

The solution developed by EPTA (https://www.carel.com/emj-electronic-modulating-ejector) to handle the liquid separator level with flooded evaporators simplifies the layout of efficient CO_2 booster systems (Fig. 7.9 right) very much. However, there is still space for improvement in this solution, since the stability of the LT evaporators could be compromised when the supply to the expansion valves changes from the liquid receiver to the separator.

An idea to improve such systems is to enable an active control of the liquid level of the separator, so that the valve supplying the LT evaporators from the liquid receivers could be removed, i.e. there would be two independent liquid lines for the MT and LT evaporators. The technical solution is still under development, but it is based on the implementation of several liquid switches arranged vertically at the liquid separator, and a controller that could shift the opening degree of the MT evaporator (electronic) expansion valves is regarded as a function of this liquid level.

Low pressure lift ejectors to supply specific cabinets

As discussed along this chapter, commercial refrigeration systems usually operate with two levels of evaporation temperature, MT and LT for cooling and freezing of products, respectively. However, the air temperature requested by the different display cabinets of the same evaporation temperature level could be quite different (Table 7.1). Conventional systems select a setpoint for the evaporation temperature that fits the "worst" cabinet, i.e. the display cabinet that contains products with very low preservation temperature or the most loaded one, due to the amount of product or open configuration (high air infiltration). Thus, most of the cabinets will operate with a lower evaporation temperature than what they actually need, which has a negative impact on the performance of the whole refrigeration system.

A solution to increase the general evaporation temperature increase without compromising the preservation in those specific cabinets is to implement a low pressure lift ejector as shown in Fig. 7.19 left (https://www.enex.it/en/refrigeration/yukon). This puts in an additional evaporation temperature within the same level but using only the highest evaporation pressure as suction conditions for the corresponding compressors. In the sketch, the liquid refrigerant supply for the expansion valve of the ejector supported display cabinet comes from the liquid receiver, but it could be supplied also from the liquid separator, analogously to the previous case. This could reduce the need for liquid ejector to control the liquid level at the separator.

LT evaporation with high pressure lift ejector

There is usually a low share of the LT load to the total load in supermarkets in Europe. The MT and LT loads in a typical food retail application could account for 120 kW and 25 kW, respectively. The LT load share could be even lower in CO_2 booster systems integrating AC production. Nevertheless, significant investments must be made to cover this LT demand, including dedicated compressors, inverter, desuperheater, etc.

The strategy behind this development is to eliminate the LT compressors and their auxiliaries, and substitute them with a high pressure lift ejector (Fig. 7.19 right) (https://www.enex.it/en/refrigeration/yukon). The ejector would use the high pressure stream at the motive nozzle to entrain the flow coming from the LT cabinets and discharge it at the liquid separator. Since ejectors can handle liquid–vapor mixtures at their suction, overfeed evaporation can be easily implemented after this solution, avoiding any additional suction accumulator. Finally, fixed size ejectors with simple control could be used, due to the stability that usually LT loads have throughout the day and the year. The implementation of liquid supply to the LT cabinets from the liquid separator could also be performed here.

7.6 Conclusions

Throughout this chapter, it has been shown how CO_2 could be used as refrigerant for commercial refrigeration applications, starting as secondary fluid or low stage in a cascade and reaching all-CO_2 units. Actually, CO_2 refrigeration systems have overcome the challenges concerning their efficiency in warm climates, and they have become the preferred solution for commercial refrigeration in many parts of the world. Parallel compression and overfed evaporators have unquestionably contributed to this change. The enhancement in efficiency due to some other technologies, such as ejectors or mechanical subcooling, is also clear and they can be found in systems supplied by different manufacturers. The chapter has been ended with some of the research efforts that are under development to give the last push to this refrigerant and produce simpler, more reliable, and more efficient compressor packs.

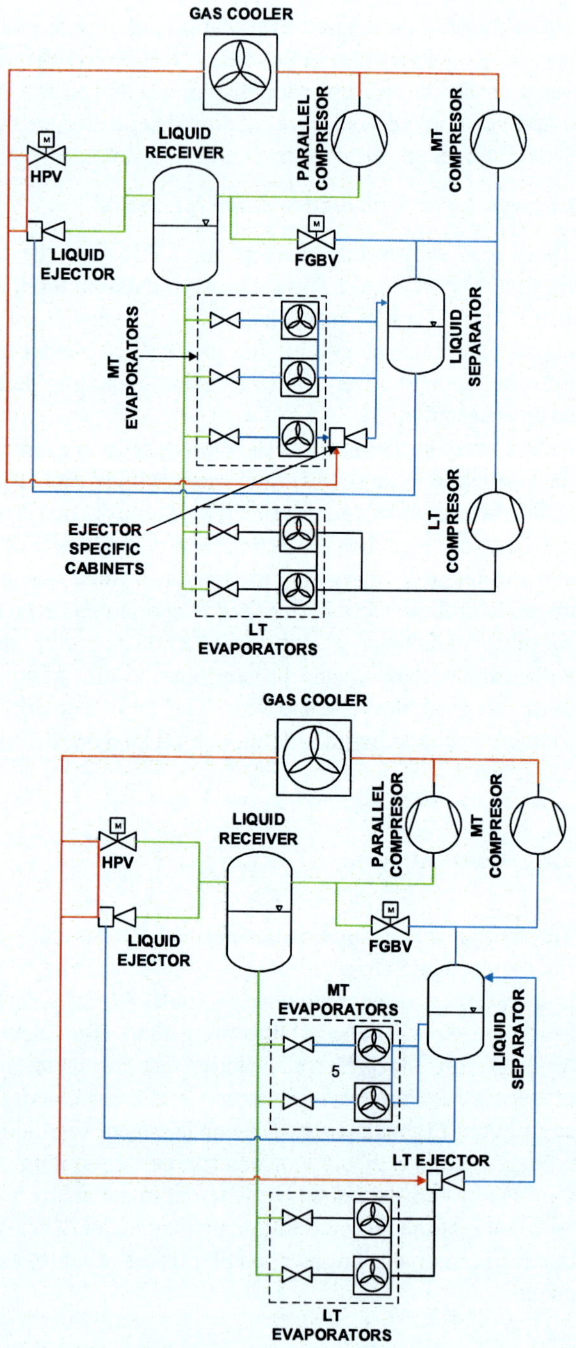

Fig. 7.19 Development of CO_2 booster system implementing ejectors. Left, ejector for specific cabinets within an evaporation temperature level; right, LT ejector

References

Devin E, Michineau T, Moulins F et al (2015) Etude sur le confinement des fluides frigorigènes, [online]. Rapport Final, N° de convention: 1481C0048. Paris, Cemafroid, IRSTEA and ADEME, 2015, 81 pages. http://afce.asso.fr/wp-content/uploads/2015/09/Rapport-AFCE-final.pdf. Accessed 14/02/2018

Directive 2006/40/EC of the European Parliament and of the Council of 17 May 2006 relating to emissions from air conditioning systems in motor vehicles and amending Council Directive 70/156/EEC (Text with EEA relevance)

GEA Refrigeration Technologies (2016) Refrigerating compressors for industrial refrigeration GEA Grasso V

Girotto S (2016) Direct space heating and cooling with refrigerant CO_2. ATMOsphere Europe, Barcelona

Hafner A (2013) CREATIV Energy efficient supermarket—overview and status. CREATIV Consortium Day

Hafner A (2018) Latest CO_2 refrigeration trends. ATMOsphere Europe, Lago di Garda

Hägg C (2005) Ice slurry as secondary fluid in refrigeration systems. Department of Energy Technology, Royal Institute of Technology, Stockholm, 71 p

https://www.unenvironment.org/

https://eur-lex.europa.eu/legal-content/EN/TXT/?uri=uriserv:OJ.L_.2014.150.01.0195.01.ENG

https://webstore.iec.ch/publication/22446

https://www.danfoss.com/en/products/valves/dcs/electric-expansion-valves/multi-ejector-solution/#tab-overview

https://www.enex.it/en/enjector/

https://www.carel.com/emj-electronic-modulating-ejector

https://www.eptarefrigeration.com/en/node/739

https://www.guntnerus.com/company/news/guentner-press-center/detail/guentner-v-shape-compact-powerful-and-smart/?tx_guentnerfe_news[action]=show&tx_guentnerfe_news[controller]=News&cHash=36ec96f22ec9b2e0812e6ec275a3ce92.

https://www.enex.it/en/refrigeration/yukon/

Karampour M, Sawalha S, Arias J (2016) Eco-friendly supermarkets: an overview. SuperSmart EU Project 696076. www.supersmart-supermarket.info

Kauko H, Kvalsvik K, Hafner A (2016) How to build a new eco-friendly supermarket. SuperSmart EU Project 696076. www.supersmart-supermarket.info

Lazzarin R, Ge Y, Zilio C (2018) Advancements in supermarket refrigeration. In: 37th informatory note on refrigeration technologies. IIR international institute of refrigeration

Munters (2011) Munters DryCoolTM dehumidification systems. http://webdh.munters.com/webdh/BrochureUploads/Engineering%20Catalog-%20DDS.pdf

Standard 15–2016 (packaged w/ Standard 34–2016)—Safety Standard for Refrigeration Systems and Designation and Classification of Refrigerants (ANSI Approved)

Pardiñas ÁÁ, Hafner A, Banasiak K, Kvalsik KH, Larsen L (2018) Strategy for the control of two groups of ejectors operating in parallel in integrated R744 refrigeration systems. Refrigeration Science and Technology, 2018-June, pp 1069–1076

Purohit N, Sharma V, Sawalha S, Fricke B, Llopis R, Dasgupta MS (2018) Integrated supermarket refrigeration for very high ambient temperature. Energy 165:572–590

Chapter 8
CO_2 Refrigeration Cycles and Systems for Ice Rinks and Snowmaking

Guan-Bang Wang and Xin-Rong Zhang

Skating and skiing have become more and more popular with people all over the world. However, the natural ice and snow are not always available, and the period of cold seasons for skating and skiing is being shortened by the serious problems of global warming. Researchers are then prompted to develop science and technologies about producing ice and snow because of the gap between the limited natural resources and the growing demand for winter sports.

CO_2 refrigeration cycles and systems for ice rinks and snowmaking are extensively presented and analyzed in this chapter. The background and state-of-the-art of the ice rink systems and snowmaking systems are firstly given. The applications of CO_2 cycles and systems are then respectively discussed with respect to ice rink systems and snowmaking systems based on the literature review and previous research by the authors. The system configuration and performance evaluation are shown by comparative studies according to the technology development.

8.1 Introduction

8.1.1 Ice Characteristics and Applications

Ice forms by the solidification of liquid water at the normal freezing point of 0 °C with a heat rejection of about 334 J/g. This latent heat is comparable to the sensible heat rejection of 80 °C temperature reduction for liquid water. Moreover, ice can be directly

G.-B. Wang · X.-R. Zhang (✉)
Department of Energy and Resources Engineering, College of Engineering, Peking University, Beijing 100871, China
e-mail: xrzhang@pku.edu.cn

X.-R. Zhang
Beijing Engineering Research Center of City Heat, Peking University, Beijing 100871, China

X.-R. Zhang and T. M. Eikevik (eds.), *CO_2 Refrigeration Cycle and Systems*, Lecture Notes in Energy 96, https://doi.org/10.1007/978-3-031-22512-3_8

obtained from water vapor in the sub-freezing air by the deposition process with more than 8 times the latent heat rejection in the solidification. Because of the high latent heat, ice generation is commonly used in different sectors for cold energy storage and cooling demands (Saito 2002; Bellas and Tassou 2005; Kauffeld et al. 2010).

Ice is a promising natural medium for the wide use of direct contact cooling in food processing and medical treatment. Ice chunks are hand-packed around the organ for conventional protective hypothermia. Flake ice is employed for heat generation removal in the bakery, rapid post-harvest cooling of fruits and vegetables, and long-term preservation in the fishery. Ice slurry is an excellent heat transport medium in the refrigeration system because of its liquidity, high heat transfer coefficient and large surface area, resulting in a substantial reduction of system size, pipeline length and pumping energy.

The development of ice generation and snowmaking technologies must be based on microscale mechanisms (Hayashi and Kasza 2001), because both snow and ice grow from single ice crystals with sizes of hundreds of micrometers (Egolf and Kauffeld 2005; Nakaya 1951). The ice crystals show various patterns under different conditions in terms of temperature and supersaturation, which have a significant influence on the physical properties (Nakaya 1951). For example, fresh natural snow can be identified as porous material with high compressibility, while the density of ice slurry is much higher than fresh snow. Therefore, ice generation and snowmaking technologies are involved with complex mechanisms and principles governing the crystal growth.

8.1.2 Snowmaking

Snowmaking was originated in the 1950s to provide enough snow for sports, recreation, and sightseeing. The snowmaking methods have been developing a lot during the last few decades, resulting in different kinds of methods including outdoor snowmaking, ice-generation-based snowmaking, and indoor snowmaking.

As presented in Fig. 8.1a, the traditional outdoor snowmaking method employs the snow guns where pumped water and compressed air mix with each other, and then the water is sprayed out through the nozzles. During the long-distance travel in the cold air, the liquid water will become small ice crystals which are called "snow". The latent heat of water crystallization is dissipated into the cold environment leading to the feasibility of the traditional snowmaking method highly dependent on the ambient temperature and relative humidity as shown in Fig. 8.2.

Consequently, techniques and commercialized products for temperature-independent snowmaking have been developing these years, most of which are mainly based on the ice generation methods shown in Fig. 8.1b. The typical method is based on the ice generation in the refrigerated vessel where the ice layer formed on the refrigerated surface, and the ice layer is then broken into smaller pieces in the shapes of plate or flake. Other methods directly produce ice slurry or small snowflakes with water evaporation under vacuum or subcooled conditions. The ice generation

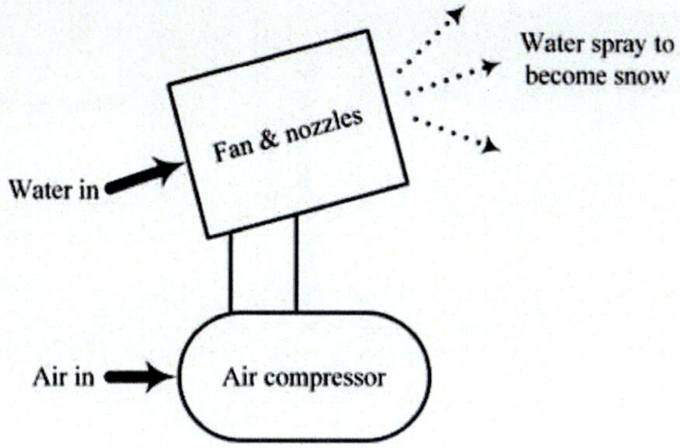

(a) Snow gun for outdoor snowmaking.

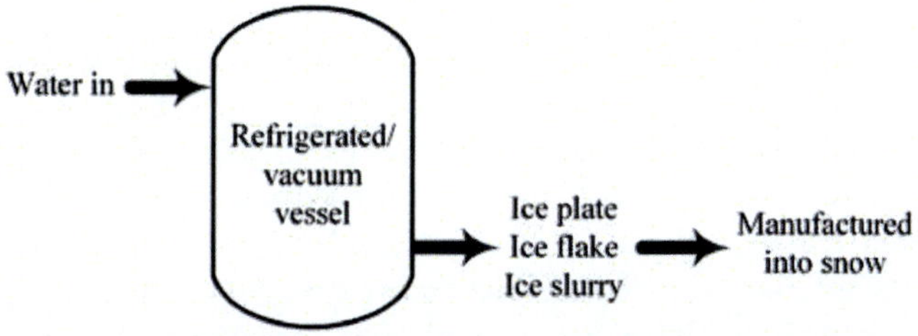

(b) Ice generation for temperature-independent snowmaking.

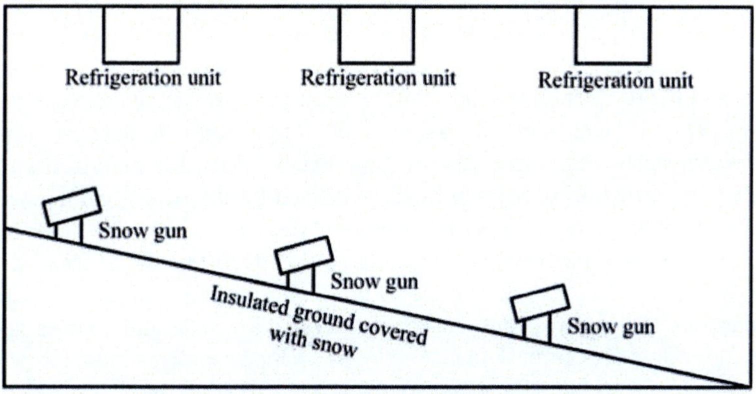

(c) Indoor snowmaking.

Fig. 8.1 Schematic illustration of different kinds of snowmaking methods

methods can be classified into four types according to the different processes and principles.

The plate ice machine is composed of refrigerated vertical plates where water is sprayed on one side of the plates and then the ice layers form hereon (Graham et al. 1993). Afterward, the ice layers are taken down by defrosting with running water

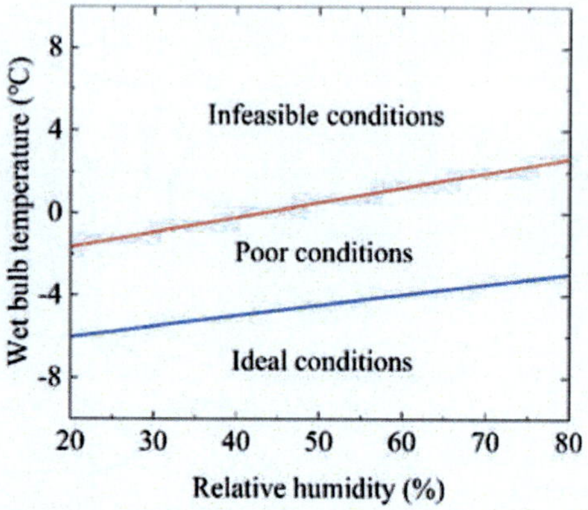

Fig. 8.2 Temperature restriction on outdoor snowmaking method

on the other side of the plates. Internal defrosting enables a large production rate by forming ice layers on both sides of the plates. The thickness of the produced ice layer is usually 10 ~ 12 mm, so an ice crusher is necessary to break the ice layer into smaller pieces with suitable sizes. The requirement of high-temperature defrosting water leads to high energy consumption and low profitability. Nevertheless, an American company SnowMagic Inc. adopted this method to develop the first equipment for temperature-independent snowmaking by the patented technology of Infinite Crystals Snowmaking.

The scraper type system is commonly used to produce flake ice (Cao et al. 2015) and ice slurry (Stamatiou et al. 2005; Martínez et al. 2014). The evaporator consists of double cylinders where water (or other solution) flows through the inner cylinder and the refrigerant at −30 ~ −20°C flows within the gap between the two cylinders. The ice layer with a thickness of 2 ~ 3 mm forms on the inner wall and is then mechanically scrapped down in the form of flake ice or mixed with the solution. The formation of ice can be also realized by the supercooling operation where the water temperature is below the normal freezing point under cooling in the evaporator without water crystallization until physically disturbed (Bédécarrats et al. 2010). An Italian company TechnoAlpin AG developed commercialized machines "SnowFactory" for snowmaking based on the flake ice generation, while another manufacturer SnowTek from Finland employed scrapped surface method to produce both flake ice and ice slurry for snowmaking by different models of the machine called "SnowGen".

The vacuum freezing type system gets rid of the refrigerators and the latent heat is removed by water evaporation under vacuum conditions (Asaoka et al. 2009). The combination of refrigeration and dehumidification is used for ice generation based on the same principle of lowering the water vapor pressure (Li et al. 2012). Israeli company IDE Technologies released their products VIM series for snowmaking by the generation of ice slurry using a vacuum system. The dehumidification process also results in water evaporation without the need for vacuum conditions, which is considered to be more energy-efficient (Li et al. 2012, 2009).

Direct contact heat transfer can be also used for ice generation by injecting refrigerant into water (Hawlader and Wahid 2009; Wijeysundera et al. 2004). Similarly, Polar Technologies utilized the mixture of compressed air, water and liquid nitrogen injected into a large insulated space for snowmaking in consideration of the required quality. However, the application of this method is restricted by the high operation cost due to the expensive liquid nitrogen.

A detailed comparison of the commercialized products above for temperature-independent snowmaking based on the ice generation methods is given in Table 8.1. Most of the temperature-independent snowmaking products adopt ice-based methods with a higher snow density of above 450 kg/m^3 compared with that of fresh natural snow resulting in less compressibility and higher hardness, which is unfavorable for sports and entertainment. Additionally, high operation cost is also caused by the low refrigeration temperature and system design without optimization. Hence, the combination of direct contact heat transfer and refrigeration system should be used to produce snow in a refrigerated space for better snow quality and low energy consumption. Moreover, HFCs with high GWP and ammonia with toxicity and flammability are used as refrigerants in many products, although water can be directly used as a safe natural refrigerant with advanced technologies for the specific components. Consequently, carbon dioxide, as a widely used and safe natural refrigerant should be employed in the snowmaking system for environment-friendly operation.

As given in Fig. 8.1c, the indoor skiing hall is indeed a kind of temperature-independent snowmaking technology by producing snow in a refrigerated space, which has been developing for more than 30 years (Paul 2003). The cold space equipped with good insulation and refrigeration systems for air and floor cooling provides suitable conditions for operating temperature dependent snow machines and snow storage (Clulow 2006). There are numerous indoor skiing halls in the world, which have large floor areas with skiing slopes to build similar conditions to outdoor skiing. Whereas, in contrast to the indoor skiing hall, snowmaking inside more limited indoor space for outdoor use of snow requires a more delicate design for the efficient integration of refrigeration systems and snow machines by the synergy of fields of flow, temperature and humidity. In addition, snow preservation before use and the design of transport to the outdoor venues are also important to lower the snow loss. Furthermore, the integrated energy system should be designed and employed to improve the overall energy efficiency in consideration of the simultaneous demands of cooling and heating energy with different temperatures by various kinds of facilities.

8.1.3 Ice Rink

Various kinds of ice rinks with different sizes and shapes are used for such sports and recreation activities on ice like ice hockey, curling, figure skating, speed skating, and so on. As shown in Fig. 8.3, the arena with an indoor ice rink requires a complex

Table 8.1 Detailed comparison of the commercialized products for temperature-independent snowmaking based on the ice generation methods (Paul 2003; Clulow 2006)

Company	Model	Principle	Operating air temperature (°C)	Capacity (m^3/day)	Refrigerant	Operation cost (USD/m^3)[a]	Snow density (kg/m^3)
TechnoAlpin AG	SF100	Flake ice	−5 ~ 25	100	R404a	3.12	450
	SF220	Flake ice	−5 ~ 25	220	Ammonia	2.51	455
SnowMagic Inc	UMF-50 T/D	Plate ice	−6 ~ 30	102	–	3.62	490
	UMF-100 T/D	Plate ice	–	204	–	2.98	490
	UMF-150 T/D	Plate ice	–	300	–	2.90	500
	UMF-200 T/D	Plate ice	–	408	–	3.27	490
SnowTek	SnowGen 100SI	Ice slurry (scrapped surface)	−10 ~ 30	181	R404a, R507, ammonia	ca. 4.4	552
	SnowGen 100FI	Flake ice	−10 ~ 30	222	R404a, R507, ammonia	ca. 4.5	450
IDE Technologies	VIM100	Ice slurry (vacuum)	−40 ~ 35	200	Water	2.04	560
	VIM400	Ice slurry (vacuum)	−40 ~ 35	860	Water	0.66	651
	VIM850	Ice slurry (vacuum)	−40 ~ 35	1720	Water	0.55	651
Polar Technologies and AGA	–	Direct heat transfer	<30	550	Liquid nitrogen	Ca. 65	–

[a]The electricity price is assumed to be \$0.1/kWh, while the nitrogen price is approximately \$0.18/kg

energy system with the demands of refrigeration, heating, lighting, ventilation, dehumidification, and so on.

Refrigeration is the most important energy demand accounting for 43% of the total energy demands due to the necessity of low-temperature of −7 ~ −6 °C for ice hockey and −2 ~ −1 °C for recreation activities to keep the ideal hardness and quality of the ice layer (IIHF 2018; ASHRAE. Ice rinks 2006). Heating is another essential energy demand following refrigeration. Typical heating demands include ground heating for freeze protection (10 ~ 30 °C (Rogstam and Bolteau 2015)), space heating by radiators (30 ~ 50 °C (Rogstam and Bolteau 2015)) to maintain the temperature of 10 °C in the stands and 20 °C in other public areas (Nguyen 2012), resurfacing water (10 °C (Karampour 2011), 30 ~ 50 °C (Rogstam and Bolteau 2015), 55 ~ 80 °C (ASHRAE. Ice rinks 2006)) and domestic hot water (50 ~ 70 °C (Rogstam and Bolteau 2015)). The refrigeration and heating demands contribute to more than two-thirds of the total energy demands. Therefore, the refrigeration and heating systems

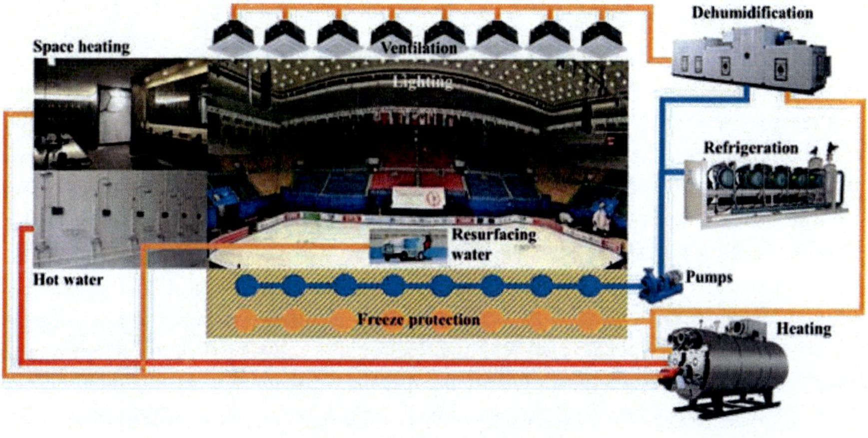

(a) Energy system configuration.

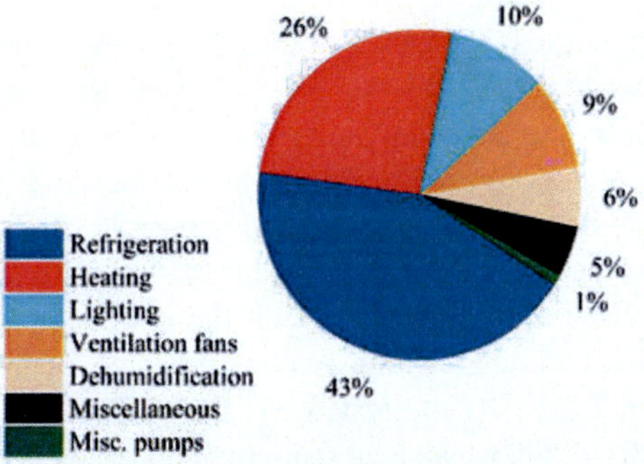

(b) Energy demands share [19].

Fig. 8.3 Energy system and demands in the arena with an ice rink should be integrally designed and operated. Humidity generates from the ice surface, resurfacing process, and people, hence dehumidification is more important than that in common buildings. The humidity of air can be removed either by the cooling coils to condensate the water vapor or by the adsorption process in the desiccant wheel (Rogstam and Mazzotti 2014). The energy consumption of the ventilation system is mainly caused by the fans, and this system is often combined with refrigeration, heating and dehumidification systems to provide comfortable indoor air, which is integrally termed by HVAC system.

It was reported that the skating season of an ice rink ranged from 8 to 12 months per year with the operation hours of 18 h per day and 7 days a week (Erol et al. 2017). Moreover, the ice rink arenas are also used for energy-intensive sports or recreational facilities with annual energy consumption of 750 kWh/m^2 which is around 3 times compared with that of office buildings (Krarti 2010). In the case of the ice rink arena for ice hockey, the annual energy consumption ranges from 900 to 1,500 MWh corresponding to the average energy bill of $50,000 to $90,000, more than half of which is contributed by the refrigeration system (IIHF. TECHNICAL GUIDELINES OF AN ICE RINK 2016).

The refrigeration systems can be generally classified into two types as direct and indirect systems. A refrigeration system is comprised of a mechanical vapor compression system to meet the cooling demand, a distribution system to transport the coolant between the ice layer and evaporator, and maybe a heat recovery system to recover the waste heat from the condenser. The distribution system is used as the evaporator in the direct system, so there is one loop with one type of refrigerant in the whole refrigeration system. Because of the pipe network of the distribution system with the primary refrigerant, the direct system requires high capital cost caused by high operating pressure as well as a large amount of refrigerant charge resulting in the potential hazard of refrigerant leakage. Hence, the indirect system is commonly used with the heat exchange between the primary refrigerant in the mechanical vapor compression system and the secondary refrigerant in the distribution system. In consideration of the heat transfer performance, safety and compatibility with the plastic pipelines, brine and glycol are used as typical secondary refrigerants. Because of less refrigerant charge in the indirect system, both the refrigeration unit size and risk of refrigerant leakage can be reduced. However, the larger temperature difference is also caused by the heat exchange between the two refrigerants in the evaporator. For example, the secondary refrigerant in the distribution system only can be cooled to $-5 \sim -2$ °C provided that the refrigeration system generates a temperature of $-12 \sim -10$ °C in the evaporator (Nguyen 2012). Moreover, additional pumping power is also needed in the distribution system. The fully indirect system also contains a secondary loop connected with the condenser in comparison with the above partly indirect system, which is considered to be inferior to the direct condensation (Nguyen 2012). As suggested by IIHF. TECHNICAL GUIDELINES OF AN ICE RINK (2016), the waste heat from the condenser is collected by the dry cooler and distributed to different heat exchangers for space heating, water heating, dehumidification, freeze protection, and so on. The heat recovery is considered to be more suitable for space heating, and 20% of annual energy use can be saved by utilizing warm air from the condenser (Piché and Galanis 2010).

R22 was used as the traditional primary refrigerant in the refrigeration systems of many old ice rinks, but the use of R22 is against the Montreal Protocol, and people have turned to natural refrigerants in consideration of the low global warming potential. It was reported that ammonia is mostly used as the refrigerant for ice rinks in Sweden with the remaining use of such HFC refrigerants as R404a or R134a (Makhnatch 2011). The revival of carbon dioxide as the refrigerant was proposed in the later years of the last century and the use has been spreading to refrigeration applications in sports facilities. The use of carbon dioxide is thus highlighted in this work in the case of refrigeration cycle and system for the ice rink and snowmaking respectively.

8.2 CO_2 Refrigeration Cycles and Systems for Ice Rinks

8.2.1 Development and State-of-The-Art of CO_2 Applications in Ice Rinks

The use of natural refrigerants such as carbon dioxide can date back to the nineteenth century, but they were replaced by chemically synthetic refrigerants with more suitable characteristics in the 1950s (Bodinus 1999). The revival of using carbon dioxide as the refrigerant with transcritical solutions was proposed by Norwegian professor Gustav Lorentzen later in the last century (Lorentzen 1994), where the significance of heat reclaim was highlighted. Afterward, the applications of carbon dioxide in supermarket refrigeration are considered to be the earliest commercial cases with the new technology of carbon dioxide refrigeration (Sawalha 2008). As shown in Fig. 8.4, the potential of high-temperature heat recovery is one of the main advantages of the newly proposed transcritical CO_2 cycle, which can meet the various heating demands within different temperature ranges by the ice rink. The integrated concept of cooling and heating is thus widely used based on the energy system with CO_2 as the working fluid.

In 1999, carbon dioxide was first used in a pumped secondary refrigerant circulation combined with ammonia as the primary refrigerant in Messestadion (an arena in Dornbirn). The ice rink system with carbon dioxide as the secondary refrigerant in combination with any primary refrigerant is considered to be the first generation of CO_2 application in ice rinks (Rogstam 2016a). Accordingly, the second-generation of CO_2 ice rink system refers to the refrigeration system with carbon dioxide as the primary refrigerant. The use of CO_2 as the primary refrigerant in ice rink systems was first proposed by the International Ice Hockey Federation in 2002 (IIHF 2002). Nevertheless, the development of the second-generation of CO_2 ice rink system was impeded by the limited cooling capacity of the CO_2 compressor. In 2006, a CO_2 transcritical ice rink with 12 compressors was designed for the Katrineholm city ice rink

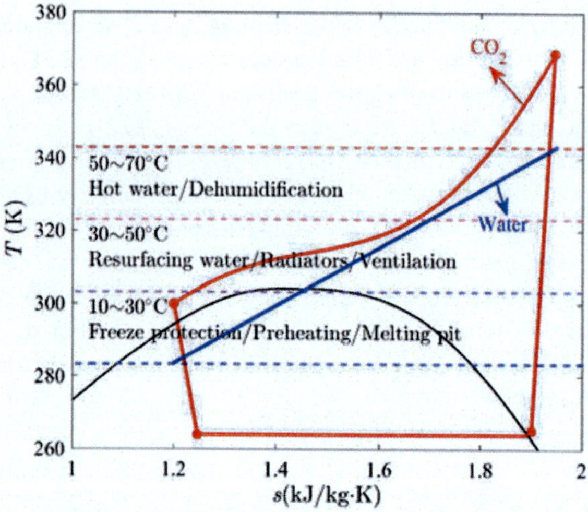

Fig. 8.4 Integrated concept of energy utilization for the ice rink based on transcritical CO_2 cycle (with temperature ranges referred to in Rogstam and Bolteau (2015))

to meet the total cooling capacity of 300 kW (Larsson 2006), but the ice rink industry was not used to the high investment cost and the complex operation resulted by the numerous compressors. Therefore, the second-generation CO_2 ice rink system could not be realized until the full development of CO_2 compressors in terms of the cooling capacity, and it was first built as a direct system with 7 compressors for the cooling capacity of 317 kW at Arena Marcel Dutil in 2010 (Simard 2012). The main challenge of the direct system is the high cost of metal pipes required by high-pressure carbon dioxide. Consequently, the first indirect second-generation CO_2 ice rink system with brine as the secondary refrigerant was built at Dollar des Ormeaux in 2012 with the heat recovery to heat a swimming pool (Heon and Guerra 2015). Additionally, a new type of secondary refrigerant, ammonium hydroxide, was recently used in the indirect second-generation CO_2 ice rink system (McLaughlin 2017).

The development and state-of-the-art of CO_2 refrigerant and applications in ice rinks are shown in Fig. 8.5. It was estimated that carbon dioxide had been used in at least 100 ice rink systems in the world by 2018, about 60 of which employed carbon dioxide as the primary refrigerant. Most of these second-generation types of carbon dioxide ice rink systems have been built since 2010, about 20 of which are located in Europe and 40 in North America (McLaughlin 2018). Moreover, the use of carbon dioxide as the refrigerant in ice rink systems, especially those systems using carbon dioxide as the primary refrigerant is still rapidly growing these years.

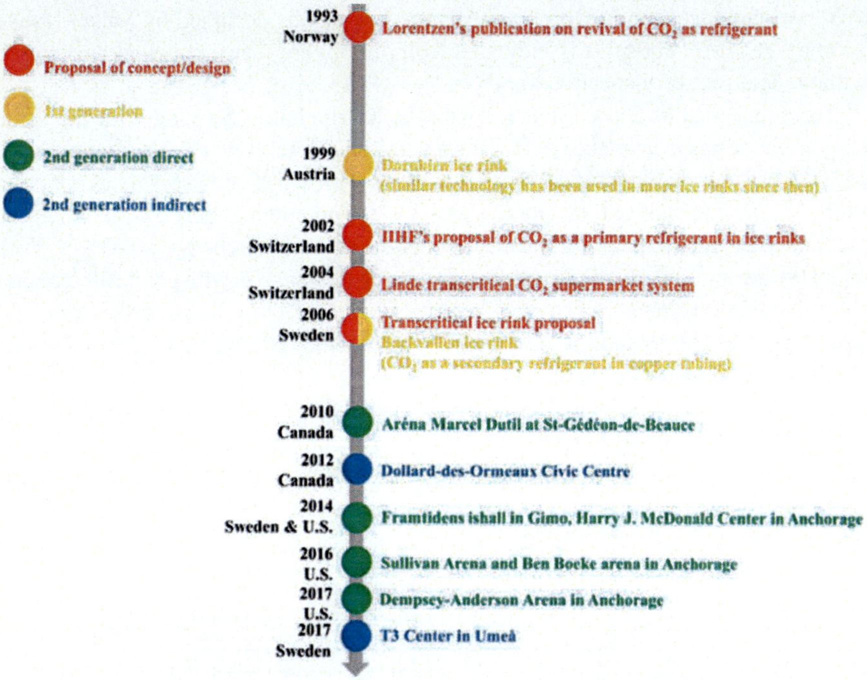

Fig. 8.5 The development and state-of-the-art of CO_2 refrigerant and applications in ice rinks (modified from Rogstam (2016b) with additional data from McLaughlin (2017); Garry (2017))

8.2.2 Description and Case Study of CO_2 Refrigeration Cycles in Ice Rink Systems

Typical ice rink systems with carbon dioxide applications are composed of the subsystems of mechanical vapor compression, distribution and heat recovery, which is similar to typical ice rink systems. Carbon dioxide was successively used in the secondary loop of the indirect system, the direct system, and the primary loop of the indirect system.

First-generation system

Carbon dioxide is used as the secondary refrigerant in the first generation type of CO_2 ice rink systems as shown in Fig. 8.6 with ammonia as the primary refrigerant in most of the cases especially in Europe, while HCFCs and HFCs can be also used as the primary refrigerants. Liquid carbon dioxide is pumped to the ice rink from the liquid receiver and then evaporates by absorbing heat from the ice layer. The liquid and vapor of carbon dioxide go back to the liquid receiver and separate from each other. The carbon dioxide vapor also goes to the evaporator to reject heat to the primary refrigerant and the condensed carbon dioxide goes back to the liquid receiver. Figure 8.7 presents the first NH_3/CO_2 ice rink system in the world built at Messestadion Arena in Dornbirn, Austria, which was designed by Sulzer (today Axima) (AXIMA 2004). It is indicated from this figure that the wet cooler is used without heat recovery specification.

Technical data of Backvallen Ice Rink in Katrineholm, Sweden with the same CO_2 applications is available in Rogstam and Prakash (2007). The cooling capacity of 300 kW is supplied by two refrigeration units with scroll compressors. Ammonia, as the primary refrigerant, evaporates at −11 °C and condenses at 38 °C, while the secondary refrigerant, carbon dioxide, undergoes the phase change at −10 °C both under the ice layer and in the evaporator. Heat recovery system with gas desuperheater and other heat exchangers provides heat for ice resurfacing, freeze protection, and space heating. The remaining low-temperature heat is rejected by a dry cooler with ethylene glycol as the coolant.

The phase change is the main process of carbon dioxide in the secondary loop, so both pipelines and vessels capable to withstand the highpressure are required by the high saturation pressure of carbon dioxide within the operating temperature range.

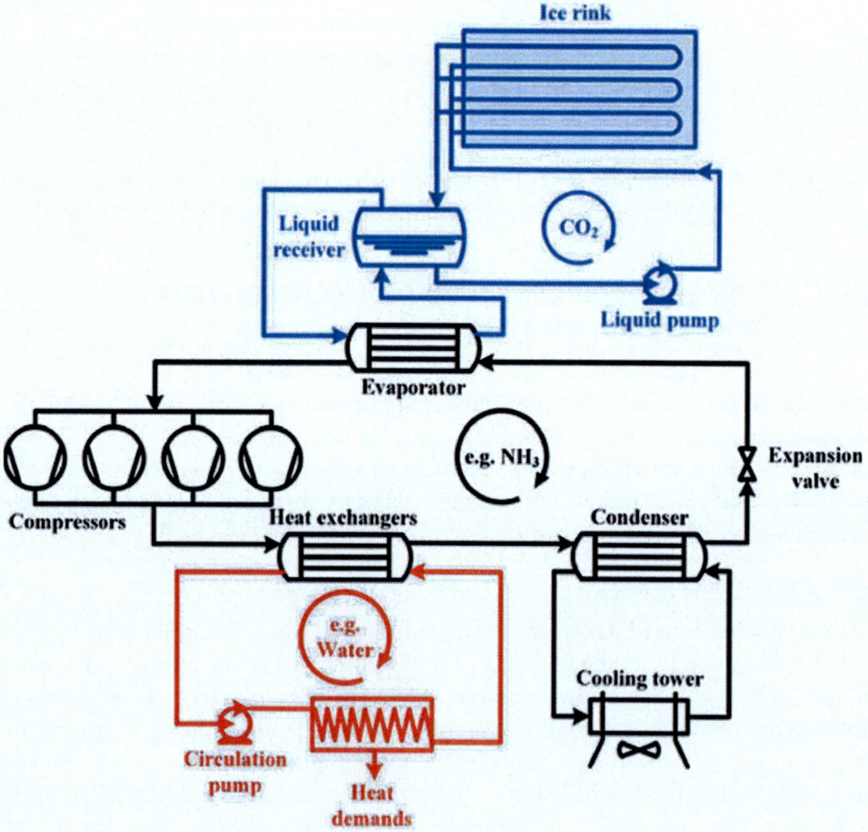

Fig. 8.6 Schematic illustration of the first generation of CO_2 ice rink system (with CO_2 as the secondary refrigerant)

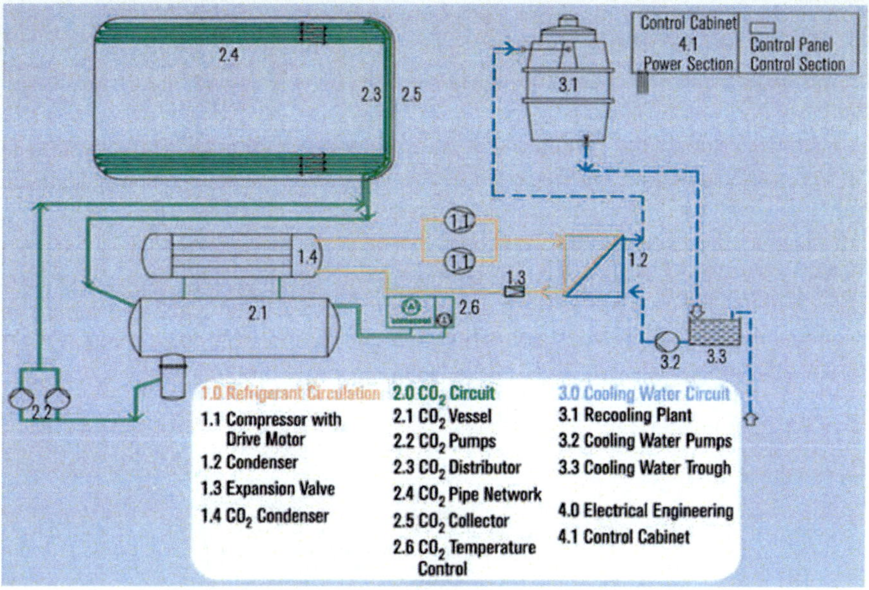

Fig. 8.7 The first NH_3/CO_2 ice rink system in the world (Messestadion, an arena in Dornbirn) (Weltweit erste eislaufbahn mit CO2 als kälteträger. 1999)

Based on the comprehensive studies on pressure drop and heat transfer performance (Shahzad 2006; Nilsson et al. 2006), the 1/2 in. (12.7 mm) copper tubing with plastic foil is commonly used as such pipelines to transport high-pressure carbon dioxide. The tubes are typically installed along the longer side of the ice rink with specific tube spacing, e.g. of 4 in. (102 mm), and the headers are installed on one of the shorter sides with return bends on the other side, so each pass of the copper tubing will be 400 ft. (122 m) for two-pass configuration in NHL-size ice rinks (Simard 2012).

The performance of the mechanical vapor compression system is similar to that of traditional ice rink systems, while the improvements of this type of ice rink system are mainly because of the new secondary refrigerant, carbon dioxide. In comparison with the traditional secondary refrigerants, such as brine and glycol, the use of carbon dioxide results in the advantages of up to 90% decreased pump power (Rostam 2010) (e.g. 3 kW for CO_2 vs. 15 kW for brine in Backvallen Ice Rink (Rogstam and Prakash 2007)), reduced corrosion problems and constant temperature under the ice layer as well as disadvantages of increased investment due to the metal pipelines required by high-pressure carbon dioxide.

Direct second-generation system

In the second-generation type of CO_2 ice rink systems, carbon dioxide becomes the primary refrigerant in both the direct and indirect systems. As presented in Fig. 8.8, the direct system eliminates the use of the evaporator and replaces the condenser with the gas cooler. The liquid carbon dioxide from the liquid receiver is pumped through the distribution pipelines and partly evaporates under the ice layer, while the vapor of carbon dioxide from the liquid receiver undergoes a mechanical vapor compression cycle and returns to the liquid receiver with a mixture of liquid and vapor.

Figure 8.9 shows the patented first transcritical CO_2 direct ice rink system in the world built at Marcel Dutil Arena in Saint-Gédéon-de-Beauce, Québec, Canada, which was designed by CSC/CIMCO (CanmetENERGY 2013). This system won first place as the existing industrial facilities or processes for the 2012 ASHRAE Technology Award and the case study is available in Simard (2012). The refrigeration capacity of 317 kW is supplied by four compressors and a 3 kW variable-speed pump maintains the CO_2 circulation at 1.9 L/s. A higher evaporating temperature of −7 °C is viable to keep the ice layer at −5 °C compared with other ice rink systems.

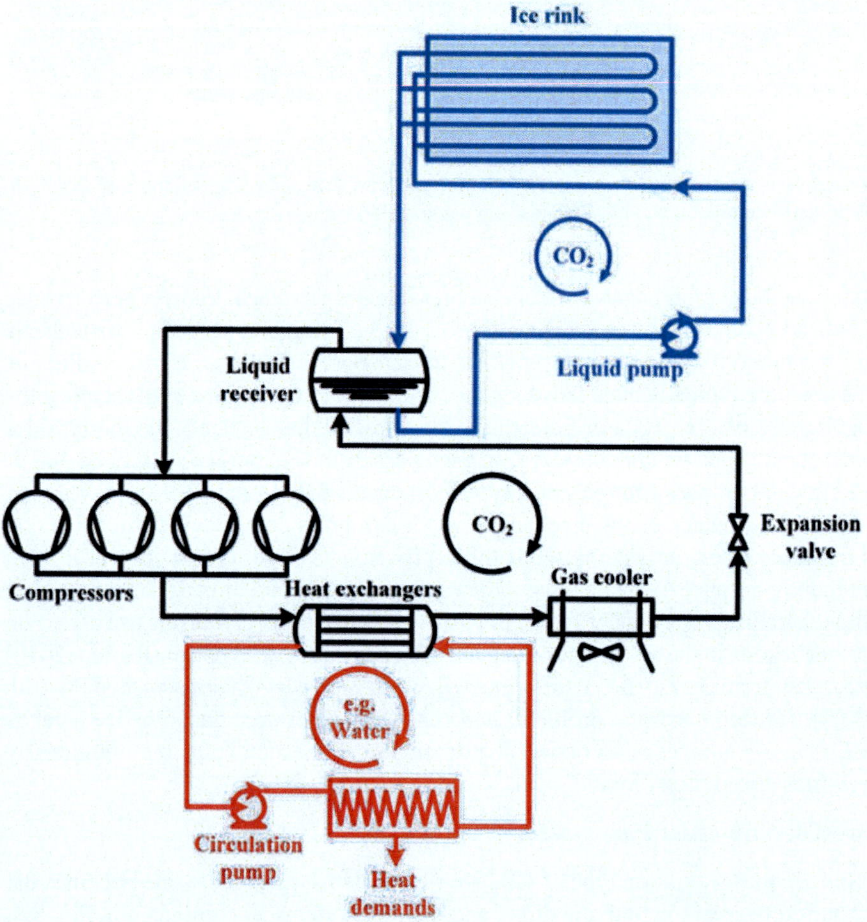

Fig. 8.8 Schematic illustration of the second-generation of direct CO_2 ice rink system (with CO_2 as the primary refrigerant)

The high-temperature heat at the outlet of compressors is used to produce hot water at 75 °C which covers all hot water needs. Further heat recovery is conducted by connecting with a warm glycol loop, which delivers 45 ~ 55 °C heating to the fan coils at the main entrance, locker rooms, and central air unit with a return temperature of 30 ~ 35 °C. Due to the original trench for the distribution manifolds on the longer side of the ice rink, a four-pass tubing configuration is adopted with plastic-coated soft copper tubes installed along the shorter side of the ice rink, which is different from the aforementioned typical tubing configuration. In comparison with the reference case with the NH_3/brine system, above 20% reduction of energy consumption is resulted from the annual refrigeration COP of 3.35 because of higher evaporating temperature, self-sufficient heat supply by heat recovery system, and 90% saving of pumping power.

Indirect second-generation system

Carbon dioxide can be also used as the primary refrigerant with common secondary refrigerants. The system configuration shown in Fig. 8.10 is similar to the traditional ice rink systems except for the gas cooler used by the primary refrigerant of CO_2. A fully indirect system is unnecessary due to the compatibility of CO_2 with a dry cooler, which saves water and maintenance costs of the cooling tower.

The ice rink system of Dollard-des-Ormeaux Civic Center in Dollard-des-Ormeaux, Québec, Canada is referred to as the first transcritical CO_2 indirect ice rink system in the world, which is designed by Carnot Refrigeration (CanmetENERGY 2013). This system also won first place as the existing commercial and institutional buildings for the 2015 ASHRAE Technology Award and the case study is available in Heon and Guerra (2015). The total refrigeration capacity of 738 kW for three NHL-size ice rinks is supplied by two groups of compressors. A shared brine distribution system is used for the three ice rinks, which require different operation periods and alternate use. Due to the high critical pressure and low critical temperature of carbon dioxide, poor compressor performance and low COP will be caused by the high air temperature in summer. At the ambient temperature of 35 °C, CO_2 refrigeration systems can reach a lower COP of 1.6 in comparison with the ammonia refrigeration systems with a COP of 3.45. However, the high operating pressure of CO_2 refrigeration systems also provides high-temperature heat rejection and large heat recovery potential. More than 40% of the heat rejection is from the desuperheater with the highest temperature of 104 °C and is used for heating a high-temperature water loop from 71 to 82 °C, which is then delivered to produce domestic hot water and heat the swimming pools and two small glycol loops for the heating of players' locker rooms, the main pool area and changing rooms. Afterward, the excess heat at around 49°C is rejected into the coils for the dehumidification and stands heating. The low capital cost of the brine distribution tubing is the advantage of the indirect transcritical CO_2 ice rink system. It was also reported that an above 50% reduction of pumping power was realized by replacing the old two-pass brine distribution with the four-pass distribution. In general, the retrofitting of the ice rink system with the indirect transcritical CO_2 system reduces the annual energy costs

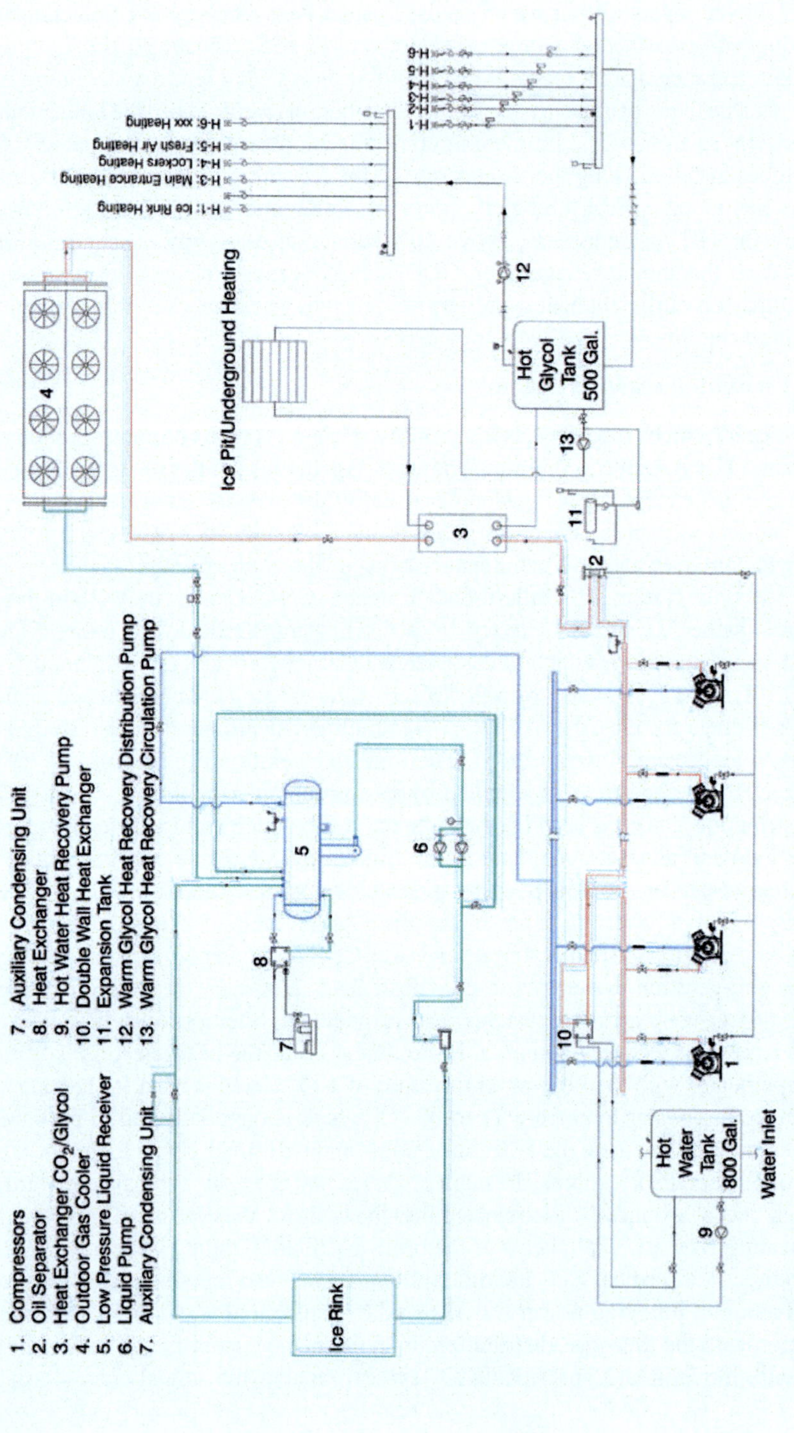

Fig. 8.9 The first transcritical CO_2 direct ice rink system in the world (Marcel Dutil Arena) (Simard 2012)

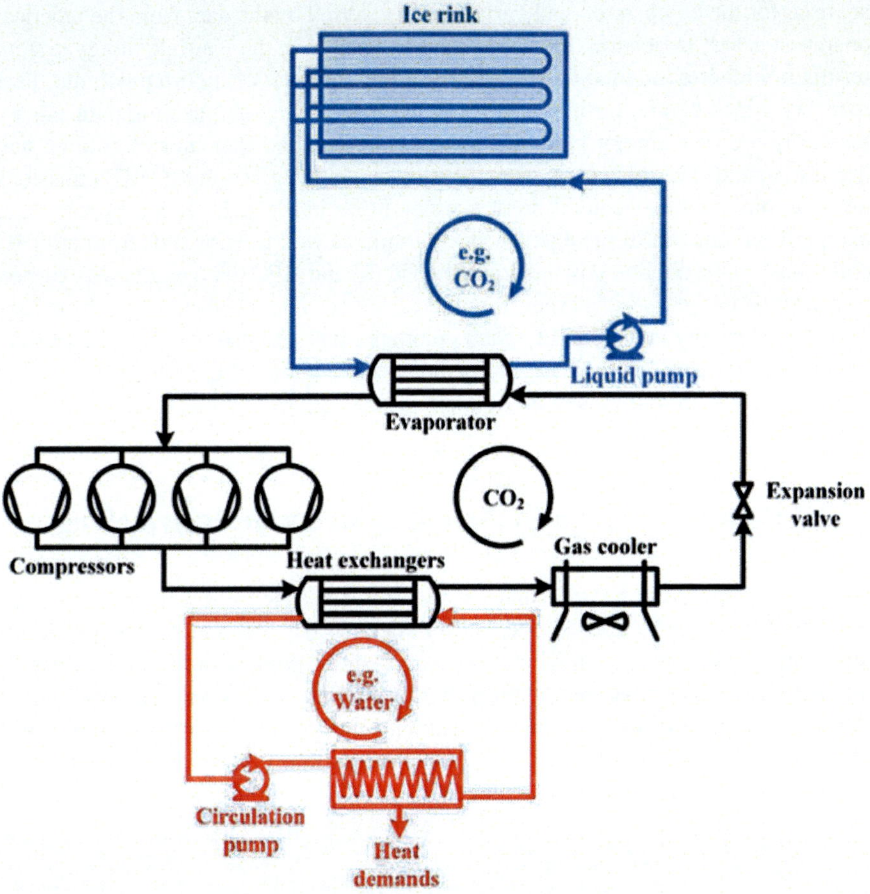

Fig. 8.10 Schematic illustration of the second-generation of indirect CO_2 ice rink system (with CO_2 as the primary refrigerant and a kind of secondary refrigerant)

by 31% resulting in a return on investment of about 8.3 years without subsidies and 3.9 years in consideration of the subsidies.

8.2.3 Comparative Remarks on Integrated Ice Rink Energy Systems

The working fluids used in ice rink energy systems have been developing rapidly these years due to the strictly restricted use of working fluids with high ODP and GWP. In Table 8.2, the ice rink energy systems with different working fluids and features are compared with each other in terms of the amount of refrigerant, cooling capacity, heating capacity, and combined COP. Similar cooling capacities of different cases are used for the comparison, but variations are caused by the data from the practical operation cases. In general, the use of carbon dioxide in the ice rink energy system results in higher combined COP because of the potential of high-temperature heat recovery which can be further used for integrated cooling and heating demands by the entire ice rink energy system. Moreover, normalized (per unit of cooling and heating capacity) annual energy consumption and greenhouse-gas (GHG) emissions are also compared between different working fluids in Fig. 8.11. It is concluded that about 50% reduction in the energy consumption as well as 95% reduction in GHG emissions can be obtained by replacing the HCFC and HFC refrigerants with carbon dioxide, while the use of carbon dioxide as the refrigerant also results in a 25% decrease of energy consumption when compared with the use of ammonia. Hence, carbon dioxide is one of the most promising working fluids of the integrated ice rink energy systems for simultaneous cooling and heating.

8.3 CO_2 Refrigeration Cycles and Systems for Snowmaking

Temperature-independent snowmaking requires a refrigeration system, while the heating energy can be also needed for hot water supply and space heating. Therefore, the heat recovery potential of CO_2 cycles can be used in snowmaking systems for simultaneous cooling and heating. In this section, the integration between the CO_2 cycles and the aforementioned temperature-independent snowmaking methods, including ice generation (vacuum and flake ice making) and indoor snowmaking methods, are respectively discussed as follows.

8.3.1 Ice Generation

In order to integrate the use of carbon dioxide with the existing ice generation methods, different systems combined with subsystems of mechanical vapor compression and ice generation were proposed in Dieseth (2016) as well as theoretical performance evaluation in terms of COP and energy consumption.

Vacuum ice making

Single-stage or double-stage water vapor compression cycles are used in the vacuum ice generation systems, which have been commercialized by IDE Technologies with the VIM series of products. However, such systems require high investment costs for the special steam compressors. Moreover, due to the limited compression ratio, a single-stage steam compressor has to be replaced with the double-stage one at higher ambient temperatures, which is considered to be 50% more expensive (Lachner et al. 2007).

A cascade system with a single-stage water vapor compression cycle at lower temperature and a single-stage CO_2 vapor compression cycle at higher temperature was then proposed as shown in Fig. 8.12a. The water vapor compression cycle in the ice generation subsystem operates with a constant condensation temperature (e.g. 5 °C in Dieseth (2016)) at different ambient temperatures, while the CO_2 vapor compression cycle in the mechanical vapor compression subsystem operates between a constant evaporating temperature (e.g. −5 °C in Dieseth (2016)) and a variable condensation temperature according to the ambient temperature. *P–h* and *T-s*

Table 8.2 Comparison of ice rink energy systems with different working fluids and features (CanmetENERGY. 2013)

Refrigerant	Rink slab fluid	Features	Amount of refrigerant (kg/TR)	Cooling capacity (kW)	Heating Capacity (kW)	Combined COP
R22	$CaCl_2$	Traditional open system assembled with electric heating	4.6	310	317	1.6
R507A	$CaCl_2$	Packaged semi-hermetic system with economizer port on compressors	1.4	287	363	2.7
R507A	Ethylene glycol	Modular semi-hermetic system with a stratified thermal buffer system	1.4	254	275	2.5
R410A	Methanol	Hermetic packaged heat pump system	0.6	267	275	2.3
R134A	$CaCl_2$	Packaged semi-hermetic system with heat pumps for water heating	2.0	289	354	2.7
R717	$CaCl_2$	Open packaged system with stratified thermal storage	0.6	262	317	3.2
R717	$CaCl_2$	Open system assembled with heat pump	1.6	247	508	2.9
R717	$CaCl_2$	Open system with integrated heat pump	3.8	254	517	2.5
R717	Ethylene glycol	Open packaged system with economizer port on the compressors	4.1	273	363	3.2

(continued)

Table 8.2 (continued)

Refrigerant	Rink slab fluid	Features	Amount of refrigerant (kg/TR)	Cooling capacity (kW)	Heating Capacity (kW)	Combined COP
R717	Ethylene glycol	Open packaged system with efficient reciprocating compressors	3.3	247	363	3.3
R744	CO_2	Semi-hermetic split packaged system with direct evaporation in the rink slab (direct second-generation system)	20.7	270	410	3.9
R744	$CaCl_2$	Semi-hermetic split-package with CO_2 space heating (indirect second-generation system)	6.5	271	365	3.8

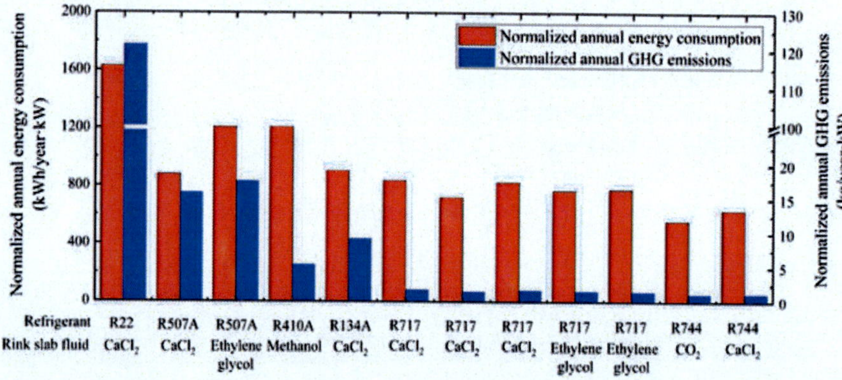

Fig. 8.11 Comparison of normalized annual energy consumption and GHG emissions between different working fluids (with data from CanmetENERGY (2013))

diagrams of the above water and CO_2 cycles are respectively presented in Figs. 8.12b and c by the data given in Dieseth (2016).

The performances of different vacuum ice making systems are compared with each other in terms of COP as given in Fig. 8.13 (Dieseth 2016). The cooling COPs of the systems without the CO_2 cycles are much higher than that of the cascade system because of the lower mass flow rate of the refrigerant and thus less compression

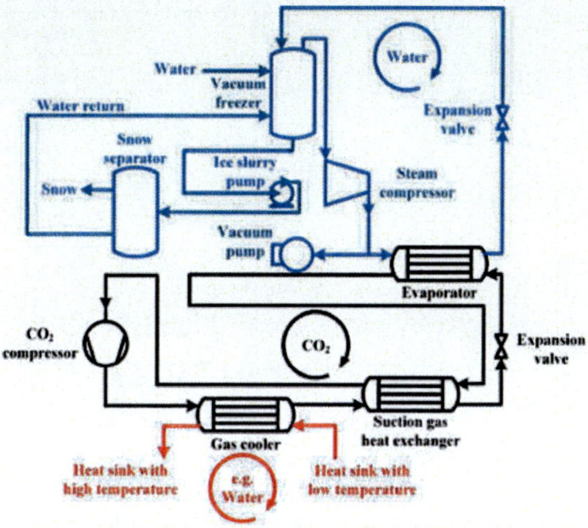

(a) Schematic illustration of the whole system.

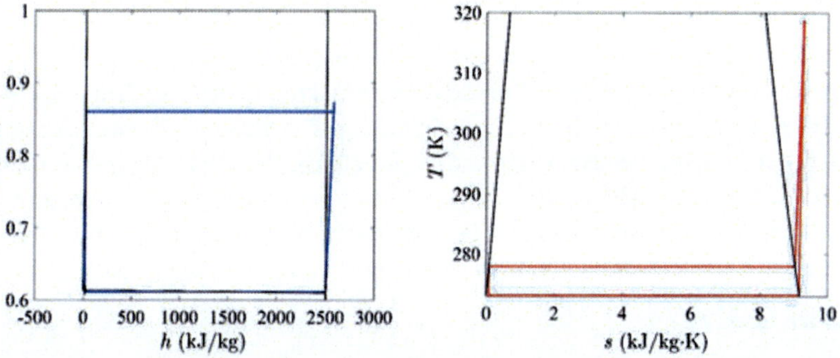

(b) *P-h* and *T-s* diagrams of the water cycle for the subsystem of ice generation.

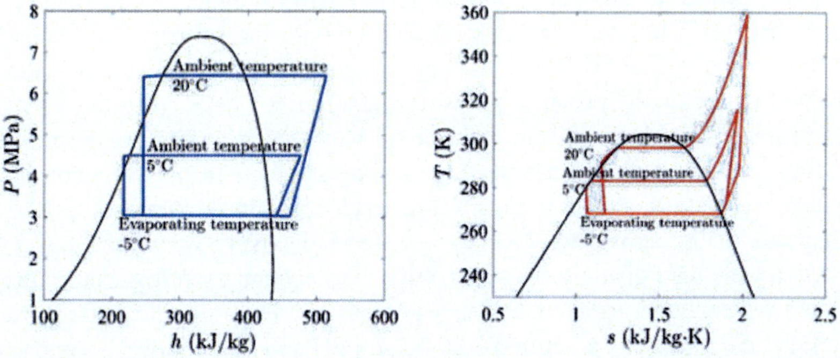

(c) *P-h* and *T-s* diagrams of the CO_2 cycle for the subsystem of mechanical vapor compression.

Fig. 8.12 Cascade system for vacuum ice making with single-stage CO_2 and water vapor compression cycles

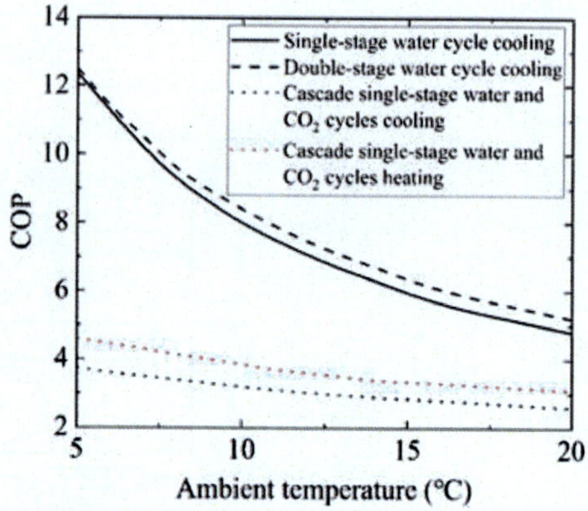

Fig. 8.13 Performance comparison between different vacuum ice making systems

work. Additionally, the temperature difference between the CO_2 and water cycles also results in extra energy consumption. It should be noted that the differences in COPs between the cascade system and the other two systems become smaller at higher ambient temperatures. Namely, the cascade system is somewhat competitive at higher ambient temperatures in consideration of the lower investment cost. The average energy consumptions of the vacuum ice making systems with the single-stage and double-stage water cycle are 5.64 kWh/m^3 and 5.8 kWh/m^3 respectively, while 12.8 kWh/m^3 for the cascade system (Dieseth 2016). Heat recovery potential can be also expected in the cascade system, but the subcritical CO_2 cycle is used to lower the compression work in the present system. Therefore, the cooling performance and use of recovered heat should be balanced in practice.

Flake ice making

The Flake ice making system requires extra refrigerants other than water to supply the cold energy to the flake ice maker. For example, ammonia and R404a have already been used in different models of the commercialized product "SnowFactory" by TechnoAlpin AG. The use of carbon dioxide as both the primary and secondary refrigerants in an indirect flake ice making system was proposed and evaluated in comparison with existing products (Dieseth 2016).

The primary CO_2 loops in the mechanical vapor compression subsystem with single-stage compression and double-stage compression are respectively considered in the flake ice systems shown in Figs. 8.14a and 8.15a. The corresponding *P–h* and *T-s* diagrams of the above two CO_2 cycles are respectively shown in Figs. 8.14b and 8.15b by the data given in Dieseth (2016). The secondary refrigerant in the ice generation system undergoes phase change processes without significant temperature and pressure changes (e.g. at around −30 °C, 1.4 MPa in Dieseth (2016)) and is just used as the heat transfer medium between the water flowing through the flake ice maker and the carbon dioxide in the primary loop.

The performances of different flake ice making systems are compared with each other in terms of COP as given in Fig. 8.16 (Dieseth 2016). The system with single-stage compression shows better performances at lower ambient temperatures and

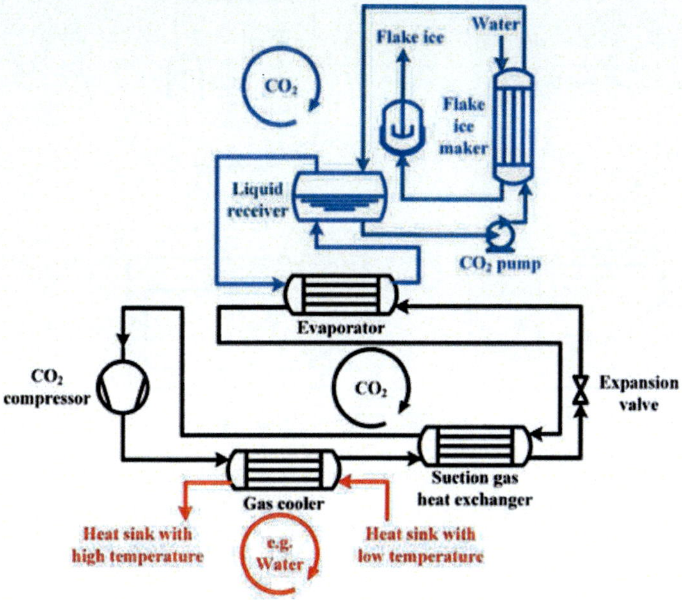

(a) Schematic illustration of the whole system.

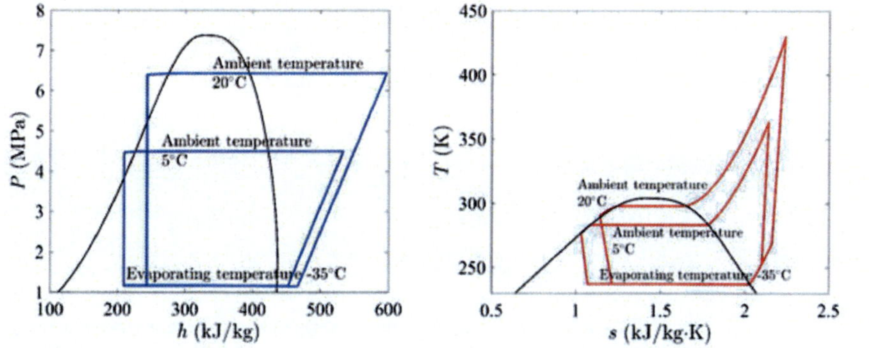

(b) *P-h* and *T-s* diagrams of the CO_2 cycle for the subsystem of mechanical vapor compression.

Fig. 8.14 Indirect system for flake ice making with a single-stage CO_2 vapor compression cycle

higher sensitivity to the ambient temperature variation. At ambient temperatures higher than 14 °C, the system with double-stage compression should be adopted because of the better cooling performance and limited maximum compressor outlet temperature. The average energy consumptions of the indirect flake ice making systems with single-stage and double-stage CO_2 compression are 24.47 kWh/m^3 and 27.1 kWh/m^3, respectively (Dieseth 2016), which are somewhat lower than that of the commercialized product using R404a.

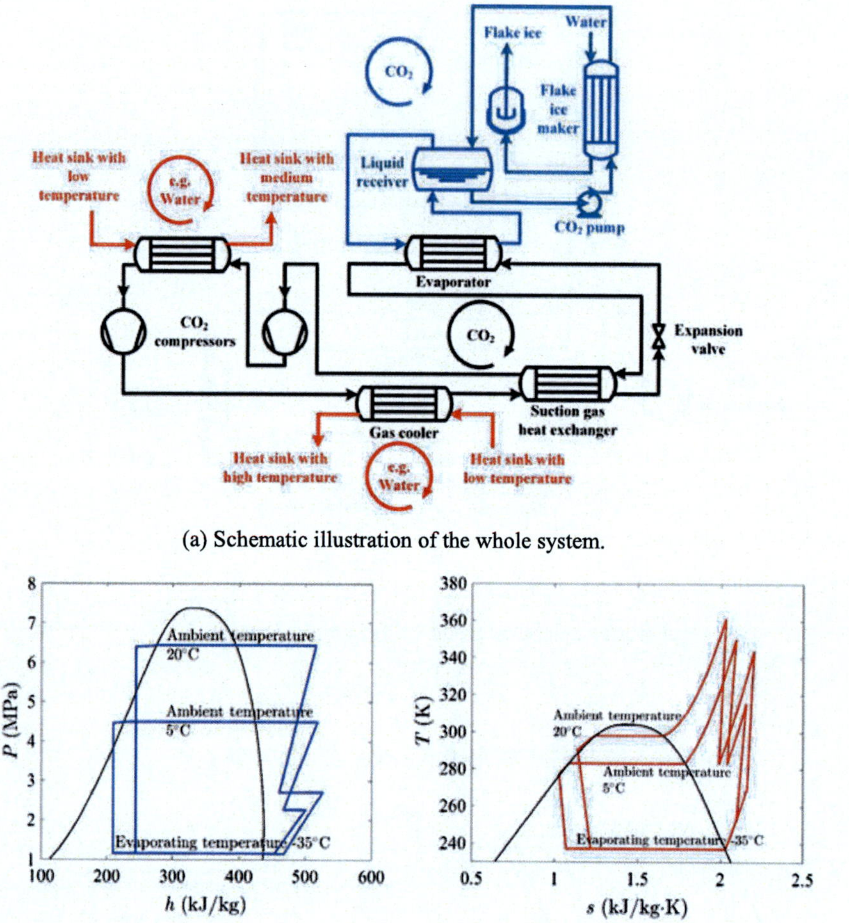

(a) Schematic illustration of the whole system.

(b) *P-h* and *T-s* diagrams of the CO_2 cycle for the subsystem of mechanical vapor compression.

Fig. 8.15 Indirect system for flake ice making with a double-stage CO_2 vapor compression cycle

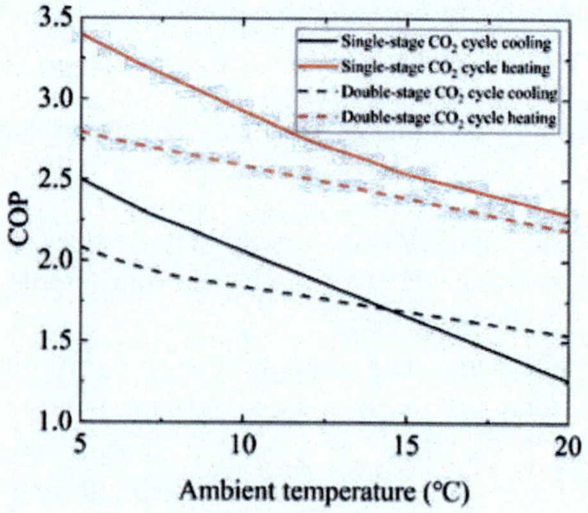

Fig. 8.16 Performance comparison between different flake ice making systems

8.3.2 *Indoor Snowmaking*

Cold rooms tend to be built for indoor snowmaking, however, in this section, they are not that large to accommodate lots of people skiing inside. The method and system of snowmaking by ejecting water into the cold air within a limited space have been developed by the authors. As shown in Fig. 8.17a, two units of fan and snowmaker are respectively installed at each end of the cold room, while the evaporators connected with cooling air ducts at the bottom are installed below the room top with six units arranged along the room length. The outdoor refrigeration units, water pump and air compressor are placed outside the cold room and respectively connected with the evaporators and snowmakers. During the operation, the water droplets will be ejected from the two snowmakers and transported to the room center by the airflow from the two fans. Meanwhile, cold air will be supplied through the cooling air ducts from the bottom outlet of the evaporators. Since enough flight time of the water droplets and efficient heat exchange between the water droplets and cold air are essential conditions for snow formation (Bergwitz-Larsen 2017), the minimum distance between the two snowmakers and the minimum height of the snowmaker nozzle are respectively set as 13 and 2.5 m, and the crossflow is also designed between the cold air with a lower velocity of 1.5 m/s and water droplets with a higher velocity of around 10 m/s. Carbon dioxide is used as the refrigerant in the refrigeration units operating with evaporating temperatures of −20 ~ −10 °C and hot water supply with 50 ~ 90 °C as presented in Fig. 8.17b.

The above indoor snowmaking system was tested at an ambient temperature of 5 °C as shown in Fig. 8.18a. The temperature inside the cold room was first lowered to about −19 °C, and then two snowmakers with 5 nozzles for each were turned on. At around 10 min, most of the nozzles were turned off with only three remaining because of the rapid temperature growth, but the temperature continued to increase. A steady indoor temperature of around −11 °C was then obtained with only two nozzles on. During the test lasted for 100 min, 966 kg of water and 79 kWh of electricity were consumed, thus the average energy consumption of the above indoor snowmaking system at an ambient temperature of 5 °C was 14.5 kWh/m^3 according to the tested snow density of 177 kg/m^3. Theoretical evaluation of the system performances in terms of cooling COP, average energy consumption and average hot water supply are shown in Fig. 8.18b. Despite the cooling COPs lower than 3.0 even at an ambient temperature of 0°C, the average energy consumption ranges from 13.5 kWh/m^3 to 20.5 kWh/m^3 at ambient temperatures of 0 ~ 20 °C, which is lower than that using flake ice making methods because of higher evaporating temperature and limited space. Moreover, the heat recovery from the gas cooler in the outdoor refrigeration units can meet different needs of water heating from 10 to 80 °C.

The above laboratory-scale system was also developed into two types of large-scale systems. As indicated in Fig. 8.18a, the red dash line separates the indoor space into two identical parts, either of which is the basic component consisting of the whole large-scale system. In the central type of large-scale system shown in Fig. 8.19a, the aforementioned basic components surrounding the room center are installed in a

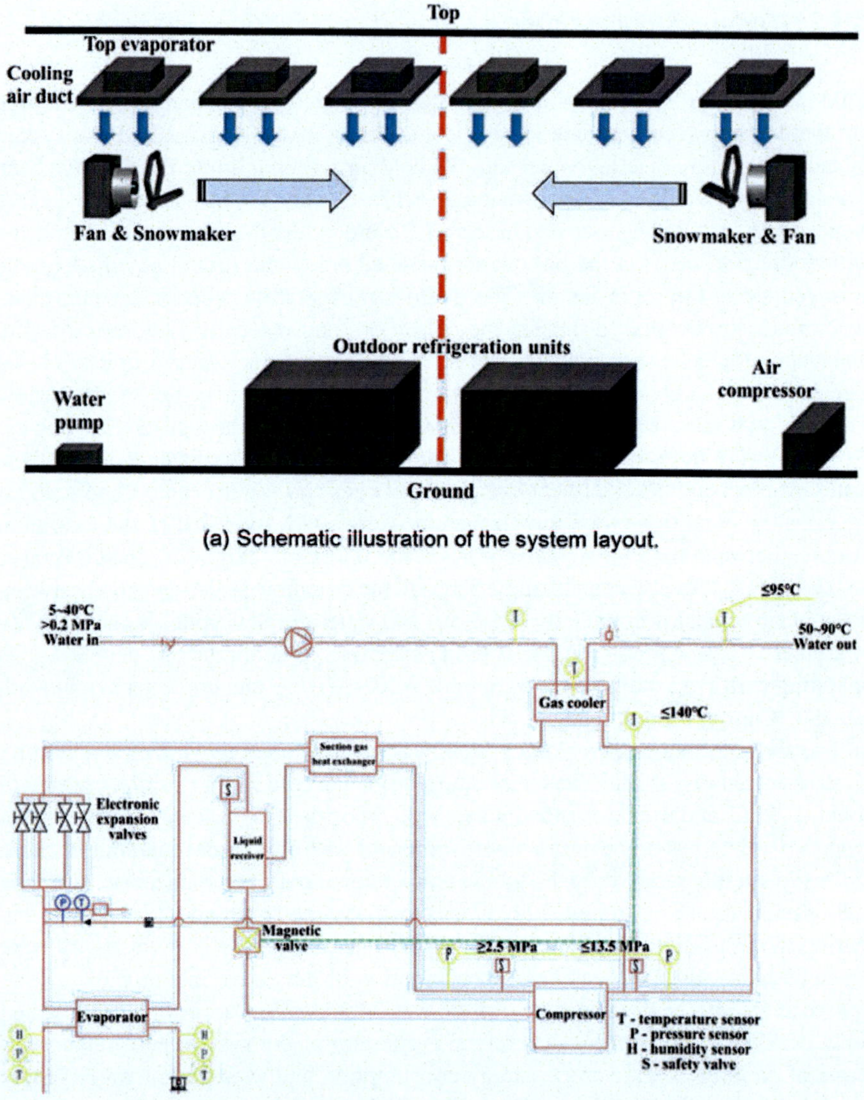

(a) Schematic illustration of the system layout.

(b) Specific illustration of the CO_2 refrigeration unit.

Fig. 8.17 Indoor snowmaking system using CO_2 refrigeration units

cylindrical cold room with water droplets ejected from the surrounding room wall and snow formed at the room center. The cold air will still be supplied through the cooling air ducts from the bottom outlet of the evaporators, and the returned air flows through the duct at the top center. Distributed type of large-scale system shown in Fig. 8.19b is more suitable for snow sports in mountainous areas, where the same

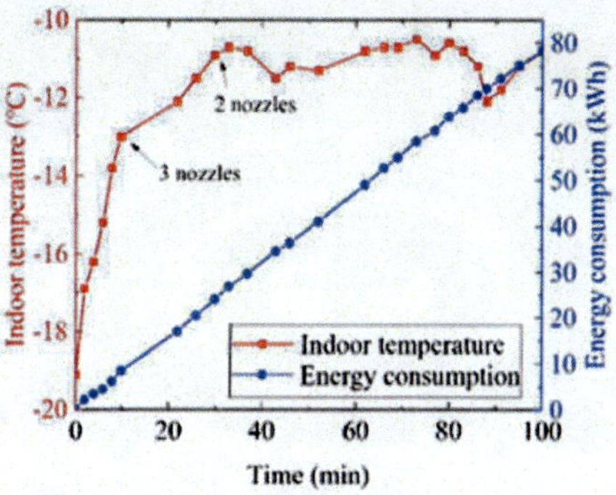

(a) Test data at the ambient temperature of 5°C.

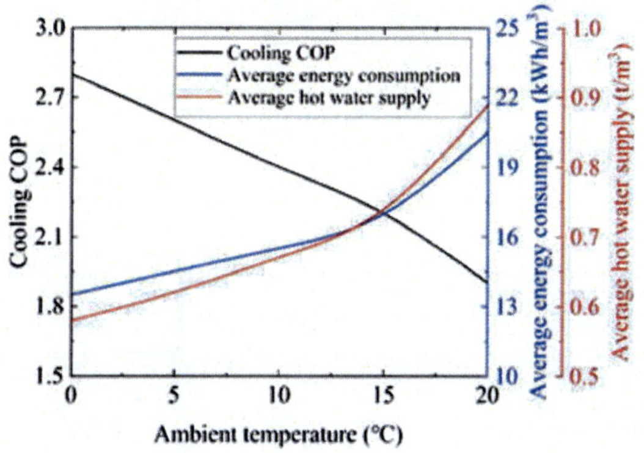

(b) Performance sensitivity to the ambient temperature variation.

Fig. 8.18 Performance test and evaluation of the indoor snowmaking system using CO_2 refrigeration units

basic components are symmetrically placed beside the track resulting in convenient snow transport from production to use.

8.3.3 Comparative Remarks on Snowmaking Methods with Heat Recovery

A comparative study on the system performances between different snowmaking methods and refrigeration cycles was performed in Fig. 8.20. The results indicated 70% lower cooling COP of the snowmaking systems with CO_2 cycles in comparison with that using water cycles, so the energy consumption of the snowmaking systems using CO_2 is at least two times that with water cycles. However, the use of CO_2 cycles results in an available hot water supply, whose profit will compensate for the higher energy cost. In fact, the energy consumptions of the vacuum ice making system using

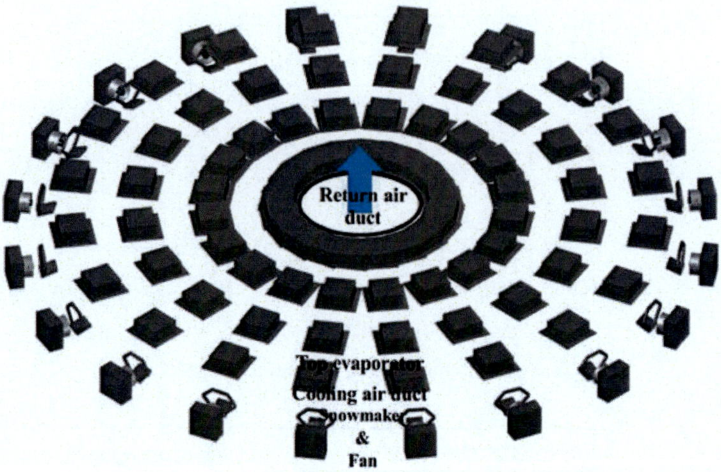

(a) Central type.

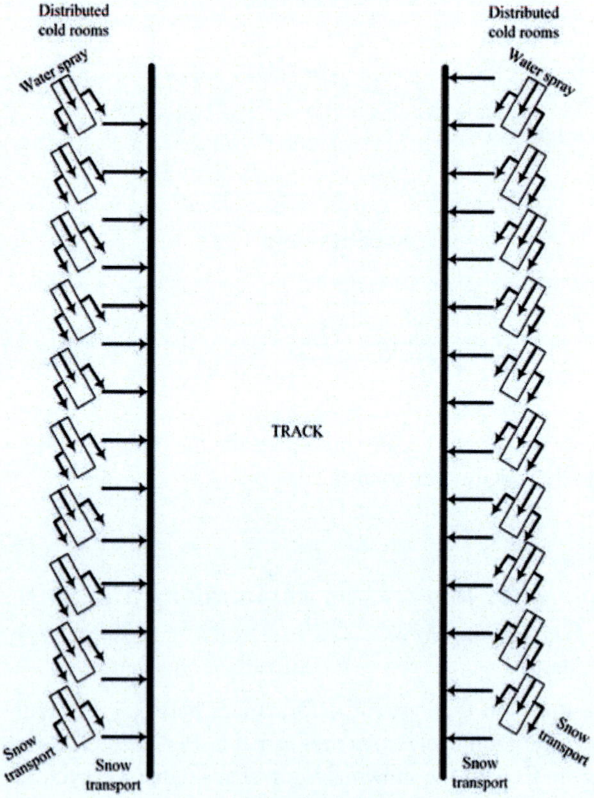

(b) Distributed type.

Fig. 8.19 Design of large-scale indoor snowmaking system

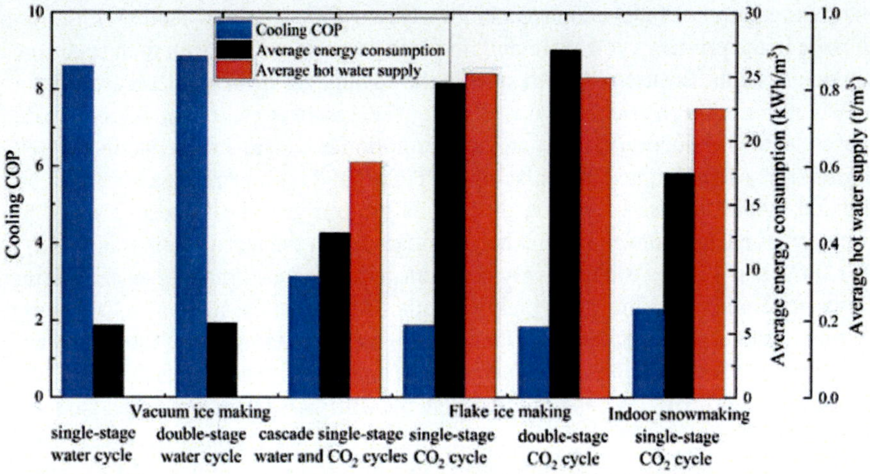

Fig. 8.20 Comparison of system performances between different snowmaking methods and refrigeration cycles

cascade water and CO_2 cycles and the indoor snowmaking system using single-stage CO_2 cycle are respectively 60% and 40% lower than that of other ice-generation-based snowmaking systems without water cycles. With the traditional snow gun, the indoor snowmaking methods are also expected to produce more snow within the same time in comparison with the ice generation methods, which is important in the case of supplementary snow production due to large snow loss under warm weather conditions. To maximize energy efficiency, a good match between heating and cooling demands should be achieved. In particular, the demands of cooling and heating vary with seasons, so thermal energy storage may be employed to accumulate the surplus heat production for peak demands.

8.4 Conclusions

The CO_2 refrigeration cycle is promising to be used in ice rink systems and snowmaking systems because of the potential for high-temperature heat recovery. Integrated energy systems with simultaneous cooling and heating can be then developed to meet different energy demands in ice rink systems and snowmaking systems. The use of carbon dioxide in these energy systems is extensively discussed and compared with the use of other refrigerants in terms of different system performances. The main concluding remarks are given as follows.

(1) The applications of CO_2 cycles and systems for ice rink refrigeration have been systematically developed over two generations in the last twenty years. Carbon dioxide was first used in the second cycle for ice rink refrigeration and then was further employed in the primary cycle for simultaneous cooling and heating. In particular, an

increasing number of the second-generation type of CO_2 ice rink systems with carbon dioxide in the primary cycle were built in these years. In comparison with traditional ice rink systems, the use of CO_2 results in the advantages of safety decreased energy consumption and operation cost as well as high heat recovery potential. It is concluded that about 50% reduction in the energy consumption as well as 95% reduction in GHG emissions can be obtained by replacing the HCFC and HFC refrigerants with carbon dioxide, while the use of carbon dioxide as the refrigerant also results in a 25% decrease of energy consumption when compared with the use of ammonia.

(2) Carbon dioxide is a relatively new refrigerant for snowmaking systems either by ice generation or indoor snowmaking methods. Theoretical studies on CO_2 refrigeration cycles and experimental studies on CO_2 refrigeration systems integrated with the above two snowmaking methods have been conducted. The results indicated 70% lower cooling COP of the snowmaking systems with CO_2 cycles in comparison with that using water cycles, so the energy consumption of the snowmaking systems using CO_2 is at least two times that with water cycles. However, the use of CO_2 cycles results in an available hot water supply, whose profit will compensate for the higher energy cost. In fact, the energy consumptions of the vacuum ice making system using cascade water and CO_2 cycles and the indoor snowmaking system using single-stage CO_2 cycle are respectively 60% and 40% lower than that of other ice generation-based snowmaking systems without water cycles.

Acknowledgements The supports of Prof. Trygve Magne Eikevik from the Norwegian University of Science and Technology, Beijing Municipal Commission of Science and Technology, the National Key Research and Development Program of China (2021YFF030680302) and Beijing Organising Committee for the 2022 Olympic and Paralympic Winter Games are gratefully acknowledged.

References

Asaoka T, Saito A, Okawa S, Kumano H, Hozumi T (2009) Vacuum freezing type ice slurry production using ethanol solution 2nd report: investigation on evaporation characteristics of ice slurry in ice production. Int J Refrig 32:394–401

ASHRAE (2006) Ice rinks. ASHRAE handbook-fundamentals. American Society of Heating, Refrigerating, and Air-Conditioning Engineers Inc., Atlanta

AXIMA (2004) Most Modern Ice Technology with CO_2 in the Messestadion Dornbirn. www.aximaref.com

Bergwitz-Larsen KW (2017) Energy efficient and environmental friendly snow production by refrigeration systems. Norwegian University of Science and Technology, Trondheim

Bellas I, Tassou SA (2005) Present and future applications of ice slurries. Int J Refrig 28:115–121

Bédécarrats J-P, David T, Castaing-Lasvignottes J (2010) Ice slurry production using supercooling phenomenon. Int J Refrig 33:196–204

Bodinus WS (1999) The rise and fall of carbon dioxide systems: the first century of air conditioning. ASHRAE J 41:37–42

CanmetENERGY (2013) Comparative Study of Refrigerations Systems for Ice Rinks. Natural Resources Canada, Varennes

Cao W, Beggs C, Mujtaba IM (2015) The theoretical approach of freeze seawater desalination on flake ice maker utilizing LNG cold energy. Desalination 355:22–32

Clulow MG (2006) Indoor snowmaking. ASHRAE J 48:18

Dieseth J-BR (2016) Snow production equipment at ambient temperatures above zero degrees celsius. Norwegian University of Science and Technology, Trondheim

Erol GO, Açıkkalp E, Hepbasli A (2017) Performance assessment of an ice rink refrigeration system through advanced exergoeconomic analysis method. Energy and Build 138:118–126

Egolf PW, Kauffeld M (2005) From physical properties of ice slurries to industrial ice slurry applications. Int J Refrig 28:4–12

Garry M (2017) Hillphoenix installs fourth CO_2 ice rink system in Alaska. http://r744.com/articles/7778/hillphoenix_installs_fourth_CO2_ice_rink_system_in_alaska

Graham J, Johnston W, Nicholson F (1993) Ice in fisheries. Food and Agriculture Organization of the United Nations, Rome

Hayashi K, Kasza KE (2001) Ice slurry cooling research: effects of microscale ice particle characteristics and freezing-point-depressant additives on ice slurry fluidity/discussion. ASHRAE Trans 107:346

Heon K, Guerra P (2015) CO_2 showcase for ice rinks, pools. ASHRAE J 57:62–65

Hawlader MNA, Wahid MA (2009) Analyses of ice slurry formation using direct contact heat transfer. Appl Energy 86:1170–1178

IIHF (2002) Ever thought of building an ice rink? IIHF Ice Rink Manual. International Ice Hockey Federation, Zürich

IIHF (2016) TECHNICAL GUIDELINES OF AN ICE RINK - CHAPTER 3. IIHF ICE RINK GUIDE. International Ice Hockey Federation, Zürich

IIHF (2018) IIHF Official Rule Book 2018–2022. International Ice Hockey Federation, Zürich

Kauffeld M, Wang MJ, Goldstein V, Kasza KE (2010) Ice slurry applications. Int J Refrig 33:1491–1505

Krarti M (2010) Energy audit of building systems: an engineering approach. CRC press

Karampour M (2011) Measurement and modelling of ice rink heat loads. KTH Royal Institute of Technology, Stockholm

Lachner BF Jr, Nellis GF, Reindl DT (2007) The commercial feasibility of the use of water vapor as a refrigerant. Int J Refrig 30:699–708

Larsson H (2006) Anbud Ishall Katrineholm-Kyla för ispist. Katrineholms Kommun, Anbud

Li X-W, Zhang X-S, Quan S (2012) Evaporative supercooling method for ice production. Appl Therm Eng 37:120–128

Li X-W, Zhang X-S, Cao R-Q, Fu X-Z (2009) A novel ice slurry producing system: producing ice by utilizing inner waste heat. Energy Convers Manage 50:2893–2904

Lorentzen G (1994) Revival of carbon dioxide as a refrigerant. Int J Refrig 17:292–301

Makhnatch P (2011) Technology and energy inventory of ice rinks. KTH Royal Institute of Technology, Stockholm

Martínez DS, Solano JP, Illán F, Viedma A (2014) Analysis of heat transfer phenomena during ice slurry production in scraped surface plate heat exchangers. Int J Refrig 48:221–232

McLaughlin C (2017) CO_2-ammonia ice rink installed in Sweden. http://www.r744.com/articles/7722/CO2_ammonia_ice_rink_installed_in_sweden

McLaughlin C (2018) EKA responds to Winter Olympics' HFC ice-rink use. http://r744.com/articles/8131/eka_responds_to_winter_olympics_hfc_ice_rink_use

Nakaya U (1951) The formation of ice crystals. Compendium of Meteorology. American Meteorological Society, Boston, pp 207–220

Nilsson P-O, Rogstam J, Sawalha S, Shahzad K (2006) Ice rink refrigeration system with carbon dioxide as secondary fluid in copper tubes. In: 7th IIR gustavlorentzen conference on natural working fluids, Trondheim, Norway

Nguyen T (2012) Carbon dioxide in ice rink refrigeration. KTH Royal Institute of Technology, Stockholm

Paul J (2003) Concept of operating indoor skiing halls with "Binary Snow" as snow substitute. In: 21st international congress of refrigeration

Piché O, Galanis N (2010) Thermal and economic evaluation of heat recovery measures for indoor ice rinks. Appl Therm Eng 30:2103–2108

Rostam J (2010) Energy usage statistics and saving potential in ice rinks. In: The conference on sustainable refrigeration and heat pump technology, Stockholm

Rogstam J, Mazzotti W (2014). Ice rink dehumidification systems energy usage and saving measures. In: 11th IEA heat pump conference, Montréal (Québec)

J. Rogstam S (2015) Bolteau. ICE RINK OF THE FUTURE—evaluation of energy and system solutions

Rogstam J (2016a) EVOLUTION OF CO_2 AS REFRIGERANT IN ICE RINK APPLICATIONS. 12TH IIR Gustav Lorentzen Natural Working Fluids Conference, Edinburg

Rogstam J (2016b) CO_2 refrigeration systems evolution for ice rinks. ASHRAE J 58:34–48

Rogstam J, Prakash M (2007) Energy analysis of Backavallen ice rink refrigeration system with CO_2 as heat transfer liquid in copper tubes. Sveriges Energi & Kylcentrum

Saito A (2002) Recent advances in research on cold thermal energy storage. Int J Refrig 25:177–189

Stamatiou E, Meewisse JW, Kawaji M (2005) Ice slurry generation involving moving parts. Int J Refrig 28:60–72

Sawalha S (2008) Carbon dioxide in supermarket refrigeration. KTH Royal Institute of Technology, Stockholm

Simard L (2012) Ice rink uses CO_2 system. ASHRAE J 54:38–44

Shahzad K (2006) An ice rink refrigeration system based on CO_2 as secondary fluid in copper tubes. KTH Royal Institute of Technology, Stockholm

Weltweit erste eislaufbahn mit CO_2 als kälteträger. Die Klima und Kältetechnik. 4 (1999)

Wijeysundera N, Hawlader M, Andy CWB, Hossain MK (2004) Ice-slurry production using direct contact heat transfer. Int J Refrig 27:511–519

Chapter 9
CO_2 Mobile Air Conditioning

Jiangping Chen, Junye Shi, and Dandong Wang

This chapter highlights the application of CO_2 refrigerant in automotive, bus and train air conditioning system. Natural refrigerant CO_2 is considered as a promising alternative to high GWP refrigerant such as HFC-134a because of its great environmental properties and cooling performance. Since Professor Gustav Lorentze proposed the novel trans-critical CO_2 cycle system in the 1990s, the CO_2 MAC system has made great progress and development in academia and industry. Before, research studies mainly focus on the performance evaluation and component development for the CO_2 MAC system. The optimum high pressure, related to gas cooler outlet temperature, is an important parameter affecting the COP of system. The trans-critical high pressure of CO_2 brings big challenges to development of key components. In recent years, CO_2 mobile heat pump has attracted more attention. At cold climate, it has a large heating capacity and a high heating COP, which could save electricity energy consumption and extend the driving range of electric vehicles, electric buses. Therefore, CO_2 heat pump has become more competitive in terms of CO_2 emission and system costs, which will promote the mass production application of CO_2 refrigerant in mobile vehicle, bus and train area in the future.

9.1 Background of CO_2 MAC

9.1.1 History of CO_2 MAC

As a refrigerant, CO_2 began in the mid-nineteenth century (1850) and continued to the twentieth century. At that time, CO_2 refrigerant was mainly applied in marine refrigeration systems. With the emergence of low cost Freon refrigerant, CO_2 refrigerant

J. Chen (✉) · J. Shi · D. Wang
Shanghai Jiaotong University, Shanghai, China
e-mail: jpchen@sjtu.edu.cn

X.-R. Zhang and T. M. Eikevik (eds.), *CO_2 Refrigeration Cycle and Systems*, Lecture Notes in Energy 96, https://doi.org/10.1007/978-3-031-22512-3_9

was gradually replaced by Freon refrigerants. In the 1970s and 1980s, researchers discovered that Freon refrigerant was the main cause of ozone depletion and greenhouse effect, so they began to pay renewed attention to natural refrigerants, such as CO_2 refrigerant. Professor Gustav Lorentze was the pioneer of the revival of CO_2 refrigerant. He submitted a patent in 1989 for a novel trans-critical CO_2 cycle system, in which the super-critical pressure of the system was controlled by the throttling valve. It was proposed that one of the applications of this trans-critical cycle is mobile air conditioning (MAC). Lorentzen and Pettern (1993) published the first experimental data on a CO_2 MAC laboratory prototype system in 1993, demonstrating COP data of that were competitive to baseline CFC-12 system performance, which is shown in Fig. 9.1 (Mu 2003; Pettersen et al. 1998). Petterson et al. (1998) reported Aluminum microchannel heat exchangers (gas cooler and evaporator) could be used for CO_2 gas coolers and evaporator with high pressure resistance in the MAC system. Figures 9.2 and 9.3 show their detailed configuration information (Mu 2003; Kim et al. 2004).

Since 1990s, the industries in Europe, Japan and the United States started to study the MAC using CO_2 natural refrigerant. The European RACE project from 1994 to 1997 included development and testing of car-installed prototype systems, with results confirming the potential for CO_2 MAC (Yang 2006). Members in the RACE project included car manufacturers (BMW, Daimler-Benz, Rover, Volvo, Volkswagen), system supplier (Behr, Valeo), and a compressor manufacturer (Danfoss). The European Union-sponsored B-COOL Project from 1995 to 1998 was devoted to the development of low-cost and efficient CO_2 systems for small cars (Malvicino et al. 2009). The project was carried out by a consortium constituted by 2 major OEMs, four supplier, and three acknowledged excellence centers. Figure 9.4 shows a Fiat Panda CO_2 system and the consortium in this project (Malvicino et al. 2009). From 2001 to 2003, SAE ARCRP (Automobile Refrigerant Cooperative Research Program) was conducted by 25 major automobile manufacturers and air conditioning component suppliers in the United States, Europe and Asia. The top-level design components for R134a and CO_2 systems were selected, and comparison experiments were conducted in the ACRC laboratory of University of Illinois, whose results represent the best MAC performance level at that time. Figure 9.5 shows the schematic diagram of CO_2 MAC system tested by ACRC (Bullard et al. 2000). In addition, a number of international conferences were held to discuss and promote the application of the CO_2 refrigerant, including Gustav Lorentzen Natural Working Fluids Conference, VDA alternative refrigerant winter meeting and SAE automotive Alternate Refrigerant System Symposium.

In the fifteen years after 1990, CO_2 MAC achieved great progress. Figure 9.6a shows a Daimler-Benz vehicle equipped with a CO_2 MAC system was tested in a wind tunnel (Armin and Petter 2005). Figure 9.6b shows a CO_2 MAC system installed in a Toyota FCHV-4 vehicle in December 2002 (Armin and Petter 2005). Figure 9.6c shows the vehicle measurements with Obrist-Ford Galaxy at the air-conditioned roller dynamometer test rig at Graz University of Technology (Wolf et al. 2007). Figure 9.6d, e and f show the test vehicle of Toyota Yaris in 2006, which is a B-class vehicle for Asian and European market (Wieschollek and Heckt

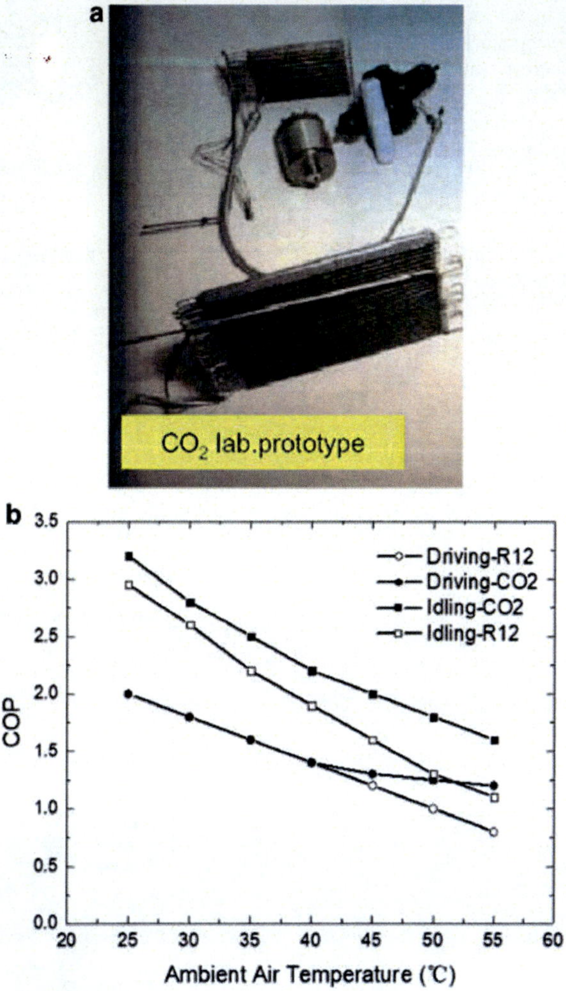

Fig. 9.1 **a** First prototype of CO_2 MAC (Mu 2003) **b** Performance comparison of CO_2 and R12 system (Pettersen et al. 1998)

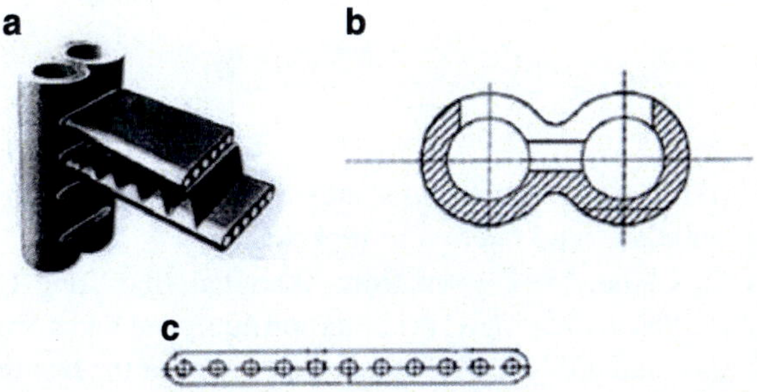

Fig. 9.2 A prototype microchannel CO_2 gas cooler for MAC system: **a** geometry of heat exchanger (Mu 2003), **b** cross-section of header (Kim et al. 2004), **c** microchannel type flat tube (Kim et al. 2004)

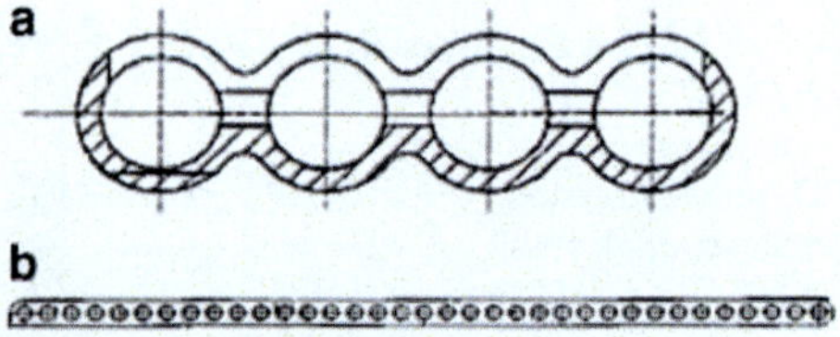

Fig. 9.3 A prototype CO_2 evaporator for MAC system **a** cross-section of the header (Kim et al. 2004) and **b** microchannel type flat tube (Kim et al. 2004)

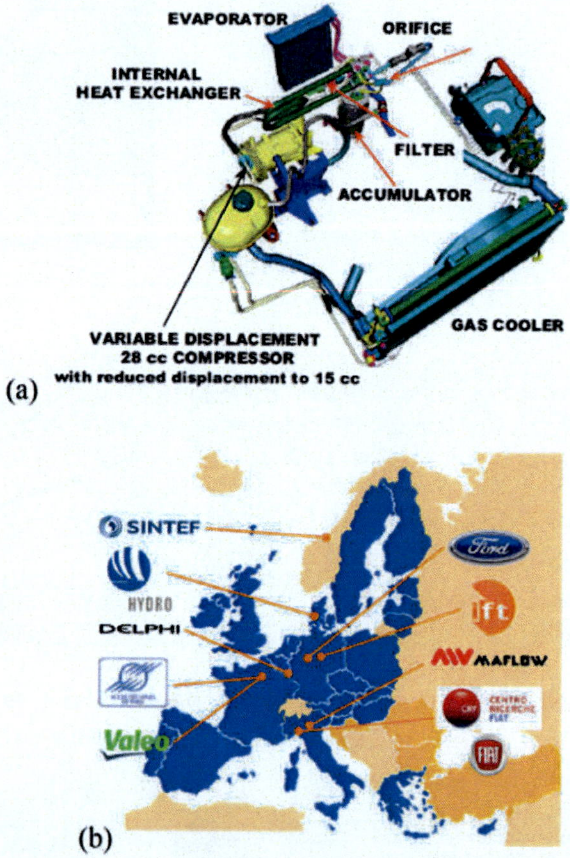

Fig. 9.4 **a** First vision of Fiat Panda CO_2 system (Malvicino et al. 2009) **b** the consortium of B-COOL project (Malvicino et al. 2009)

2007). The test vehicle has standardized architecture for CO_2 systems with combined accumulator and internal heat exchanger.

In China, MAC team from Shanghai Jiao Tong University developed the first prototype of CO_2 MAC in China and equipped it in a Shanghai Volkswagen vehicle in 2005 (Liu 2005). In 2005, the expert group of the National Development and Reform Commission evaluated that this system has reached to the international advanced level. In 2010, the MAC team developed its second CO_2 MAC and tested it on a Changan vehicle, as shown in Fig. 9.7c and d (Hu 2011; Jin 2011). In addition, the

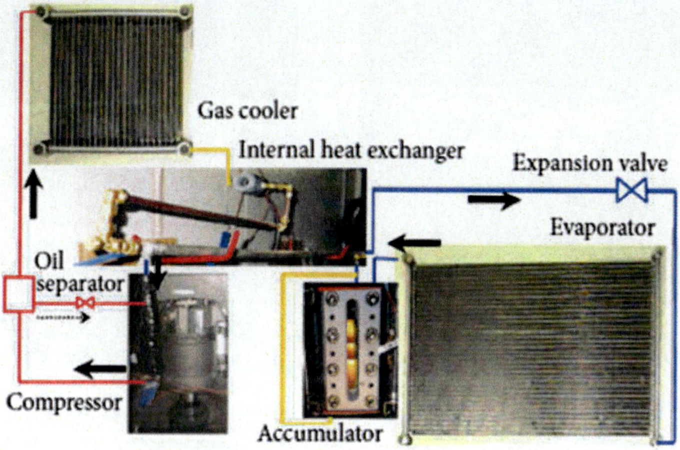

Fig. 9.5 Schematic diagram of CO_2 MAC system tested by ACRC (Bullard et al. 2000)

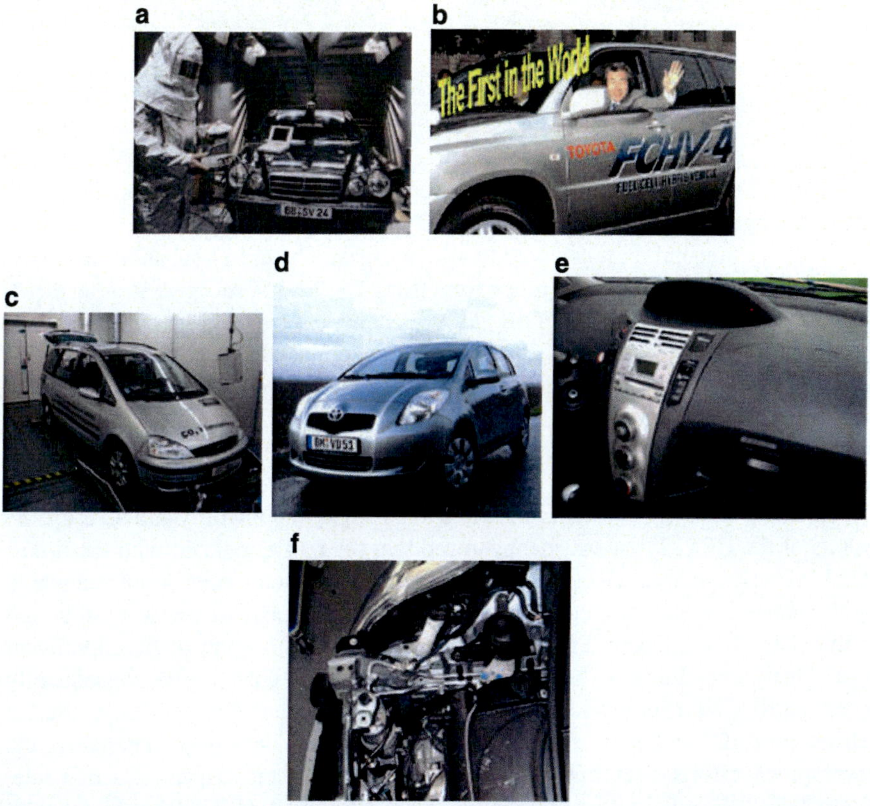

Fig. 9.6 Vehicles equipped with CO_2 MAC system **a** Daimler Benz in 1998 (Armin and Petter 2005) **b** Toyota FCHV-4 in 2002 (Armin and Petter 2005) **c** Ford Galaxy tested by Obrist (Wolf et al. 2007) **d–f** Tyota Yaris in 2006 (Wieschollek and Heckt 2007)

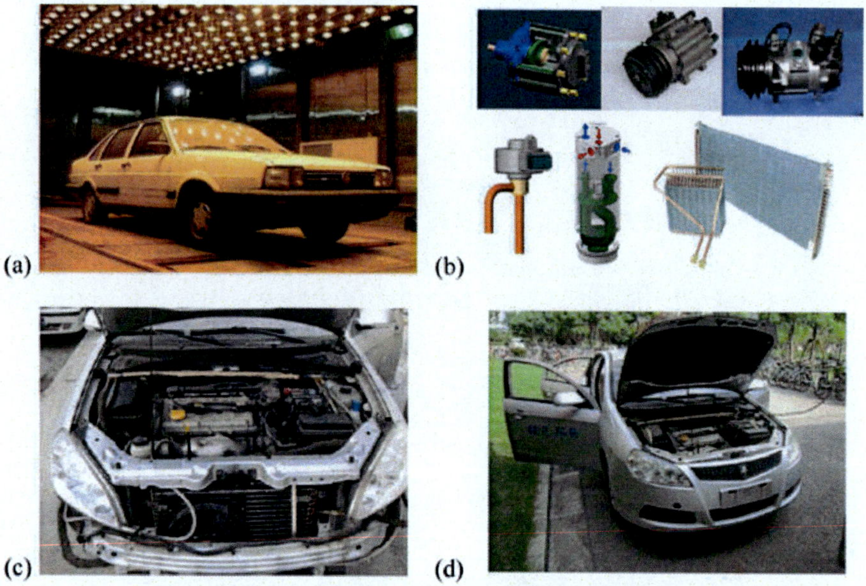

Fig. 9.7 **a**, **b** Shanghai Volkswagen vehicle (Liu 2005) **c**, **d** Chang'an vehicle with CO_2 MAC system (Hu 2011; Jin 2011)

MAC team developed the simulation models for calculating the performance of CO_2 heat exchanger, valves, compressor, and CO_2 system.

For environmental reasons, in 2006, the European Commission banned all MAC systems on all vehicles sold in Europe from using GWP >150 refrigerants after 2017. Natural refrigerant CO_2 and HFO-R1234yf become the two refrigerant alternatives to replace R134a. From 2006 to 2009 SAE International administered a cooperative research project (CRP1234) which evaluated R1234yf, which is proposed by Honeywell and Dupont. As a synthetic refrigerant, R1234yf has similar physical properties to R134a. Therefore, the system which used R1234yf as a refrigerant does not have to change a lot based on the R134a system. Though there is a security threat of the MAC system using R1234yf as a refrigerant due to its slight flammability and toxicity, the results of the CRP1234 assessment showed that the risk associated with the use of R1234yf in automotive vehicles is well below those commonly considered acceptable by the general public and regulatory agencies. The risk assessment was submitted to the U.S. Environmental Protection Agency (US EPA) as part of the Significant New Alternatives Policy (SNAP) approval process, and the US EPA subsequently approved R-1234yf for use in US vehicles. In 2009, the industry decided to use the refrigerant R1234yf and only these MAC systems were developed. Therefore, the development efforts regarding MAC systems using CO_2 were suspended at that date. Up to July 2016, about 35 manufactures have sold about 8,000,000 vehicles with R1234yf. By 2017, it is expected that 18,000,000 vehicles with the R1234yf MAC system will be sold in the EU (Andersen et al. 2017).

9.1.2 *Low GWP Refrigerant Policy*

As a device to provide cabin climate control, MAC has become a standard equipment used for most vehicles. In 2017, the world's production of vehicles exceeded 90 million. More than three of four new vehicles have been equipped with MAC systems in Europe in 2003 (Ayad et al. 2012). However, HFC-134a is the main refrigerant for MAC, with zero ozone depletion potential (ODP) and 1300 global warming potential (GWP). The use and the leakage of the R134a refrigerant caused high greenhouse emission. In order to reduce the environmental effect of MAC refrigerant, several policies have been made to eliminate the use of R134a.

In 2006, the European Commission (Directive 2006/40/EC) banned all MAC systems on all vehicles sold in Europe from using GWP >150 refrigerants after 2017. The California Air Resources Board (CARB) started to make rules to reduce the use of HFC-134a in MAC, and aimed to reduce greenhouse gas (GHG) emissions to 1990 levels by 2020. The US EPA announced that HFC-134a will be removed from the list of acceptable MAC refrigerants by MY 2021 (Andersen et al. 2017). In 2013, Japan decided to reduce annually allocations and emissions of HFCs and require low-GWP refrigerants after MY 2023.

In addition, the Kigali HFC Amendment passed by 197 countries in 2016 limits production and consumption of HFCs. According to the amendment, developed countries will begin reducing HFC production and consumption starting in 2019, with some developed countries starting one year later. Some developing countries including China will freeze HFC production and consumption in 2024, while other developing countries will freeze in 2028.

9.1.3 *Recent Development*

With the development of CO_2 MAC system technology, the German Institute for Standardisation (DIN) has issued first standards relating to CO_2 components and MAC systems in 2015, which was strongly supported by German Association of the Automotive Industry (VDA). The standards are designed to promote the CO_2 MAC system and reduce initial system cost between HFC and HFO systems. In addition to the requirements on components of CO_2 MAC systems, these specifications also describe the process requirements for their development and qualification. The standards include specifications for the heat exchanger, compressor, valve and connector. The series of specifications will make an important contribution to safety, quality and cost optimization for the manufacture and operation of CO_2 MAC systems (https://www.vda.de/en/topics/safety-and-standards/standardization/preparation-of-a-din-specification-series-for-CO2-air-conditioning-systems.html).

The European automotive manufacture Daimler-Benz has announced its first-ever plan to offer a CO_2 MAC system in a production vehicle. From 2017, it will offer in Europe the S-class and E-class as the first production passenger cars equipped with CO_2 MAC systems. And these completely newly developed systems are based on the DIN specifications defined by the automotive standards committee of the German Association of the Automotive Industry (VDA) (https://www.daimler.com/sustainability/product/further-environmental-technologies/CO2-air-conditioning-system.html).

What's more, electrical vehicles (EV) are becoming increasingly popular due to the benefits of low vehicle emissions. However, additional sources for cabin heating in the winter are needed due to the lack of engine exhaust waste heat in EV. The power consumption for this greatly affects the driving range in cold temperature conditions, which makes the heat pump system a good choice to supply for cabin heating. Compared to conventional R134a heat pump, CO_2 heat pump can provide sufficient heating capacity at lower ambient temperatures. Recently, the CO_2 mobile heat pump system has attracted much attention for EV and HEV application.

9.2 Characteristics of CO_2 MAC

9.2.1 CO_2 MAC Structure

A basic CO_2 MAC system consists of the compressor, gas cooler, internal heat exchanger (IHX, also called suction line heat exchanger), throttling valve, evaporator and accumulator. The configuration of the CO_2 MAC system is quite similar to the R134a MAC system. One additional component of the CO_2 MAC system is the IHX, which is quite important to increase the capacity and COP of the CO_2 system. A typical schematic of the CO_2 MAC system is shown in Fig. 9.8.

The CO_2 MAC cycle mainly consists of four major process, which is similar to the conventional vapor compression cycle (Martin et al. 2005). The changes of temperature and entropy of CO_2 during the working process of the CO_2 MAC system is depicted in the temperature-entropy diagram shown in Fig. 9.9. The compressor sucks the superheated vapor from the IHX and then compresses it to the gas cooler (1–2). Then the refrigerant rejects heat in the gas cooler (2–3) and then has a process of cooling in the IHX with the low-pressure refrigerant from the evaporator (3–4).

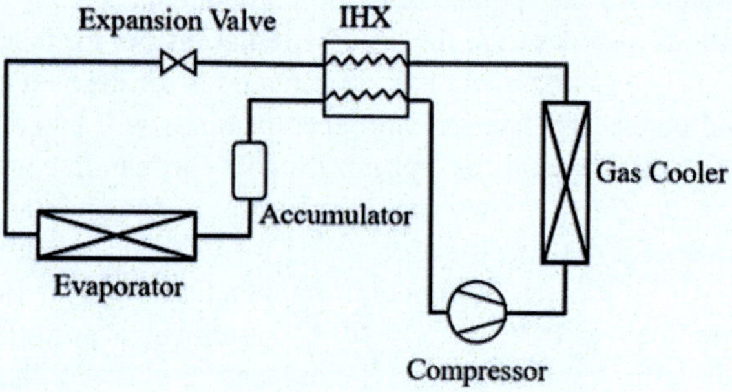

Fig. 9.8 Schematic diagram of CO_2 MAC system

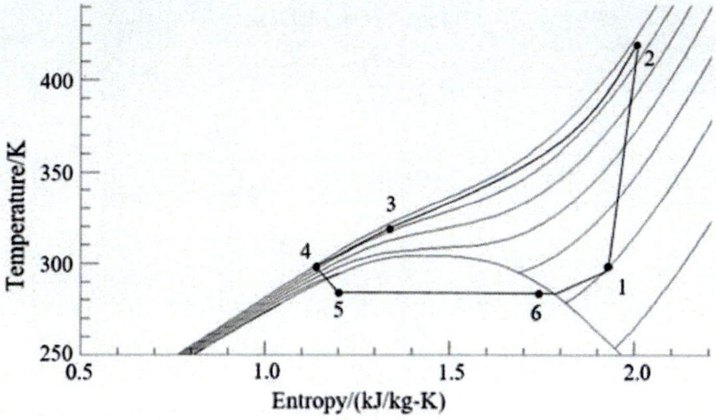

Fig. 9.9 CO_2 MAC system working process in T-s diagram (Martin et al. 2005)

When the high-pressure refrigerant goes through the throttling device (expansion valve in the figure), its pressure goes down and enters the two-phase state (4–5). In the evaporator, the refrigerant absorbs heat (5–6). After that, the superheated refrigerant comes back to the compressor passing the accumulator and IHX (6–1).

Compared to conventional refrigerants, CO_2 has some different property, thermal cycle and work pressure (Mu et al. 2003). This leads to three distinct features of the CO_2 MAC system:

(1) Heat is rejected at a super-critical pressure in many situations. The critical temperature of CO_2 is 31.1 °C. The CO_2 system uses a trans-critical cycle that operates partly below and partly above the critical pressure. The high-side pressure in the system is not determined by the saturation pressure but by the refrigerant charge. The performance of the CO_2 system is greatly affected by the high-side pressure. At a certain ambient temperature condition, the cooling capacity will increase with the high pressure increasing, and there will be a maximum coefficient of performance (COP). The system design thus has to consider the need for controlling high-side pressure to ensure sufficient COP and capacity.

(2) The pressure level in the CO_2 MAC system is quite high (around 30–120 bar). Therefore, the components have to be redesigned to fit this new situation. Due to smaller volumes of piping and components, the stored explosion energy in the CO_2 system is not much different from a conventional system. A benefit of high pressure is the 80–90% smaller compressor displacement needed for a given capacity. Compressor pressure ratios are low, thus giving favorable conditions for high compressor efficiency.

(3) The refrigerant temperature glide is large during heat rejection. The T-s diagram comparison of R134a and CO_2 is shown in Fig. 9.10 (Wang et al. 2018). There is usually no phase change occurring with CO_2 while R134a has a process of condensing. This results in a uniform distribution of temperature at the outlet of the CO_2 gas cooler of the system. At super-critical or near-critical pressure,

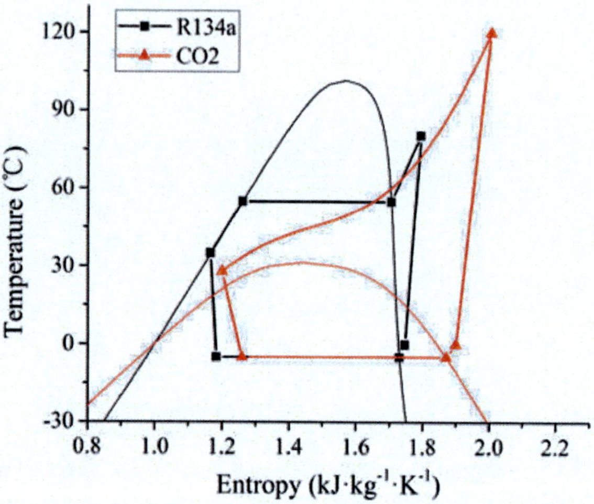

Fig. 9.10 The T-s diagram comparison of R134a and CO_2 (Wang 2018a)

all or most of the heat transfer from the refrigerant takes place by cooling the compressed gas. The heat rejecting heat exchanger is then called the gas cooler instead of the condenser. Gliding temperature can be useful in heat pumps for heating water or air. With proper heat exchanger design, the refrigerant can be cooled to a few degrees above the entering coolant temperature, and this contributes to high COP of the system.

9.2.2 *Cooling Performance*

Perterson et al. (2003) compared the COP between CO_2 and R134a MAC system. Figure 9.11 shows the comparison data from HFC data of SAE ARCRP (2002) and CO_2 data of Hrnjak (2003) (Pettersen and Nekså 2003). Figure 9.12 shows the typical efficiency of CO_2 and R134a system at different gas cooler air inlet temperatures (Pettersen and Nekså 2003). It is obvious that the CO_2 system tends to have a higher efficiency than the R134a system at lower ambient temperatures, while the situation is reverse at a higher ambient temperature.

Kim et al. (2004) found that the CO_2 system was sized to provide approximately equal capacity at the extreme high-temperature (54.4 °C) idling condition, but its COP fell 10% short of the baseline R134a system. At outdoor ambient temperatures below 40 °C, the CO_2 system COP exceeded that of the baseline R-134a system up to 40%. Figure 9.13 shows their performance comparison (Kim et al. 2004).

Visteon carried out a series of experiments on a Toyota Yaris vehicle equipped with the CO_2 MAC system. Figure 9.14a shows that the CO_2 system provided lower evaporator outlet temperature and faster cooling down performance than the conventional R134a system (Wieschollek and Heckt 2007). In addition, Visteon evaluated the impact of R134a and CO_2 MAC systems on vehicle fuel consumption. Tests were

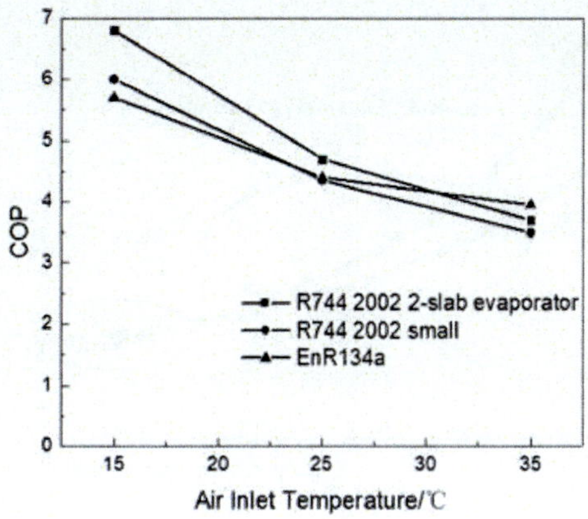

(a) 900rpm (idling, 20% of the usage)

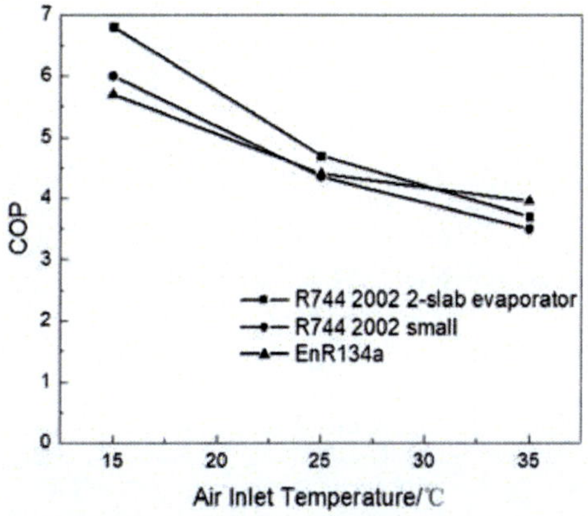

(b) 2500rpm (driving, 80% of the usage)

Fig. 9.11 COP comparison between CO_2 and R134a, 5 °C air from evaporator or equal capacity (Pettersen and Nekså 2003)

carried out with AC off, with R134a and CO_2 systems at different ambient conditions, based on the NEDC procedure. Test results shown in Fig. 9.14b show that the fuel consumption of the CO_2 MAC system was reduced by 0.3 and 0.5 L/100 km at ambient temperatures of 25 °C and 35 °C, respectively, compared to the R134a system (Wieschollek and Heckt 2007).

In the B-COOL project, a Ford Ka and a Fiat Panda vehicle were tested to investigate the impact of the CO_2 MAC system on the cooling performance and vehicle fuel consumption (Malvicino et al. 2009). Figure 9.15a shows that the two versions R744 CO_2 vehicles perform a slightly higher fuel consumption than the conventional R134a vehicle at different operating conditions (Malvicino et al. 2009). Figure 9.15b

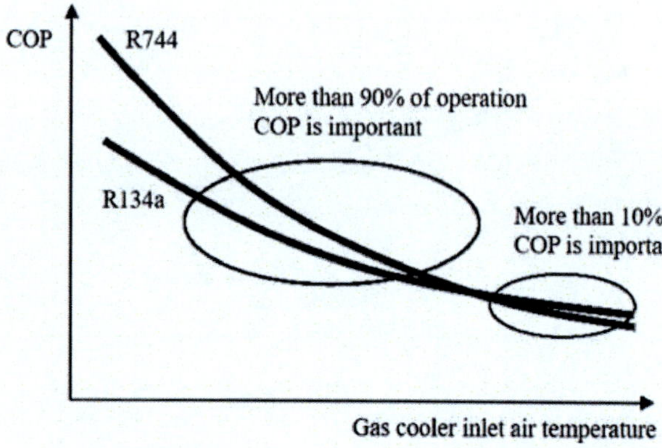

Fig. 9.12 Typical efficiency at varying gas cooler air inlet temperature (Pettersen and Nekså 2003)

Fig. 9.13 Measured performance of CO_2 and baseline R134a MAC systems (Kim et al. 2004)

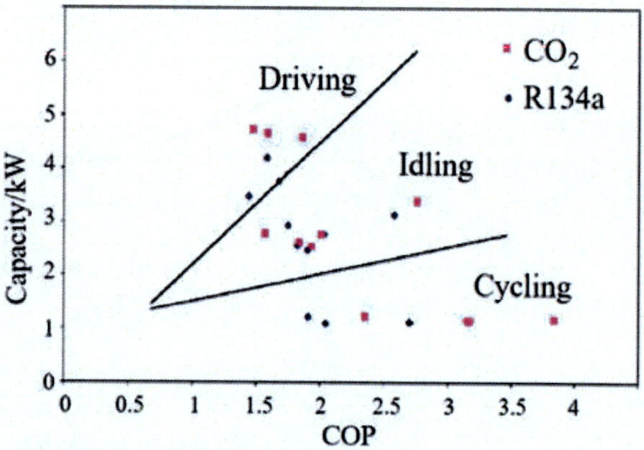

shows that the second version R744 CO_2 vehicle with increased compressor displacement could achieve better cool down performance than the baseline (Malvicino et al. 2009).

In 2016, the MAC team from Shanghai Jiao Tong University conducted comparison experiments conducted for MAC systems operating with CO_2 and R134a (Wang 2018b). Figure 9.16 shows the results of COP and cooling capacity with respect to the refrigerant and outdoor temperature (Wang 2018b). Increasing the outdoor temperature results in declination in COP and cooling capacity for both systems. When the outdoor temperature increases from 27 to 45 °C, the cooling capacity of CO_2 and R134a is decreased by 21% and 13%, respectively. At the constant outdoor temperature, the difference of the cooling capacity between the CO_2 system and R134a system is less than 5%. It indicates that the CO_2 system is able to perform a comparable cooling capacity with the R134a system regardless of the outdoor temperature. Figure 9.16 also shows that the COP of the CO_2 system reduced from 2.0 to 1.3 by 38%, and the COP of the R134a system declines from 2.8 to 1.4 by 49%, with the increasing outdoor temperature from 27 to 45 °C. It is concluded that the COP of the

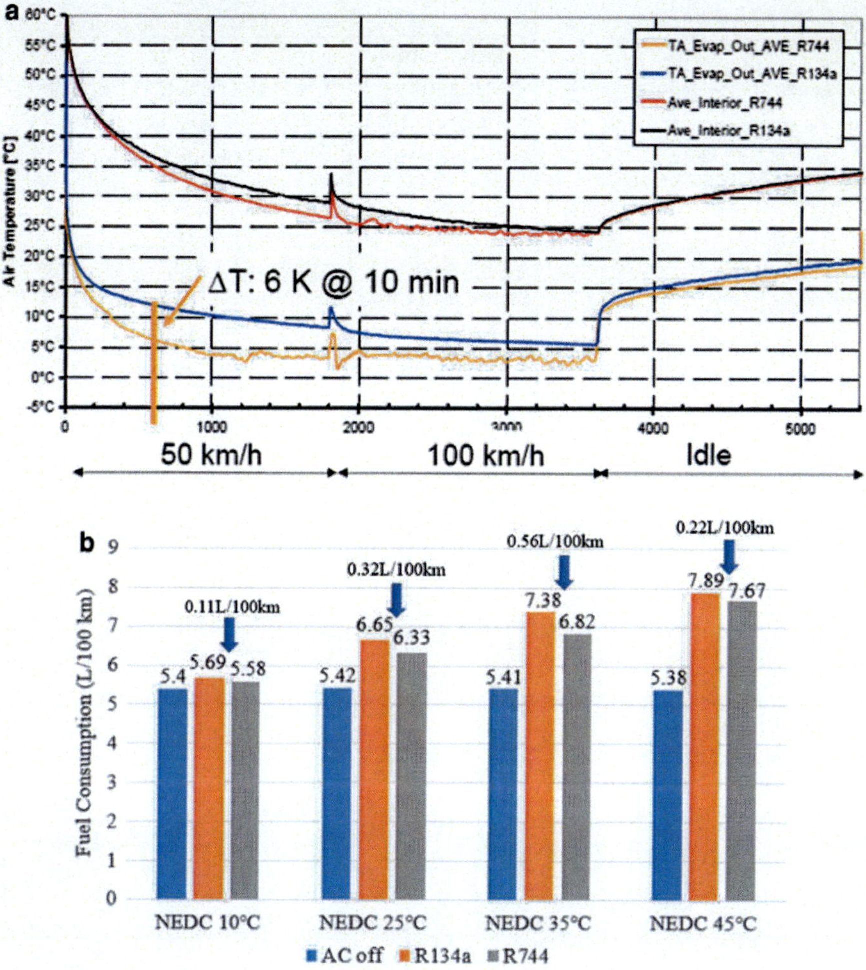

Fig. 9.14 Test results of **a** cooling performance, at 43 °C/40%RH, sun load 1000 W/m^2 (Wieschollek and Heckt 2007) **b** fuel consumption at varied ambient temperature (Wieschollek and Heckt 2007)

CO_2 system is 10–26% lower than the R134a system at same operation conditions and the COP disparity decreases with the increase of the outdoor temperature.

Figure 9.17 shows the experimental results of CO_2 gas cooler outlet temperature and R134a condensing temperature with varied outdoor temperature (Wang 2018b). The CO_2 gas cooler outlet and R134a condensing temperature are averagely 9.3 °C and 17.7 °C higher than the outdoor temperature, respectively. Figure 9.17 also shows the comparison of evaporation temperature with varied outdoor temperature (Wang 2018b). As the outdoor temperature increases from 27 to 45 °C, the CO_2 evaporation temperature improves from 1.9 to 7.2 °C, and the R134a evaporation temperature increases from 1.6 to 4.7 °C. By comparison, evaporation temperature

Fig. 9.15 **a** Measured fuel over consumption on a NEDC based testing cycle (Malvicino et al. 2009) **b** Cool down test cycle at 43 °C, 30% R.H. and 800 W/m^2 (Malvicino et al. 2009)

a

	Testing Conditions		
Temperature Humidity	15°C 70% R.H.	28°C 50% R.H.	35°C 60% R.H.
Air Enthalpy	34J/kg	59J/kg	91J/kg
	Fuel Overconsumption (% of baseline @28°C)		
Panda R134a	38	100	177
Panda R744 1st	38	112	180
Panda R744 2nd	45	126	225
Ka R134a	65	100	135
Ka R744	70	109	143

b

	T Outlet Mean/°C				
	10'	30'	60'	90'	120'
Panda R134a	9.9	9.3	7.4	10.6	16.9
Panda R744 1st	10.4	5.4	4.9	7.1	14.3
Panda R744 2nd	8	7	7	5	9
Ka R134a	14	9	7	7	17.5
Ka R744	12	5	5	5	19

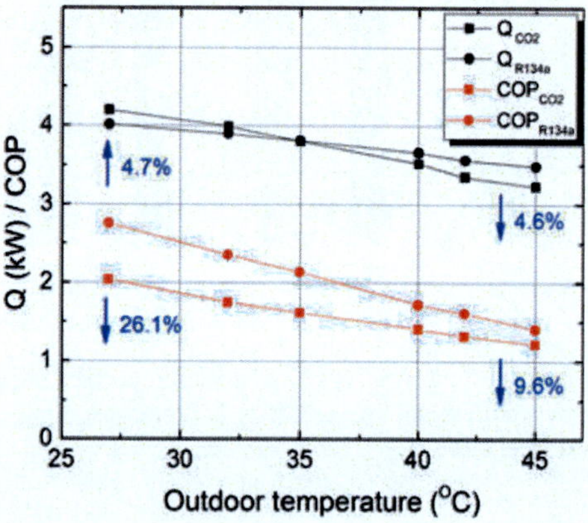

Fig. 9.16 Cooling capacity and COP for CO_2 and R134a MAC systems at varied outdoor temperature (Wang 2018b)

of CO_2 is 0.3–2.5 °C higher than that of R134a, and their difference increases with the increasing outdoor temperature. The relatively higher evaporation temperature of the CO_2 system could partly explain the experimental results of narrowed COP disparity at high outdoor temperature.

Figure 9.18 shows the compressor overall efficiency and pressure ratio with respect to the refrigerant and outdoor temperature (Wang 2018b). The pressure ratio was defined as the discharge pressure divided by the suction pressure. And the compressor overall efficiency was calculated by experimental data. It should be noted that the

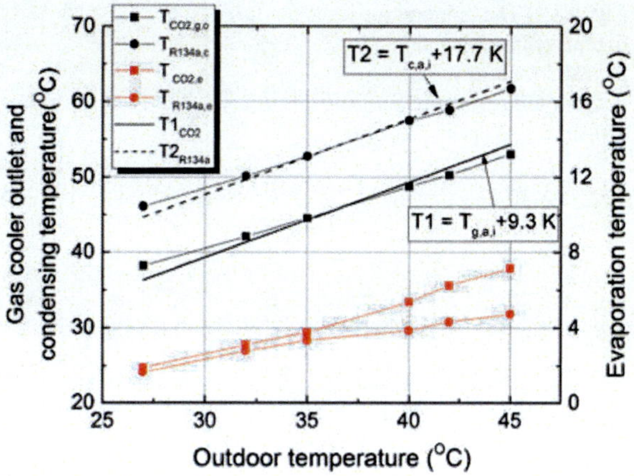

Fig. 9.17 The gas cooler outlet, condensing temperature and evaporation temperatures at varied outdoor temperature (Wang 2018b)

measured mass flow concluded the refrigerant and lubricant oil, the oil circulation rate (OCR) was normally around 3% for the MAC system, hence the calculated results of efficiency will be a few percentage higher than the actual efficiency. With the outdoor temperature increasing from 27 to 45 °C, the efficiency of the CO_2 rotary compressor changes from 0.72 to 0.69, and the efficiency of the R134a scroll compressor decreases from 0.65 to 0.47. It indicates the overall efficiency of the CO_2 compressor is higher than R134a by 11.8% and by 47.4%, at 27 and at 45 °C outdoor temperature, respectively. Since the compressor overall efficiency proportionally affects the COP of system, the 47.4% higher efficiency of the CO_2 compressor at 45 °C will improve the COP of the CO_2 system by 47.4%, which manly account for the narrowed COP disparity at 45 °C outdoor temperature. Figure 9.18 also shows that with the change of the outdoor temperature, the pressure ratio of CO_2 increased from 2.8 to 3.0 by 8%, but the pressure ratio of R134a increased from 4.5 to 6.2 by 37%. It indicates that the pressure ratio of CO_2 is smaller and less affected by the outdoor temperature, compared with that of R134a. This may account for why the CO_2 compressor is more efficient and maintains high efficiency around 70% even under high outdoor temperature condition.

9.2.3 *Influencing Parameters*

9.2.3.1 Effect of High-Side Pressure

Figure 9.19A shows the P–h diagram of the trans-critical CO_2 refrigeration cycle (Yang et al. 2005). If the high pressure drops from P to P1, the q will drop to q1 and w will drop to w1. Similarly, q will increase to q2 and w will increase to w2 when the high pressure rises from P to P2. Whether the high pressure increases or decreases, the change of COP depends on the value of ΔP and Δw to the change of ΔP. Figure 9.19b demonstrates the relationship between high pressure and COP,

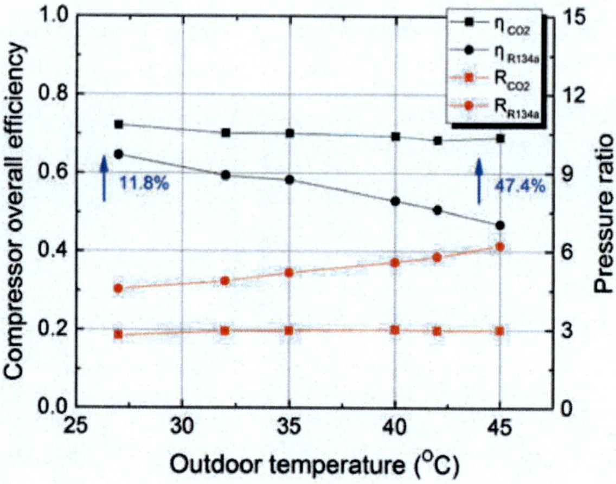

Fig. 9.18 The compressor overall efficiency and pressure ratio at varied outdoor temperature (Wang 2018b)

cooling capacity and driving power (Martin et al. 2005).When the high pressure increases, the cooling capacity will increase on a diminishing scale, and the driving power will increase linearly nearly. Therefore, the COP has a maximum value as the high pressure increases (Martin et al. 2005).

The outlet temperature of the gas cooler has a great impact on the optimum high side pressure, which is extremely important to obtain a good performance of the CO_2 MAC system. The COP-maximum pressure is sensitive to the outlet temperature of the gas cooler and it is the most important parameter to control the trans-critical CO_2 system operation. However, the sensitivity of the optimum high pressure to the outlet temperature of the gas cooler varies with the frequency of the compressor. In general, a higher compressor speed can decrease the sensitivity (Yang et al. 2005).

Yang et al. (2005) studied the relationships between the optimum high pressure P_{op} and the outlet temperature of the gas cooler $T_{ga,o}$. Under different compressor speeds, the related equations derived are:

950 RPM

$$P_{op}(\mathrm{MPa}) = 2.66 + 0.166 \times T_{ga,o}(^{\circ}\mathrm{C}) \tag{9.1}$$

1800 RPM

$$P_{op}(\mathrm{MPa}) = 3.78 + 0.140 \times T_{ga,o}(^{\circ}\mathrm{C}) \tag{9.2}$$

3000 RPM

$$P_{op}(\mathrm{MPa}) = 4.48 + 0.137 \times T_{ga,o}(^{\circ}\mathrm{C}) \tag{9.3}$$

Kim et al. (2009a) studied the optimum high pressure control algorithm for the CO_2 air conditioning system. The experiment shows a good agreement with Yang's conclusions. The related equation is given by:

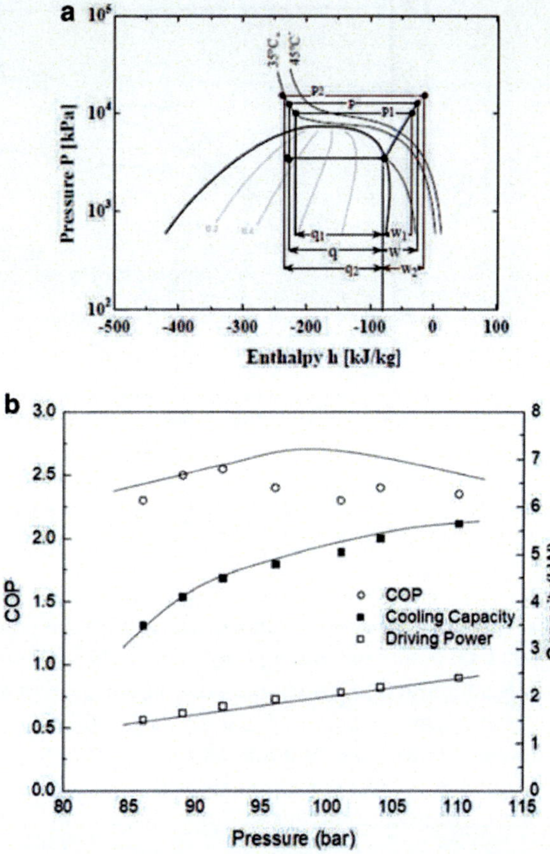

Fig. 9.19 **a** The effect of high pressure on trans-critical CO_2 cycle (Yang et al. 2005) **b** COP, cooling capacity and driving power with the change of high pressure (Martin et al. 2005)

$$P_{op}(\text{MPa}) = 0.987 + 0.194 \times T_{ga,o}(^{\circ}\text{C}) \tag{9.4}$$

MAC team from Shanghai Jiao Tong University conducted experiments to investigate the relationship between temperature and optimal pressure at the gas cooler outlet measured in the CO_2 heat pump (Wang et al. 2018). The effects of temperature on pressure are lower than those predicted by Kim et al., but higher than those predicted by Yang.

$$P_{op}(\text{MPa}) = 3.60 + 0.139 \times T_{ga,o}(^{\circ}\text{C}) \tag{9.5}$$

9.2.3.2 Effect of CO_2 Refrigerant Charge

The CO_2 system performance is strongly related to the CO_2 mass charge level, which is demonstrated in Fig. 9.20 (Liu et al. 2005). When the CO_2 mass charge increases,

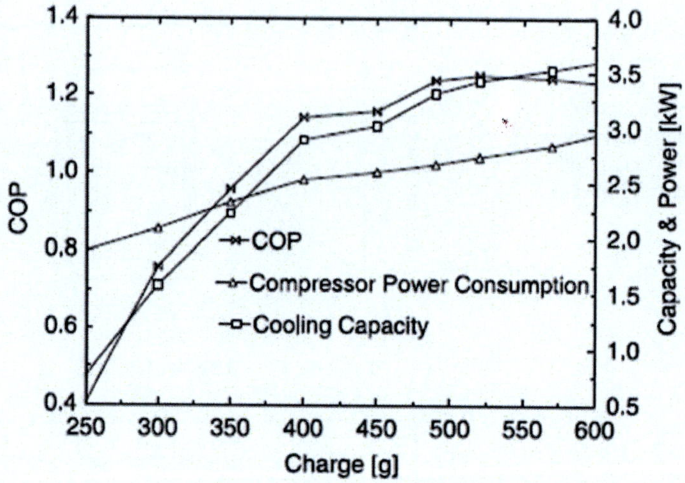

Fig. 9.20 Effect of CO_2 mass charge on the system performance (Liu et al. 2005)

the cooling capacity will increase because the CO_2 flow rate increases slightly, and the compressor input power increases because the compressor outlet pressure goes up; the COP value increases up to a point, then decreases slightly. In other words, there is an optimum CO_2 charge point at which the COP value is maximum. When CO_2 is less than the optimum charge, the cooling capacity decreases faster than the compressor power consumption and the COP will go down remarkably; however, the compressor power consumption increases faster than the cooling capacity if CO_2 is overcharged (Liu et al. 2005).

9.2.3.3 Effect of Compressor Speed

When the compressor speed increases, the compressor input power, the CO_2 flow rate and evaporator inlet quality all become larger. At higher compressor speed and discharge pressure, the capacity of heat exchangers becomes a limiting factor, so the cooling capacity increases slightly. The efficiency of the heat exchanger will decrease as the refrigerant mass flow increases. For the CO_2 system with a fixed displacement compressor, when the compressor speed goes up, the CO_2 mass flow rate will increase, resulting in decrease of the heat exchanger efficiency. For a CO_2 trans-critical cycle, the lower the gas cooler outlet temperature, the higher the COP of the system. To make a CO_2 air conditioning system more efficient, the variable displacement CO_2 compressor should be adopted in the CO_2 automotive air conditioning system.

Brown et al. (2002) evaluates performance merits of CO_2 and R134a MAC systems using semi-theoretical cycle models. Figure 9.21 shows COP as a function of ambient temperature and compressor speed (Brown et al. 2002). At 1000 RPM for an ambient condition of 32.2 °C, the R134a system's COP is approximately 21% higher than

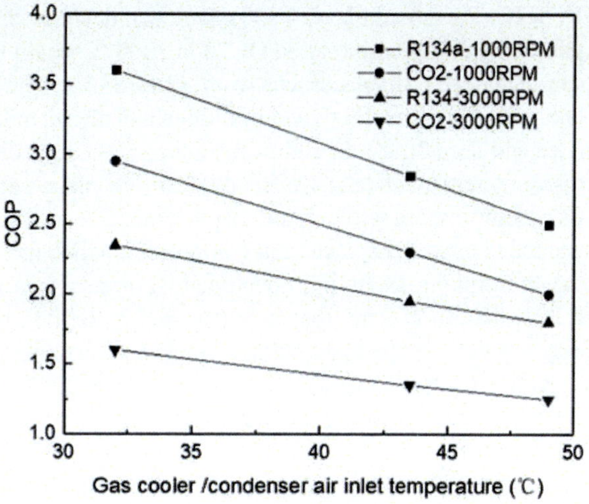

Fig. 9.21 COP comparison between CO_2 and R134a MAC systems (Brown et al. 2002)

the CO_2 system's COP. The COP gap increases to 29% at 43.3 °C and to 34% at 48.9 °C. At 3000 RPM for an ambient condition of 32.2 °C, the R134a system's COP is approximately 42% higher than that of the CO_2 system. The COP gap increases to 51% at 43.3 °C and to 60% at 48.9 °C.

9.3 Key Components

9.3.1 Compressor

The trans-critical CO_2 cycle operates at high pressure up to 15 MPa, but its pressure ratio is lower than that of conventional refrigerant, which could enhance the CO_2 compressor efficiency. Due to the higher density of CO_2 compared to the conventional refrigerant, the CO_2 compressor displacement can be much smaller. In addition, Suss and Kruse et al. reported that the leakage may have strong influence on the CO_2 compressor performance duo to the high pressure difference, but the effect of leakage on the compressor performance can be reduced to a reasonable amount with an appropriate design of the compressor (Süβ and Kruse 1998).

In the early days of the revival of CO_2 technology (1990), no small capacity CO_2 compressors were available. SINTEF Refrigeration Engineering developed a prototype of a three cylinder wobble plate CO_2 compressor with 3.1 m^3/h capacity, which is based on a Sanden R12 wobble plate compressor. Figure 9.22a shows this compressor prototype (Armin and Petter 2005). TES Wankel and Obrist developed mobile CO_2 compressor with a fixed displacement, which is shown in Fig. 9.22b (Armin 2009). Figure 9.22c shows a radial piston CO_2 compressor designed by

IXETIC (Parsch and Hotz 2008). Doowon developed an external variable displacement CO_2 compressor called DC28 in 2008 (Lee and Kim 2008). DC28 compressor can change the displacement from 22 to 33 cc. The external fin can improve the cooling ability, and the swivel Joint Mechanism can reduce friction loss and hysteresis (Lee and Kim 2008). In addition, Sanden (29.8 cm^3 displacement), Zexel (28.0 cm^3 displacement), Calsonic-Kansei (30.0 cm^3 displacement) have developed swash plate CO_2 compressors with a fixed displacement. A magnetic clutch for on–off cycles is needed in a fixed displacement compressor, which increases the cost and the mass of the compressor. Ixetic has developed a compact swashing compressor for compact cars, with 20–25 cm^3 displacement and clutchless drive. A swash-plate variable displacement compressor with 33 $cm^3$was developed by Sanden, with the collaboration with LuK, which is shown in Fig. 9.22f (Mu et al. 2003). Figure 9.22g shows the comparison of a mobile R134a compressor and CO_2 compressor, whose displacements are 155 cm^3 and 21 cm^3, respectively (Hafner 2006).

Figure 9.23a shows a family of variable displacement mobile CO_2 compressor developed by Obrist Engineering applied for CO_2 MAC systems (Wolf 2008). Figure 9.23b shows that the overall efficiency of the CO_2 compressor outperformed than the R134a compressor up to 79% (Wolf et al. 2007). Figure 9.23c shows that the vehicle fuel consumption could be reduced by 14–45% by using this updated CO_2 compressor (Wolf 2008). Variable spelling change in the figure below.

9.3.2 Gas Cooler

Microchannel heat exchangers are commonly used in the MAC systems (Boewe et al. 1999). Figure 9.24 shows a product of a micro-channel gas cooler applied in the CO_2 MAC system. A typical gas cooler configuration is shown in Fig. 9.25 (Chen and C Z 2011; Raiser et al. 2006). The flat tube port diameter is 0.79 mm, much smaller than that of the conventional micro-channel. And the maximum working pressure of the gas cooler can reach up to 15 MPa, which is about 10 times as a conventional R134a condenser.

Owing to the fact that the critical temperature of CO_2 is very low by 31 °C, the CO_2 refrigerant in the gas cooler operates at a supercritical state in most operation conditions. Its special thermophysical properties at the supercritical state make it have a high refrigerant-side heat transfer coefficient, greater than 10,000 W/(m^2*K), which is much higher than conventional refrigerant. Besides, due to the large gliding temperature of the CO_2 refrigerant, the CO_2 gas cooler exhibits a very small approach temperature (TA), which is defined as the difference between the refrigerant exit and air-inlet temperature. Figure 9.26 shows the refrigerant temperature profile comparison of a R22 condensor and a CO_2 gas cooler (Pettersen et al. 1998). Since reducing the gas cooler outlet temperature could significantly improve the COP of the CO_2 system, the small TA value of the CO_2 gas cooler is beneficial to the performance of the CO_2 system.

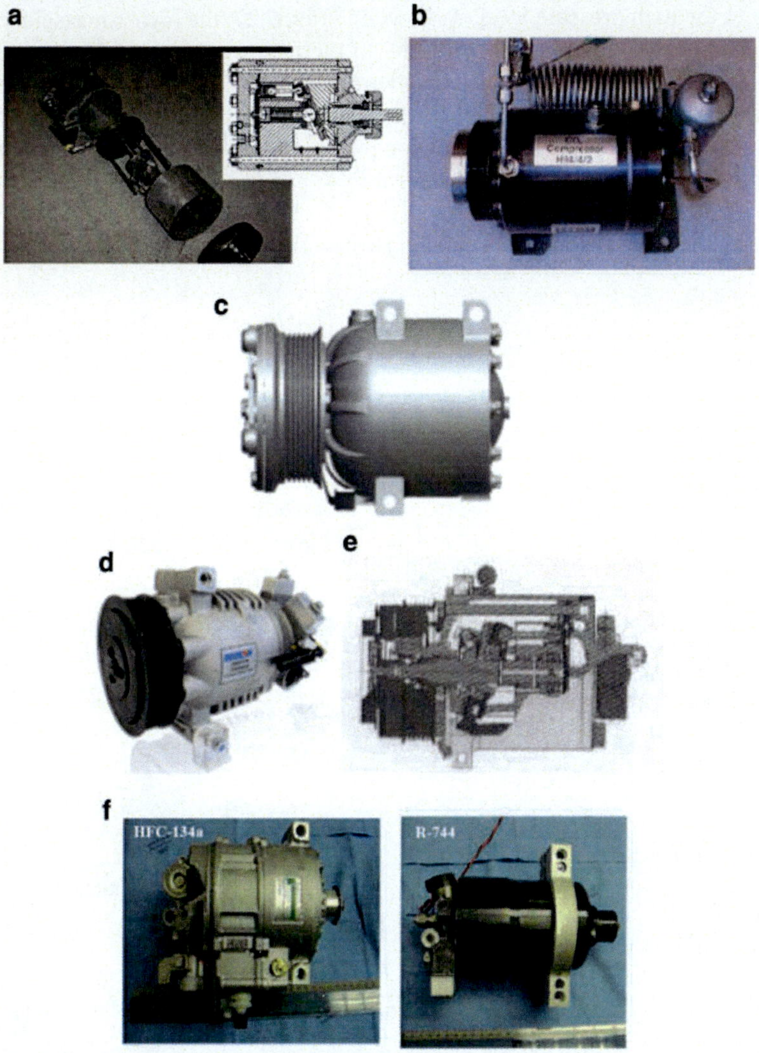

Fig. 9.22 **a** SINTEF first CO_2 MAC compressor (Armin and Petter 2005) **b** TES Wankel/OE H94 fixed displacement compressor (Armin 2009) **c** fixed displacement compressor, IXETIC (Parsch and Hotz 2008) **d** DC28 CO_2 compressor (http://www.doowonrnd.com/eng/?page_id=15) **e** Sanden-LuK swash-plate variable displacement compressor (Mu et al. 2003) **f** mobile R134a and CO_2 compressor with variable displacement (Hafner 2006)

In addition, due to the high operation pressure, the CO_2 gas cooler performance is less sensitive to pressure drop on the refrigerant-side, thus a smaller flat tube diameter and longer flow path could be used to enhance its heat transfer coefficient. Pettersen (1998) developed a microchannel gas cooler with a port diameter of 0.7 mm. The reduced port diameter can lead to higher mass flux and greater pressure drop. However, quite large pressure drops can be tolerated in the CO_2 gas cooler owing to the high pressure level. As shown in Fig. 9.27, the higher mass flux inside the microchannel tube can be beneficial to the heat transfer performance of CO_2 (Pitla et al. 2002). By introducing smaller microchannel tubes, the refrigerant-side surface areas can be reduced, giving room for more air-side surface and increasing the system compactness. The calculation procedure for the heat transfer coefficient of supercritical CO_2 can be expressed as below.

$$\xi = (0.79 \ln(\mathrm{Re}) - 1.64)^{-2} \tag{9.6}$$

$$Nu = \frac{\xi/8(Re-1000)Pr}{12.7\sqrt{\frac{\xi}{8}}\left(Pr^{\frac{2}{3}}-1\right)+1.07} \tag{9.7}$$

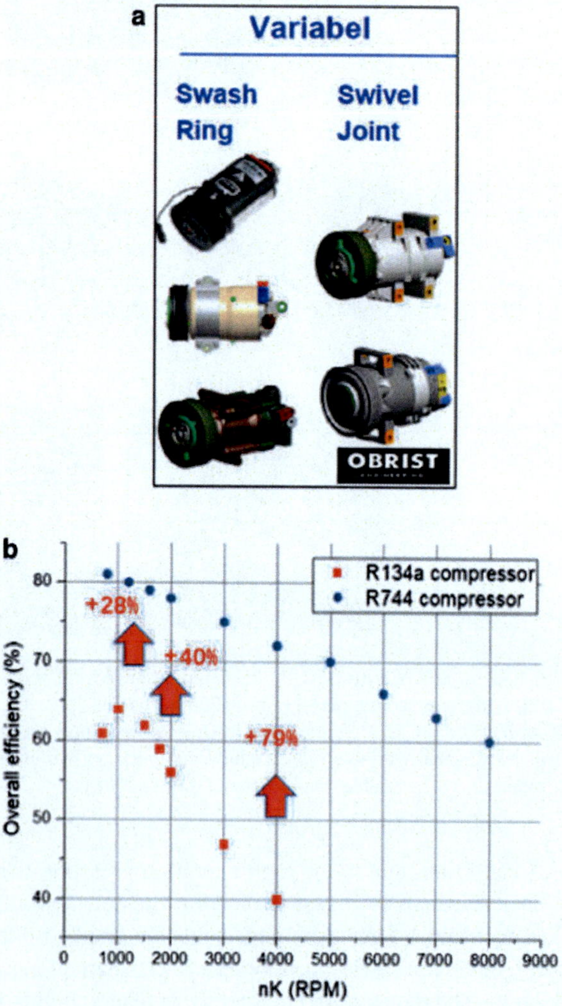

Fig. 9.23 **a** Family of variable displacement compressor developed by Obrist Engineering (Wolf 2008), **b** overall efficiency comparison of variable displacement compressor (Wieschollek and Heckt 2007), **c** fuel consumption comparison of variable displacement compressor (Wolf 2008)

Fig. 9.23 (continued)

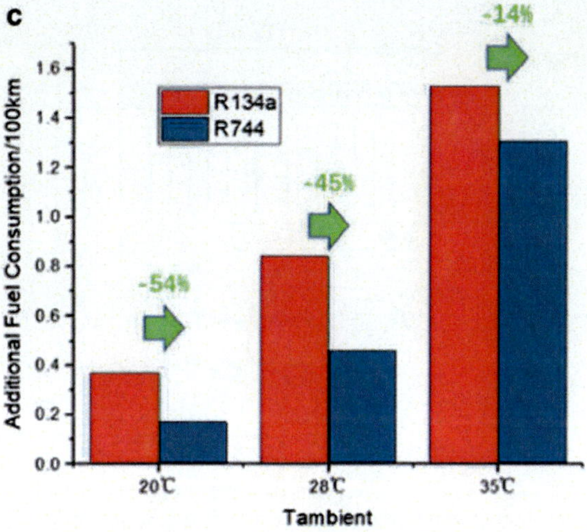

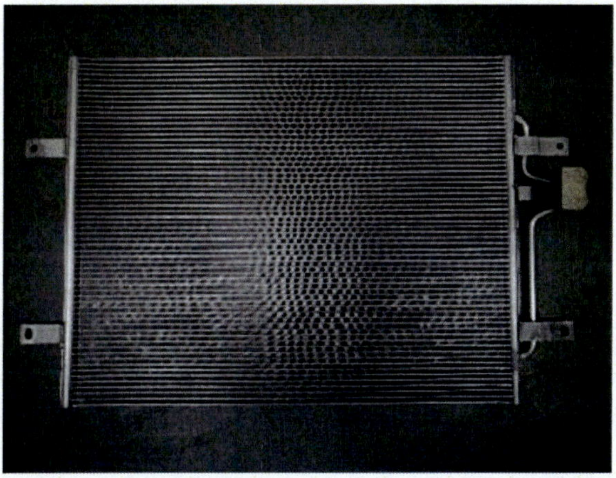

Fig. 9.24 Product of a CO_2 microchannel gas cooler

$$h = \frac{Nu}{D}k \tag{9.8}$$

ξ is the friction coefficient, Pr is the Prandtl number, Re is the Reynolds number, and Nu is the Nusselt number.

And the calculation of the pressure drop coefficient f can refer to the Churchill correlation.

$$f = 8\left[\left(\frac{8}{Re}\right)^{12} + \frac{1}{(\theta 1 + \theta 2)^{1.5}}\right]^{\frac{1}{12}} \tag{9.9}$$

$$\theta 1 = \left[-2.457\ln\left(\left(\frac{7}{Re}\right)^{0.9} + \frac{0.27\varepsilon}{D}\right)\right]^{16} \tag{9.10}$$

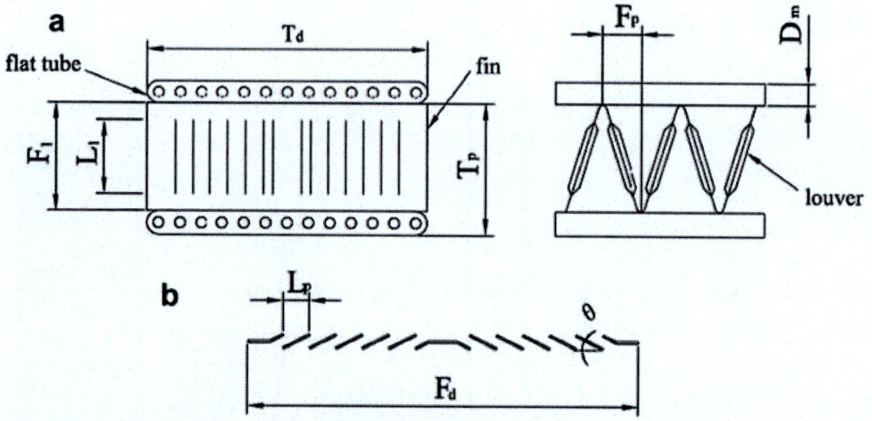

Fig. 9.25 Schematic diagram of a prototype microchannel CO_2 evaporator for MAC system: **a** definition of geometric parameters, **b** cross-section of louvered fin geometry

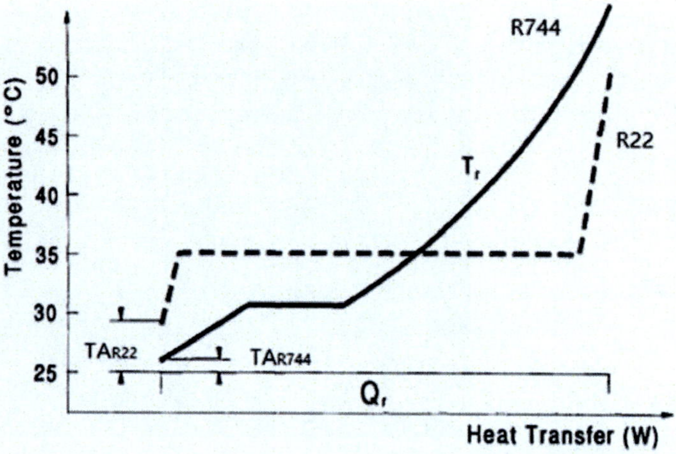

Fig. 9.26 Refrigerant temperature profile comparison of a R22 condensor and a CO_2 gas cooler (Pettersen et al. 1998)

$$\theta 2 = \left(\frac{37530}{\text{Re}}\right)^{16} \tag{9.11}$$

There are several options to enhance a gas cooler's performance. Yin et al., concluded that increasing the pass number of a gas cooler could will improve the higher refrigerant side heat transfer coefficient, but at the expense of increasing pressure drop, as shown in Fig. 9.28 (Jian et al. 2001). Besides, using multiple slabs may enhance the gas cooler performance, since the heat transfer process of a gas cooler with multiple slabs will be closer to an ideal counter flow heat exchanger. In addition, the reduction in port diameter can be considered to enhance heat transfer

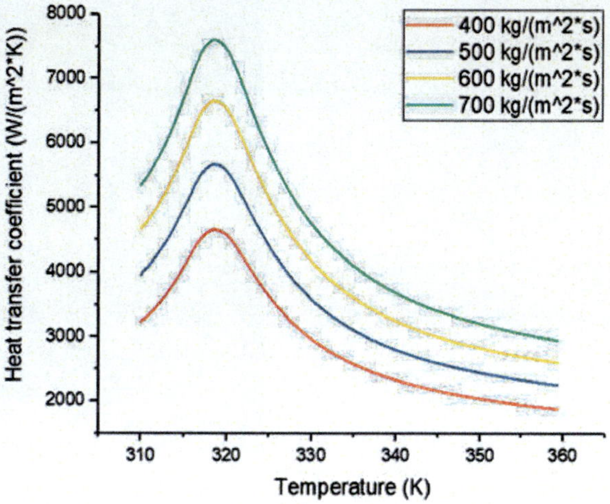

Fig. 9.27 Heat transfer coefficient of supercritical CO_2 at different mass flux (Pitla et al. 2002)

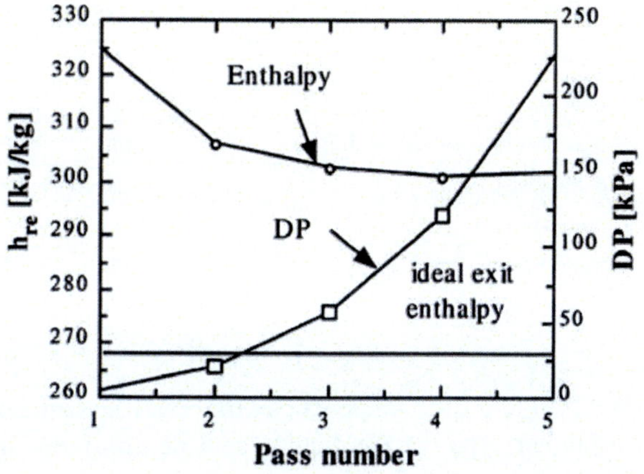

Fig. 9.28 Effect of pass numbers on a CO_2 gas cooler performance (Jian et al. 2001)

performance of a gas cooler. Due to its superior refrigerant heat transfer properties and lower mean temperature difference between the air and CO_2 refrigerant, the large ratio between the air-side surface area and refrigerant-side surface area will be beneficial to the gas cooler's performance.

9.3.3 Evaporator

Figure 9.29 shows a product of a micro-channel evaporator applied in the CO_2 MAC system.

Operation pressure of an evaporator for CO_2 MAC systems ranges from 3 to 7 MPa. Table 9.1 shows the various thermophysics properties of saturated refrigerants at 0 °C. The high thermal conductivity, low kinematic viscosity and high specific

Fig. 9.29 Prototype of microchannel evaporator

Table 9.1 Thermophysics properties of saturated refrigerants at 0 °C

	HCFC-22	HFC-134a	CO_2
Liquid thermal conductivity (W/(m*K))	0.099	0.092	0.111
Liquid kinematic viscosity (m^2/s)	0.2	0.21	0.11
Liquid specific heat (J/(kg*K))	1180	1340	2430
Vapor density (kg/m^3)	21.5	14.4	97.6
Liquid density (kg/m^3)	1285	1293.5	926.4
Density ratio	59.8	89.8	9.5
Surface tension (N/m)	13.1	11.2	4.6

heat of liquid CO_2 are beneficial to the refrigerant-side heat transfer of MAC CO_2 evaporator. And the low liquid/vapor density ratio could be beneficial to the two-phase refrigerator distribution among micro-channel tubes inside the evaporator, which could improve the heat transfer performance of the CO_2 evaporator.

Compared with conventional refrigerants, CO_2 has better heat transfer coefficient for two-phase evaporation process, but it is more affected by the dryout phenomenon, as shown in Fig. 9.30 (Cheng et al. 2008; SG, MR, MAW 2013). Contrary to conventional refrigerants, increasing the mass velocity does not significantly improve the heat transfer coefficient of CO_2 since it accelerates the appearance of the dryout. Also, increasing the mass flux causes a larger pressure drop and smaller temperature difference between the refrigerant and air, which results in the reduction of the cooling capacity of the CO_2 evaporator.

The calculation procedure for the heat transfer coefficient of the two-phase CO_2 during evaporation can be expressed as below.

$$\mathrm{h}_{tp} = \frac{\theta_{dry} h_v + (2\pi - \theta_{dry}) h_{wet}}{2\pi} \tag{9.12}$$

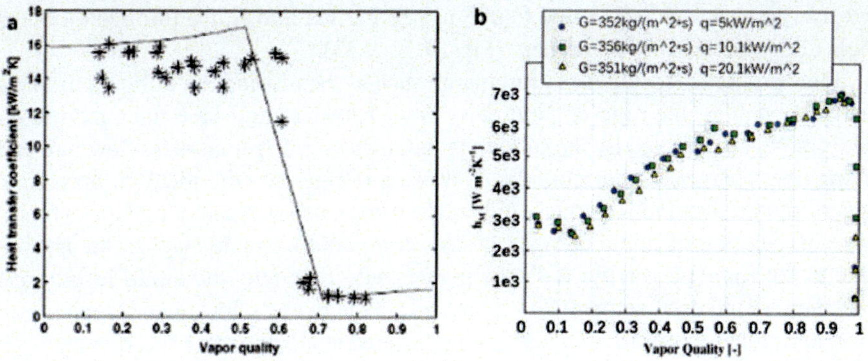

Fig. 9.30 Comparison of heat transfer coefficient between CO_2 and R134a during evaporation (Cheng et al. 2008; SG, MR and MAW 2013)

$$\mathrm{h}_v = \frac{0.023Re_v^{0.8}Pr_v^{0.4}k_v}{D} \tag{9.13}$$

$$\mathrm{h}_{wet} = \left[(S\mathrm{h}_{nb})^3 + \mathrm{h}_{cb}^3\right]^{\frac{1}{3}} \tag{9.14}$$

$$\mathrm{h}_{nb} = 131\mathrm{pr}^{-0.0063}(-log_{10}pr)^{-0.55}M^{-0.5}q^{0.58} \tag{9.15}$$

$$\mathrm{h}_{cb} = \frac{0.0133Re_\delta^{0.69}Pr_l^{0.4}k_l}{\delta} \tag{9.16}$$

The subscript v represents vapor, δ represents liquid film thickness, tp represents two-phase, nb represent nucleate boiling and cb represent convective boiling. θ is the angle of tube perimeter, and pr is reduced pressure [p/pcrit].

9.3.4 Accumulator

Accumulator plays a key role in a CO_2 MAC system. It is placed between the evaporator outlet and low pressure side IHX inlet. It is used to store excess CO_2 refrigerant, separate gas and liquid refrigerant, and feed the oil back to the compressor. This refrigerant storage in low pressure side allows the optimum high pressure adjustment by EXV controlling and could compensate for the refrigerant leakage along the life cycle. A typical accumulator structure consists of a jar, an inlet tube and a J-tube, as shown in Fig. 9.31 (Lemke and Stulgies 2005). The Two-phase fluid enters the jar from the inlet tube, and the liquid and vapor separate under the effect of gravity force and centrifugal force. The liquid accumulates at the bottom, whereas the vapor mixed with small liquid droplets floats at the top of the jar, and then exits the jar from the J-tube inlet. An oil bleed hole is installed at the bottom of the J-tube to recycle the deposited oil. Some liquid may be carried into J-tube through oil bleed hole during the oil recycle process (Figs. 9.32, 9.33).

Flow properties inside the accumulator, such as the distribution of liquid droplets and gas bubbles, the ratio of fluid mass flows between the J-tube inlet and the oil bleed hole, are affected by the following parameters: refrigerant mass flow rate and refrigerant charge in the accumulator. When a refrigerant with medium mass flow enters into the accumulator, the liquid will be broken up and transformed into a foam-type distribution of liquid and gas (Raiser et al. 2006). Liquid stays at the bottom due to the force of gravity. With the higher mass flow rate, the liquid level drops and more liquid droplets are whirled upwards. At high mass flow rate, this effect can lead to a situation where no liquid level is formed at all, and there is a homogenous

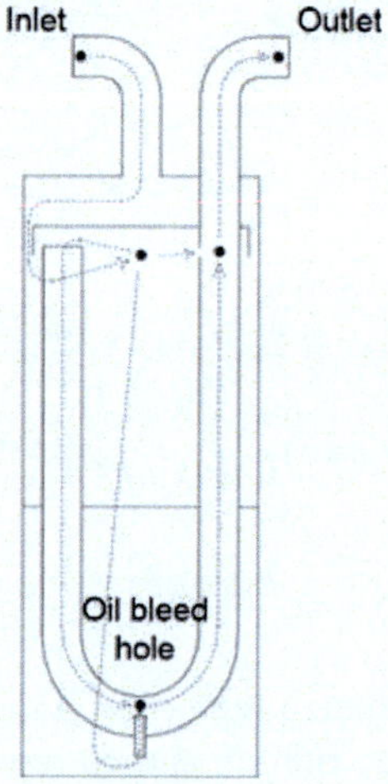

Fig. 9.31 A typical CO_2 accumulator structure (Lemke and Stulgies 2005)

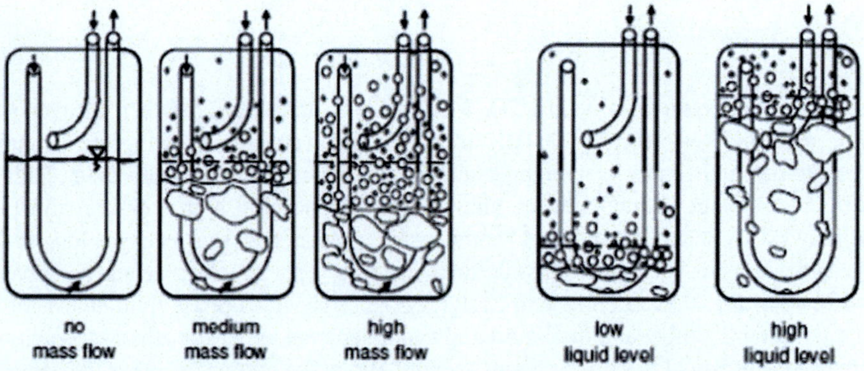

Fig. 9.32 CO_2 accumulator behavior under different mass flow rates and liquid levels (Raiser et al. 2006)

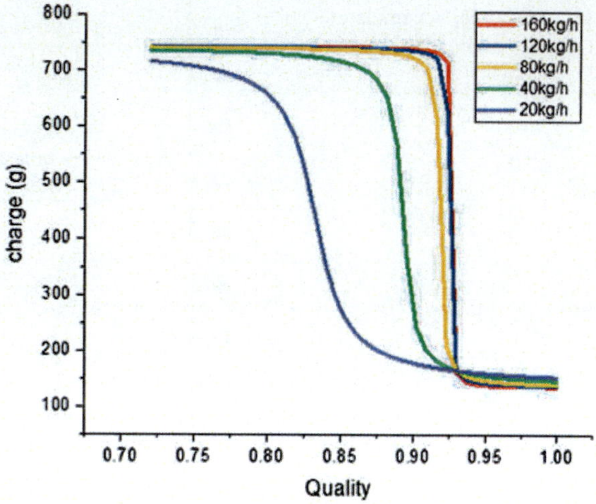

Fig. 9.33 CO_2 accumulator charge at different mass flow rates

foam-type distribution of liquid droplets and gas bubbles. When the accumulator is filled a small amount of refrigerant, there are almost no droplets in the upper part of the accumulator, resulting in a larger vapor fraction at the accumulator outlet. In contrast, when the accumulator is filled with a large amount of refrigerant, there is a very high liquid level in the upper part of the jar, which leads to a larger liquid fraction at the accumulator outlet.

The performance of the accumulator could be affected by the structure parameters, such as the length and radius of inlet tube and J-tube, the jar's volume, and the diameter of the oil bleed hole. Since liquid at the bottom of the accumulator tends to flow into the oil bleed hole with oil, the existence of liquid at the accumulator outlet is inevitable in a system with refrigerant stored in the accumulator. The smaller the diameter, the higher the flow resistance of the oil bleed hole, and the lesser will be th liquid carried by the vapor flowing through the J-tube, leading to a larger vapor fraction at the outlet. Hence, under the precondition of adequate oil recycling, smaller oil bleed hole diameter is recommended. According to Bockholt's research, the maximum charge of an accumulator seems to be irrelevant with the diameter of the oil bleed hole, as is shown in Fig. 9.34 (MB, TW and LN 2008). According to the German standard, the accumulator should be designed for mass flows from up to 300 kg/h during continuous operation. And oil circulation shall be ensured under all operational conditions, especially up to a filling level of 50% from the original filling level with the refrigerant.

9.3.5 *Internal Heat Exchanger*

In a typical CO_2 MAC system, IHX is used to exchange the heat between the high-pressure side refrigerant after the gas cooler and the low-pressure side refrigerant

Fig. 9.34 CO_2 accumulator charge at different oil bleed hole diameter

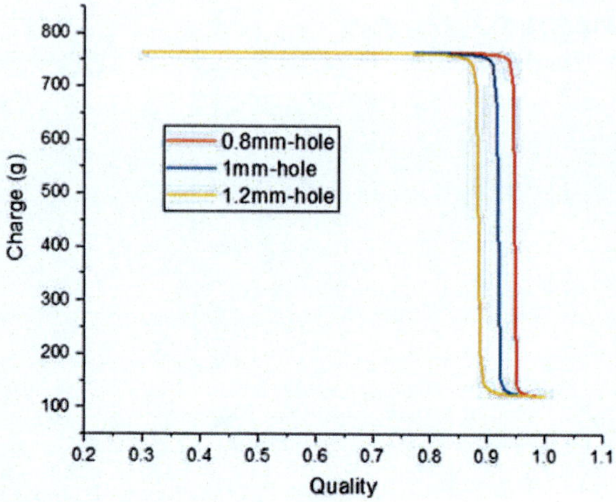

after the accumulator. It is installed after accumulator or integrated with accumulator. The utilization of IHX could decrease the temperature after the gas cooler and vapor quality at the evaporator inlet, thus improving the cooling capacity and efficiency of the CO_2 system (Kim et al. 2004). Aprea and Maiorino (2008), Torrella et al. (2011) conducted experiments and concluded that the use of the IHX increases the COP of the system by 10% (Dewa Made Cipta Santosa MS 2015).

The IHX used in a MAC system mainly adopts coaxial type and micro-channel type, both of them having counter flow configuration. The hot fluid flows in the inner tube, and the cold fluid flows in the outer tube with opposite flow direction. Figure 9.35a, b and c show Alumnium coaxial type IHX used in a CO_2 MAC system. Figure 9.35 shows the configurations of different IHXs (Kim et al. 2004, 2009a; Armin 2009; Boewe et al. 1999; Kwon et al. 2009).

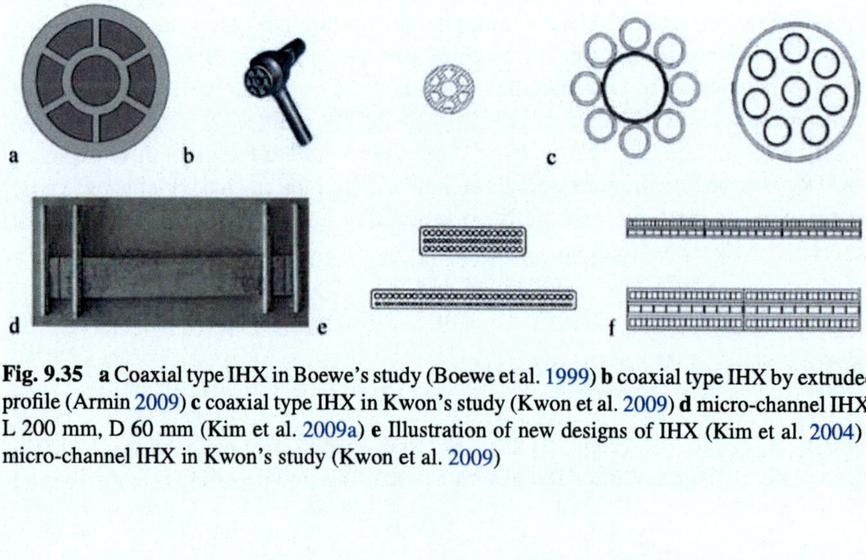

Fig. 9.35 **a** Coaxial type IHX in Boewe's study (Boewe et al. 1999) **b** coaxial type IHX by extruded profile (Armin 2009) **c** coaxial type IHX in Kwon's study (Kwon et al. 2009) **d** micro-channel IHX, L 200 mm, D 60 mm (Kim et al. 2009a) **e** Illustration of new designs of IHX (Kim et al. 2004) **f** micro-channel IHX in Kwon's study (Kwon et al. 2009)

Some parameters can affect the performance of IHX, including the effectiveness, the length the pressure drop and the capacity (Kwon et al. 2009). With the increase of IHX length, the capacity, the effectiveness and the pressure drop will increase. As the mass flow rate increases, the pressure drop and the capacity of IHX will increase, whereas the increase of the effectiveness is not significant. When the capacity of the IHX is constant, increasing the heat transfer coefficient on both sides of IHX or reducing the flow resistance can improve the performance. It is important to choose proper IHX capacity for a specific MAC system. If the capacity is too small, the undercooling degree will be small and can't significantly improve the efficiency of system. If the capacity is too large, the suction gas superheat will be too high, which results in high discharge temperature and decreases the performance of compressor.

9.3.6 Throttling Device

A CO_2 throttling device allows the CO_2 refrigerant to expand from the high pressure after the gas cooler to the low evaporating pressure, undergoing an isenthalpic process. Fixed orifice valve and electronic expansion valve (EEV) are typical throttling devices used in a CO_2 MAC system. Fixed orifice valve can't flexibly adjust the high pressure and requires a bypass device to avoid excessive high pressure. EEV could be used to control the optimum high pressure by regulating the opening degree of EEV (Li et al. 2004; Liu 2003). When the throat diameter of EEV becomes smaller, the high pressure of the system will be higher. When the throat diameter becomes larger, the high pressure of the system will be lower. As a result, the EEV can adjust the high pressure to optimum by changing the opening degree. Figure 9.36 shows a CO_2 electronic expansion valve.

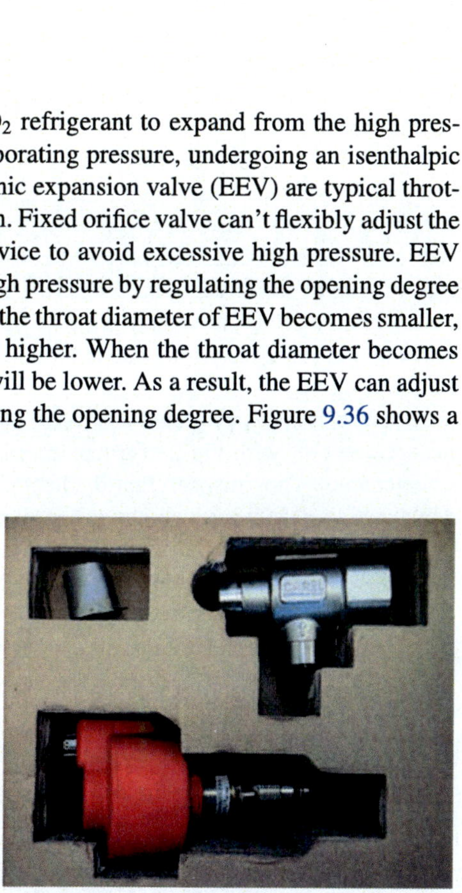

Fig. 9.36 CO_2 electronic expansion valve

9.4 CO_2 Mobile Heat Pump

Electrical vehicles (EV) are becoming increasingly popular due to the benefits of low vehicle emissions. However, the lack of engine exhaust waste heat in EV means that additional heater sources are required for cabin heating in the winter. The power consumption for cabin heating significantly affects the driving range, resulting in a 50% reduction in range in cold climate conditions (Higuchi et al. 2017). Compared with a conventional positive temperature coefficient (PTC) heater for cabin heating, an air source heat pump may be an effective method to improve heating efficiency and reduce energy consumption. However, a heat pump with the HFC-134a refrigerant suffers significantly diminished heating capacity and system efficiency as ambient temperature decreases. As a result, the HFC-134a heat pump system is unable to provide sufficient heating capacity without the assistance of a PTC heater at low ambient temperature conditions such as −20 °C (Lee 2015; Kwon et al. 2017). Therefore, for improved EV development, it is necessary to develop a novel mobile heat pump system with superior heating performance at low ambient temperature conditions.

Based on the advantage of thermos-physical properties, CO_2 refrigerant has been successfully applied in tap water heat pumps and supermarket refrigeration systems. Tamura et al. (2005) developed a prototype CO_2 automotive cooling and heating AC system for conventional vehicles and concluded that its dehumidifying efficiency performance exceeds that of the HFC-134a system. Hammer and Audi (2002) experimentally compared the vehicle warming-up time between a system with the CO_2 heat pump system and with a conventional PTC heater for an Audi car with a 1.6 L gasoline engine.

Kim et al. (2017) provided a novel CO_2 heat pump for use in FCEV (fuel cell electric vehicle) with the performance evaluation considering the heat exchanger arrangements. They insisted that the heat pump must be efficiently utilized for the cabin heating of fuel cell vehicles in the absence of exhaust heat source of relatively high temperature as in internal combustion engines. In their another published article, Kim et al. (2009b) found that use of a CO_2 heat pump system with exhaust heat recovery for fuel cell vehicles could improve the heating capacity by 100%, compared to conventional PTC heating.

9.4.1 Conventional CO_2 Mobile Heat Pump

In 2016, MAC team from Shanghai Jiao Tong University investigated the heating performance characteristics of a CO_2 heat pump system for an electrical vehicle in a cold climate.

9.4.1.1 Same Indoor Air Outlet Temperature

For vehicle heating, the indoor air outlet temperature is a main parameter affecting passenger comfort. A heat pump should be able to warm the recirculated or fresh air to above 40 °C for the comfort of cabin passengers. The impact of outdoor air outlet temperature on system performance was experimentally studied when the recirculated air (20 °C) was heated to the warm outlet temperature (40 °C). The outlet temperature varied from 0 to −20 °C, the compressor speed was manually controlled to satisfy the outlet temperature, and the EXV was adjusted to obtain the best COP range. The detailed test conditions are shown in Table 9.3.

When the outdoor air inlet temperature decreases, the refrigerant mass flow will drop due to the lower refrigerant density at the compressor suction point, so the compressor should increase its speed to meet the given air outlet temperature requirement. Figure 9.37 shows the change of compressor speed, COP, and compressor work at various outdoor temperatures when heating the 20 °C inlet air to the 40 °C outlet. When the outdoor air inlet temperature decreased from 0 to −20 °C, the compressor speed and work increased linearly by 59% and 50%. The heating capacity was almost constant at 2.2 kW because of controlled indoor air inlet and outlet conditions. Due to the constant heating capacity and increased compressor work, the COP of the system decreased from 2.5 to 1.7. The compressor power consumption for the CO_2 heat pump changed from 0.9 to 1.3 kW. If conventional PTC heating (1 kW electricity produces 1 kW heat) was adopted, at least 2.2 kW electricity power was consumed to produce equal heat 2.2 kW. Thus compared with PTC heating, the use of a CO_2 heat pump could save 40–60% of heating power consumption in full air recirculation mode (20 °C inlet and 40 °C outlet), indicating that the CO_2 heat pump has great potential to save energy and to extend the driving range for improved passenger comfort in cold climate.

In order to reveal the effect on the CO_2 system cycle, the *T-s* diagram and *P–h* diagram with the variation of the outdoor air temperature are shown in Fig. 9.38. The temperature and pressure at gas cooler outlet were about 10 MPa and 33 °C,

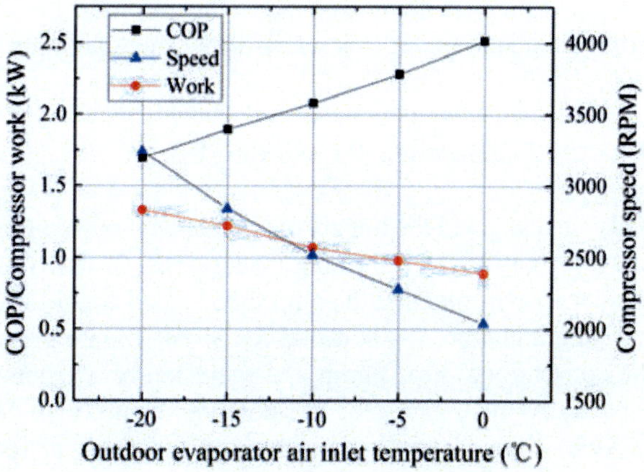

Fig. 9.37 Effects of outdoor air inlet temperatures on system performance

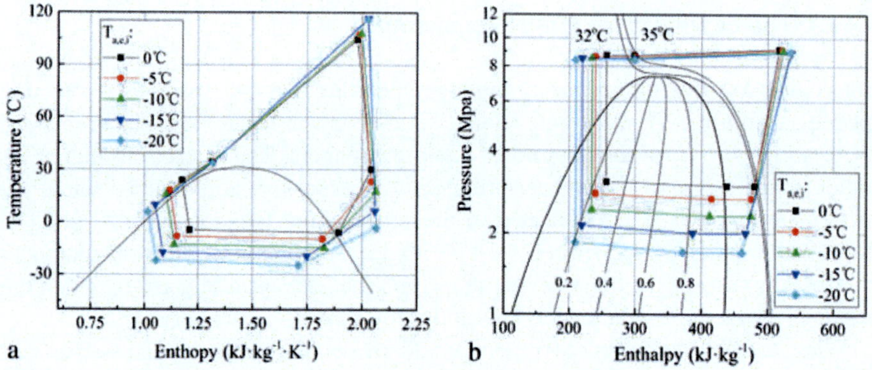

Fig. 9.38 **a** *T–s* diagram and **b** *P–h* diagram with varied outdoor air inlet temperatures

and were almost constant regardless of outdoor air temperature. This is because the indoor air inlet was controlled at 20 °C, and the compressor speed was changed to maintain 40 °C at the air outlet. Theoretical analysis reveals that the system efficiency of the CO_2 heat pump is significantly affected by the gas cooler outlet temperature, and the reduction of this temperature could increase the enthalpy difference and thus improve the heating capacity (Baek et al. 2013). Here, due to the refrigerant-to-air cross-flow configuration of the gas cooler (2-slab), the gas cooler outlet temperature was 7 °C lower than the air outlet 40 °C, but was 13 °C higher than the air inlet 20 °C. This indicated that the gas cooler can be more optimally designed to decrease the temperature difference between the air inlet and the refrigerant outlet, and the system efficiency could be further improved with a smaller approaching temperature. If the approaching temperature decreased from 13 to 0 °C, the enthalpy difference between the inlet and outlet gas cooler could increase from 221 to 272 kg/kJ, which will result in 23% heating capacity and COP improvement. As the outdoor air temperature decreased, the evaporating pressure reduced from 3.1 to 1.9 MPa, and thus the compression ratio changed from 2.4 to 3.5, accounting for the increase of compressor input work. The evaporator heat transfer temperature difference between the refrigerant and air was about 4–6 °C, indicating that the CO_2 outdoor evaporator still exhibited heat transfer characteristics during operation in cold conditions.

9.4.1.2 The Secondary Loop Results

It also investigated the impact of a secondary loop on the overall system performance of the overall CO_2 heat pump (Wang et al. 2018). The systems were tested under the same environment and compressor speed conditions, as shown in Table 9.3. The refrigerant charge was adjusted for SLHP. Figure 9.39a shows that compared with the conventional heat pump, the inclusion of a secondary loop decreased heating capacity by 6%, increased the compressor power by 15%, and reduced the COP by 19%. Thus, using a secondary loop negatively affects CO_2 system performance, especially COP. As shown in Fig. 9.39b and c, at the gas cooler outlet, the temperature increased from 34.3 to 40.5 °C and the pressure increased from 9.2 to 10.8 MPa. For the conventional heat pump, the gas cooler outlet temperature of 34.3 °C was less than the air outlet temperature 40.2 °C, due to the good transfer characteristic of the cross-flow gas cooler and direct heat rejection between the refrigerant and air. For SLHP, the coolant firstly absorbed heat from the gas cooler, and then released the heat to the air through the heater core. The heat transfer analysis of the heater core between the coolant and air showed that the coolant return temperature (40.0 °C) was a little higher than the air outlet temperature (39.0 °C). For the counter flow plate heat exchanger, the gas cooler outlet temperature (40.5 °C) must be higher than the coolant return temperature (40.0 °C). As a result, the use of the secondary loop remarkably increased the gas cooler outlet temperature and the corresponding optimal high pressure. For a CO_2 trans-critical system, the temperature at the gas cooler outlet significantly affects the system efficiency, and its increase could account for the observed COP reduction with inclusion of the secondary loop. Thus, the result suggests that the use of a secondary loop is severely harmful to system performance and should not be used in the CO_2 heat pump.

As calculated by Eqs. (9.7), (9.8), the thermal effectiveness of the gas cooler, the secondary loop plate heat exchanger, and the heater core were 80%, 98%, and 80%, respectively. Thus, the counter-flow plate heat exchanger has achieved almost maximum heat transfer capacity due to its sufficient volume and high efficiency. These values also indicate that the gas cooler and heater core could be further improved by enlarging the heat transfer area or optimizing heat transfer efficiency. In addition, a small difference in the temperature of the heat core coolant (3.8 °C) will produce a uniform distribution of outlet air temperatures in SLHP, but for the conventional heat pump, air temperature uniformity can be a big challenge due to the large temperature change on the refrigerant side (from 104.4 to 34.3 °C).

9.4.2 Novel SGC CO_2 Mobile Heat Pump

Furthermore, MAC team from Shanghai Jiao Tong University proposed a novel CO_2 heat pump system with Series Gas Cooler (SGC) configuration to improve its heating performance (Wang 2018a).

9.4.2.1 SGC CO_2 Mobile Heat Pump Structure

Figure 9.40 shows the schematic diagram of systems with (a) conventional OGC and (b) novel SGC. Both the OGC and SGC heat pump system can be switched from cooling to heating operation via stop valves and EXVs. There is no cycle difference between these two systems at the cooling mode. For OGC at the heating mode, after being discharged from the compressor, the refrigerant enters the indoor HX1 (No. 2 in Fig. 9.40a) and rejects heat to cold air, then directly passes through the

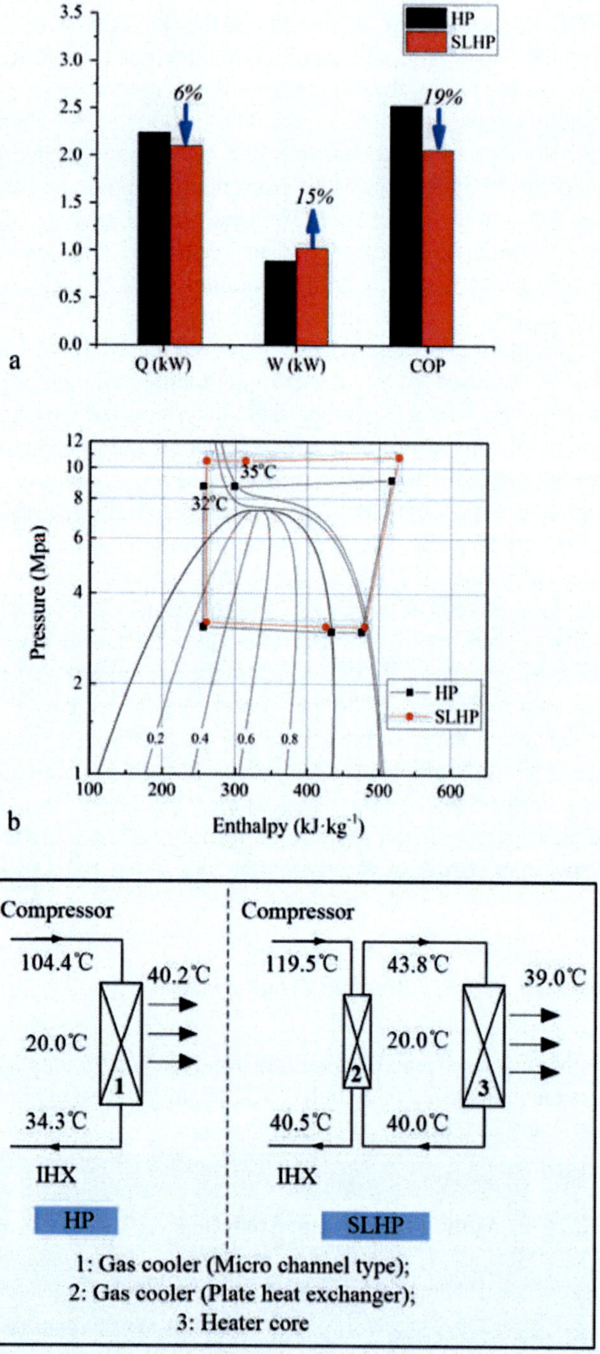

Fig. 9.39 The comparison between conventional HP and SLHP: **a** system performance, **b** P–h diagram, and **c** temperature distribution

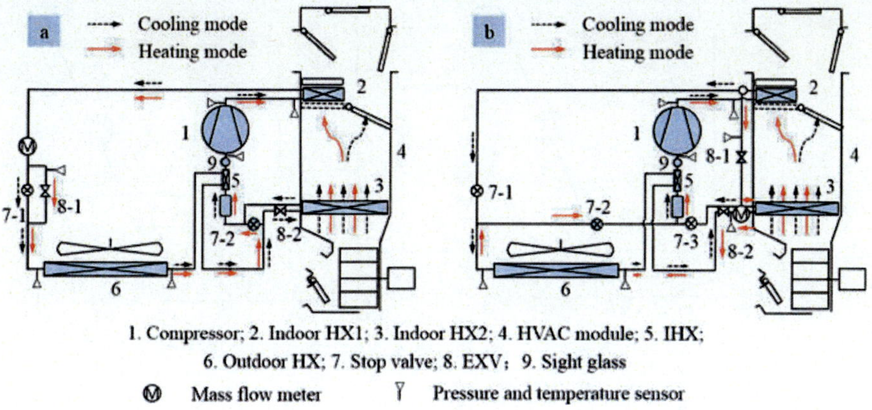

Fig. 9.40 Schematic diagram of systems with **a** conventional OGC and **b** novel SGC

EXV to the low pressure side of the cycle. The indoor HX2 (No. 3 in Fig. 9.40a) is bypassed when all refrigerant flows from the outdoor HX (No. 6 in Fig. 9.40a) to the accumulator. For SGC, however, after the heat rejection in the indoor HX1 (No. 2 in Fig. 9.40b), the refrigerant first passes through the full open EXV (No. 8–1 in Fig. 9.40b), which doesn't produce pressure drop here, then enters into the indoor HX2 (No. 3 in Fig. 9.40b) to release extra heat to the cold inlet air. Thus the indoor HX2 (No. 3 in Fig. 9.40b) is utilized as a second GC in the SGC system, and the cold inlet air is capable of absorbing the rejected heat from indoor HX1 (No. 2 in Fig. 9.40b) and indoor HX2 (No. 3 in Fig. 9.40b). After the heat rejection process in first GC and second GC, the refrigerant passes through the EXV (8–2 in Fig. 9.40b) and is throttled to a low pressure state. After throttling, the two-phase refrigerant evaporates in outdoor HX (No. 6 in Fig. 9.40b), which is used as an evaporator in the heating mode. Then the refrigerant enters into the accumulator and goes back to the compressor. When air dehumidification is required, the EXV (No. 8–1 in Fig. 9.40b) is regulated to throttle the refrigerant after indoor HX1 (No. 2 in Fig. 9.40b), then the indoor HX2 (No. 3 in Fig. 9.40b) is used as an evaporator and dehumidifies the inlet air, avoiding windshield flash fogging. Figures 9.41 and 9.42 show the *p–h* diagrams of CO_2 cycles and the psychrometric chart of indoor air for two heat pump systems, respectively.

9.4.2.2 Heating Performance of SGC System

Figure 9.43a reveals that the heating capacity of the SGC system was 20–33% higher than the OGC system, and the COP of the SGC system was 17–32% better than the OGC system. In order to illustrate the positive effect of SGC, *the p–h diagram* for results in 6000 RPM compressor speed is shown in Fig. 9.43c. A distinct difference from the OGC system is that $T_{g,r,o}$ in the SGC system was much lower. For instance, at the temperature of −20 °C, $T_{g,r,o}$ was −1.2 °C and 18.8 °C for SGC and

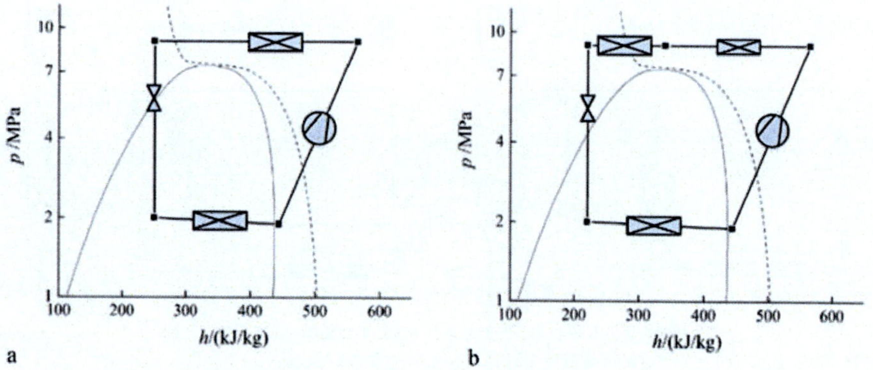

Fig. 9.41 p–h diagrams of CO_2 cycle for **a** OGC system and **b** SGC system under heating operation

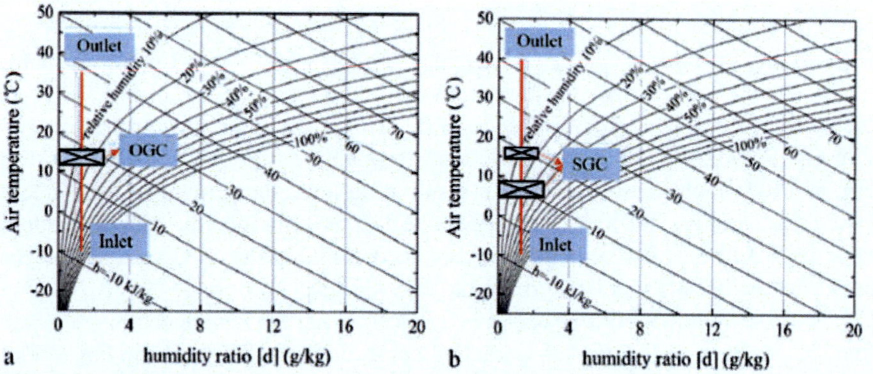

Fig. 9.42 Psychrometric charts of indoor air for **a** OGC system and **b** SGC system under heating operation

OGC, respectively. The obvious reduction of $T_{g,r,o}$ represents that the heat transfer temperature difference between the cold air and hot refrigerant was narrowed. $T_{g,r,o}$ is a significant influencing parameter in the CO_2 heat pump system, whose decrease will achieve heating capacity and COP enhancement. Thus it is concluded that the proposed method of SGC could greatly enhance GC rejection amount and thus boost the heating performance of EV CO_2 heat pump system.

What's more, at the highest compressor speed of 8000 RPM, the SGC system achieved 6.4 kW heating capacity, 59.4 °C $T_{g,a,o}$ and 1.6 COP at mild cold temperature of −10 °C, and 5.6 kW heating capacity, 40.4 °C $T_{g,a,o}$ and 1.8 COP at an extremely cold temperature of −20 °C. This indicated that the SGC CO_2 heat pump system could successfully provide comfortable warm air and sufficient heating capacity to passenger cabin even under the conditions of extreme cold weather (−20 °C) and severe heating load (100% fresh air). Based on it, the utilization of the SGC CO_2 heat pump system greatly outperforms the PTC heating supplement, which would significantly save electricity consumption and extend the driving range for EVs.

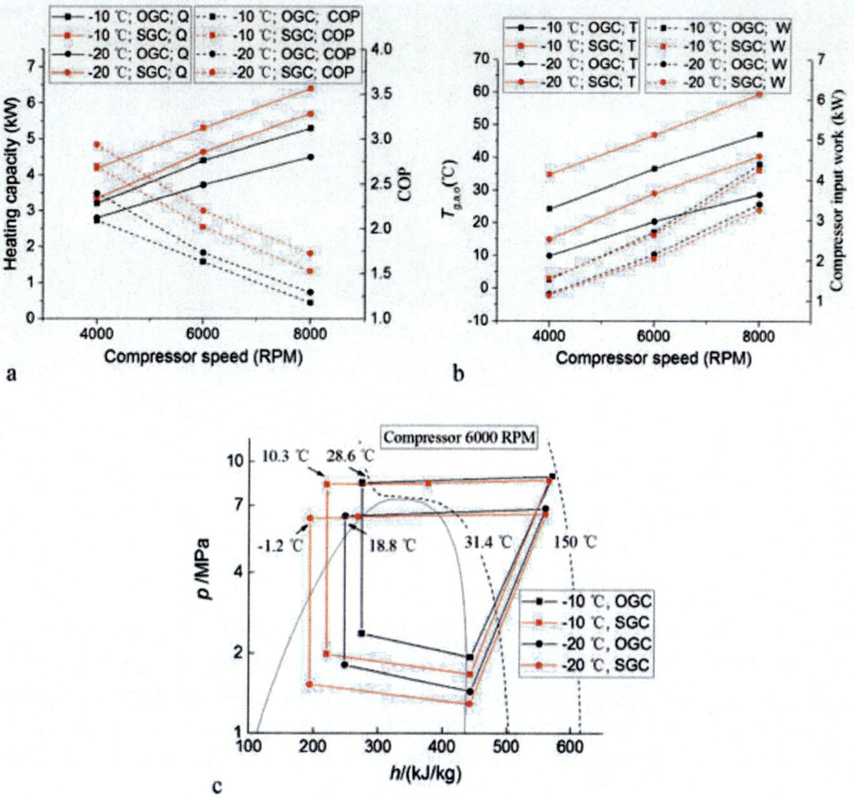

Fig. 9.43 Heating performance of EV CO_2 heat pump with different GC configurations **a** heating capacity and COP, **b** $T_{g,a,o}$ and compressor work, **c** *p*–*h* diagram

9.5 Application in Train, Bus and Container

A majority of current air conditioning systems for transport (train, bus and container) applications use high GWP HFC refrigerants (Finckh 2016). In Europe, currently available HVAC train units use R134a (75%) or R407C (25%), and most HVAC bus units use R134a as the refrigerant (Trygstad 2017). In recent years, CO_2 transport air conditioning has become one of the future options to reduce the impact of HFC refrigerant on global warming. This section introduces the research progress of CO_2 transport air-conditioning in the fields train, bus and container.

Fig. 9.44 First CO_2 train air conditioning in China

9.5.1 *CO_2 Train Air Conditioning*

9.5.1.1 Research Progress

In 2018, the MAC team from Shanghai Jiao Tong University collaborated with a Chinese train HVAC unit manufacturer to develop a CO_2 train air conditioning for high-speed train application shown in Fig. 9.44. The design capacity is 40 kW.

Europe train manufacturer Deutsche Bahn, and train HVAC unit manufacture Liebherr, Vossloh Kiepe GmbH, Merak have conducted research projects and prototype trials on CO_2 train air conditioning. Vossloh Kiepe GmbH developed its new CO_2 air conditioning unit for railway application, as shown in Fig. 9.45a and b (GmbH 2016). This unit with the type designation HKL751—CO_2 not only allows using the waste heat from the surrounding components, but also has a heat pump operation to save energy consumption. Besides, the variable frequency compressor and electronic expansion valve further improve its energy efficiency. This compact air conditioning unit has a weight of 560 kg and a size of 2997 mm × 2000 mm × 460 mm (length × width × height). Under a typical cooling operation condition (35 °C, 50%), the cooling capacity of this unit is 24 kW. At low outside temperatures, the heating capacity of this unit is greater than 24 kW, including heat output of the heat pump is 6–20 kW, and heat output of electrical heating can reach 24 kW (GmbH 2016). Liebherr has developed a CO_2 train air conditioning system with compact design and heat pump operation, as shown in Fig. 9.45c (Liebherr 2017).

9.5.1.2 CO_2 Train Air Conditioning with Ejector

With the collaboration between FTL (Faiveley Transport Leipzig GmbH & Co. KG) and SINTEF, a CO_2 train air conditioning system with an ejector cycle and reversible heat pump function was proposed. The schematic diagram is shown in

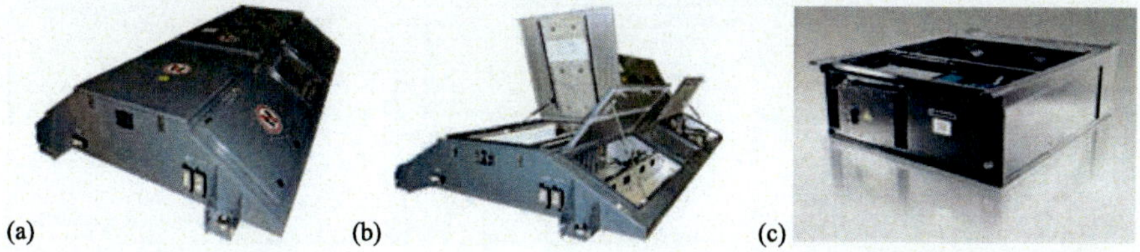

Fig. 9.45 CO_2 train air conditioning **a**, **b** by Vossloh Kiepe GmbH (GmbH 2016) **c** by Liebherr (Liebherr 2017)

Fig. 9.46 (Trygstad 2017). The system consists of two independent flow circuits, basic circuit and ejector circuit. Each circuit can independently operate in the cooling mode and reversible heating mode by switching several valves. The biggest difference between the two circuits is that in the cooling mode the basic circuit used an EXV as the throttling device, while the ejector circuit adopted a two-phase ejector to recover throttling energy loss and improve the system efficiency. In the heating mode, the ejector doesn't work, so the two circuit system have the same flow cycle. The whole CO_2 train air conditioning unit mainly consists of two compressors, two heat exchangers (interior HX1 and exterior HX2), one ejector and five expansion valves. The compressor is a two-cylinder semi-hermetic reciprocating CO_2 compressor with an integrated crankcase heater, which is manufactured by Bitzer and named 2MTE-5 K. The displacement is 3.3 m^3/h at 1450 RPM and the weight is 94 kg. The detailed technical specifications of the compressor are shown in Table 9.2 (Trygstad 2017). Besides, the compressor speed is controlled by the HVAC control unit through PWM signal connection. The heat exchanger is the copper fin and tube type heat exchanger with a tube outer diameter of 5 mm. Table 9.3 provides the detailed parameter information of these two heat exchangers (Trygstad 2017). The area of HX2 is smaller than HX1, HX2 has six tube passes, whereas HX1 has eight tube passes. Each exchanger meets the space requirement for roof-top train air conditioning installation (Trygstad 2017). As for expansion device, two EEVs are used in the basic circuit. One is for the cooling mode and the other is for the heating mode. In the ejector circuit, an ejector and three EEV are used. For EEVs, two are used for cooling mode and the other is for the heating mode. In the cooling mode, one EEV is designed to bypass the ejector to adjust the high pressure, and the other is used to maintain the evaporator pressure at the suction port of thr ejector. Figure 9.47 shows the EEVs and the two-phase ejector, which are manufactured by Carel (https://www.carel.com/emj-electronic-modulating-ejector; https://www.carel.com/electronic-expansion-valve).

The cooling performance of the CO_2 train air conditioning unit was tested in the FTL's R&D laboratory. Under the condition of 28 °C, 45% and 4000 m^3/h supply air, the test results of individual operation of the basic circuit shows that the cooling capacity and COP are 11.1 kW and 2.4, respectively. The test results of individual operation of the ejector circuit show that the cooling capacity and COP are 9.9 kW and 2.3, respectively. The test results of the two circuits running at the same time show that the cooling capacity is 18.6 kW, 5.4 kW smaller than the design capacity of 24 kW. The COP of the ejector circuit is the same as the COP of the basic circuit

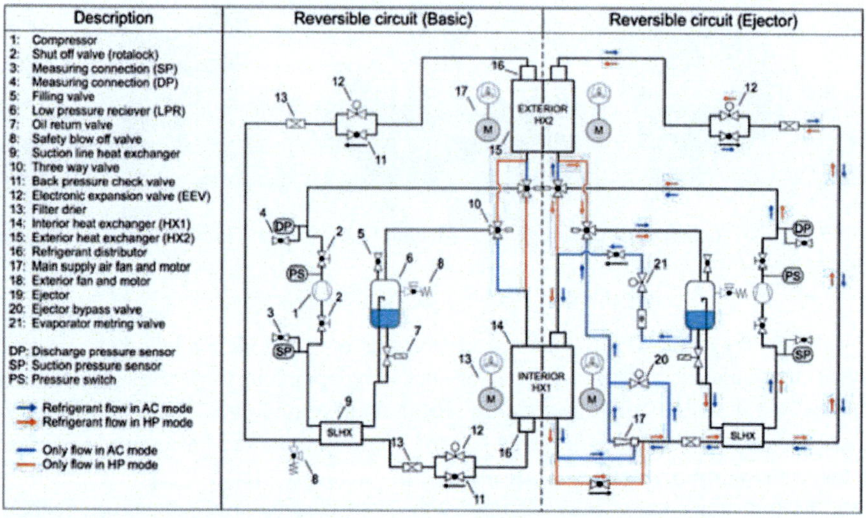

Fig. 9.46 The schematic diagram of CO_2 train air conditioning with ejector (Trygstad 2017)

Table 9.2 Technical specifications of the Bitzer 2MTE-5 K compressor

Bitzer 2MTE-5 K						
Motor	Power supply	Max operating current	Max starting current	Design load ta = 28 °C RH = 45%		Extreme load ta = 35 °C RH = 45%
	400 V–3–50 Hz AC	11.5 [A]	62.0 [A]	Approx.4.3 [kVA]		Approx. 5.7[kVA]
Compressor	Number of cylinders	Borex stroke	Displacement (1450 RPM)	Max pressure		Oil charge
				DP	SP	
	2	30 [mm] × 27 [mm]	3.3 m^3/h	100 [bar]	160 [bar]	1.20 [dm^3]

under the same conditions. In this laboratory experiment, the COP of the ejector circuit has not been improved (Trygstad 2017).

Armin Hafner et al. established the dynamic mathematical model of CO_2 train air conditioning unit with standard cycle and ejector cycle, and investigated their cooling and heating performance by numerical simulation. Figure 9.48a and b show the schematic diagrams of the standard unit in air conditioning operation and heat pump operation, respectively (Hafner 2016). Figure 9.48c and d illustrate the schematic diagrams of ejector unit in air conditioning operation and heat pump operation, respectively (Hafner 2016). The simulation model was built in Modelica software, and properties of essential components were set up by TIL libraries supported by TLK company.

Table 9.3 Parameter information of HX1 and HX2

HX1						HX2			
Fin	Material	Thickness	Length	Height	Depth	Thickness	Length	Height	Depth
	Au	0.15 mm	1350 mm	500 mm	173 mm	0.2 mm	1100 mm	495 mm	76.2 mm
Tube	Material	Diameter		Number of passes		Diameter		Number of passes	
		Outer	Inner	Horizontal	Vertical	Outer	Inner	Horizontal	Vertical
	Cu	5.0 mm	4.59 mm	8	10 (per circuit)	5.0 mm	4.4 mm	6	10 (per circuit)

Fig. 9.47 The two-phase ejector and EEV (https://www.carel.com/emj-electronic-modulating-ejector; https://www.carel.com/electronic-expansion-valve)

The system simulations were carried out under the high-load AC operating conditions (35 °C, 41%) and low load (25 °C, 45%). At the cooling load of 21.3 kW, the power consumption of the ejector unit system is 6.3 kW, which is less than the power consumption of the standard system 6.9 kW. And the COP is improved from 3.07 to 3.39. At the cooling load of 8.6 kW, the power consumption of the ejector system is 1.25 kW, which is less than the power consumption of the standard system 1.31 kW. And the COP is improved from 6.56 to 6.87.

9.5.1.3 CO_2 Train Air Conditioning with Two Evaporators and Two Stage Compression

Øystein Christensen presents a CO_2 train air conditioning with two evaporators and two stage compression. At first, the reason for using two evaporators and gas coolers is to reduce the potential exhaust air energy lost in the air stream mixing when fresh and exhaust air first is mixed together and then sent to a heat exchanger. Therefore there are separated evaporators and gas coolers for fresh and exhaust air.

Figure 9.49a, b is the schematic diagram of the CO_2 HVAC unit, with separated evaporators and gas coolers for fresh and exhaust air in the system solution (Christensen 2009). In addition, this system is able to switch cooling and heating mode by the three-way valve. Figure 9.49c is the principal sketch of the rooftop unit (Konvekta 2018). For Air stream 1, fresh air which is going to enter the system as treated air will exchange heat with evaporator 1 (summer)/gas cooler 2 (winter). For Air stream 2, fresh air which is going to exit the system will exchange heat with gas cooler 1 (summer)/evaporator 2 (winter). For Air stream 3, exhaust air which is going to be treated and recycled will exchange heat with evaporator 2 (summer)/gas cooler 1 (winter). For Air stream 4, exhaust air which is going to exit the system will exchange heat with gas cooler 2 (summer)/evaporator 1 (winter) (Christensen 2009).

In summer, the inside heat exchangers act as evaporators and the outside heat exchangers act as gas coolers. In winter, the inside heat exchangers act as gas coolers and the outside heat exchangers act as evaporators (Christensen 2009). To improve the performance of the basic system with one stage compression in Fig. 9.49 (Fig. 9.50a), Øystein Christensen gives a new system circuit with two stage compression to reduce

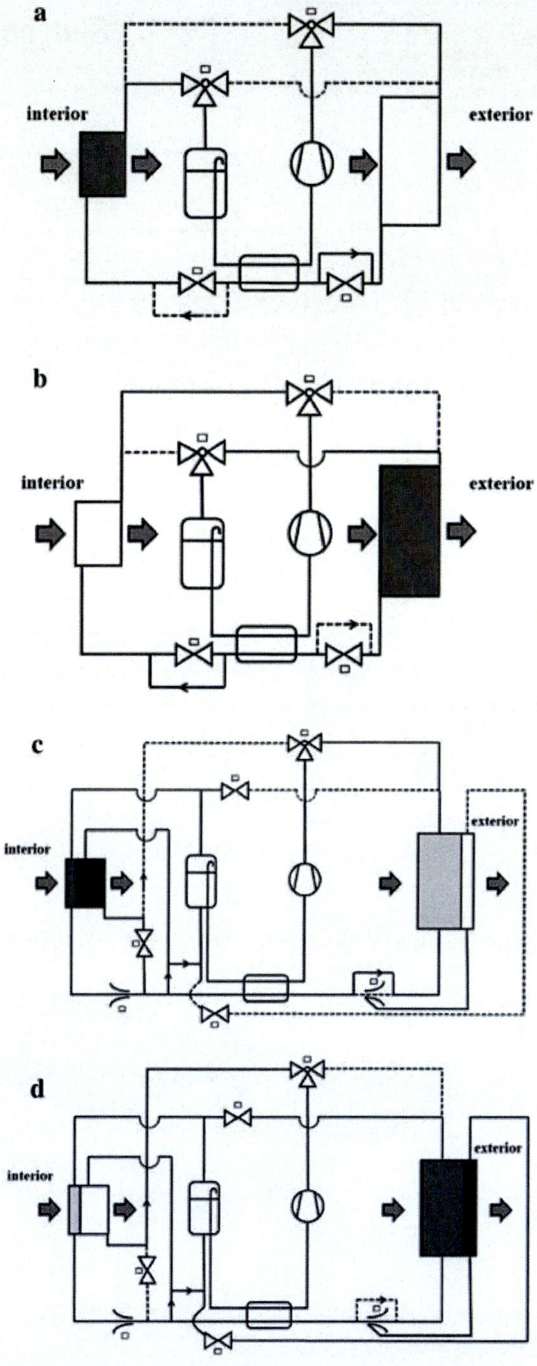

Fig. 9.48 The schematic diagram of CO_2 train air conditioning: standard unit and ejector unit (Hafner 2016)

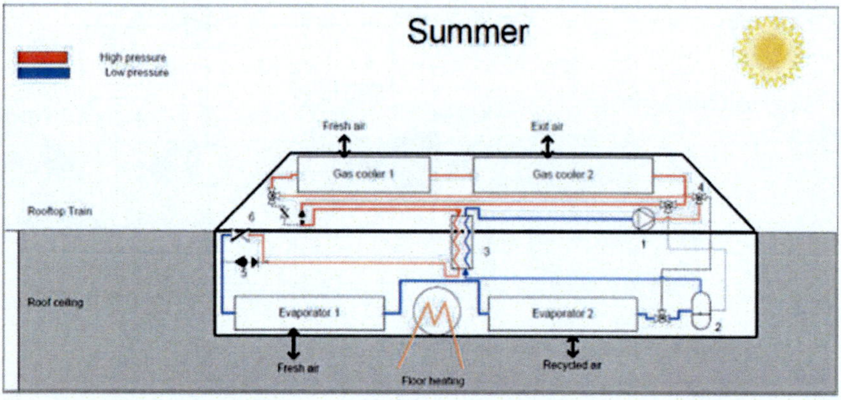

(a)

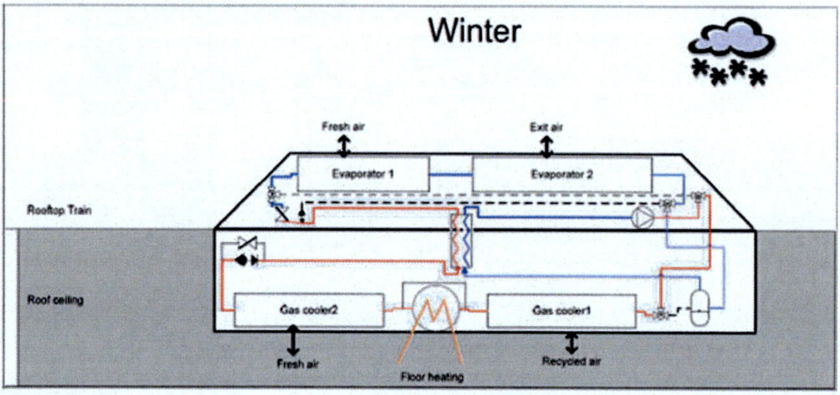

(b)

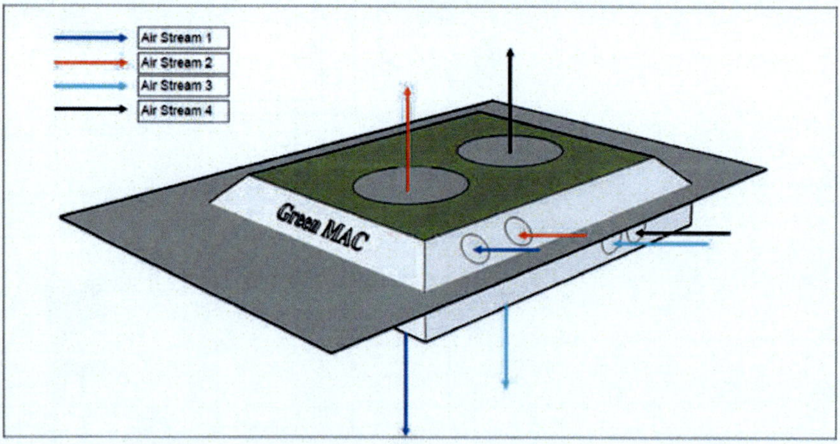

(c)

Fig. 9.49 **a**, **b** The schematic diagram of the CO_2 HVAC unit with separated evaporators and gas coolers (Christensen 2009) **c** Principal sketch of the rooftop unit (Christensen 2009) the loss caused by a high compression ratio (Fig. 9.50b). And then Øystein Christensen presents a simulation for the two system in Fig. 9.50 in the cooling mode by the Excel model and the HXsim/ProII model (Christensen 2009). Computer simulation shows the heating capacity of the basic system (Fig. 9.49a) is only about 4.7 kW, and the COP is 1.7 when the ambient temperature at −40 °C, 0%. For the improved system with a two stage compression (Fig. 9.50b), the heating capacity is up to 13.8 kW, and the COP is improved to 3.2 at same conditions (Christensen 2009).

9.5.1.4 CO_2 Train Air Conditioning with Air Reversing Configuration

A CO_2 train air conditioning with air reversing and ejector configuration was proposed by Hanne Elisabeth Bø Andreassen, as shown in Fig. 9.51 (Andreassen 2010). This unit was placed on a rotatable plate, and can be mechanically driven to rotate 180° to switch between cooling and heating mode. The advantage of this design is that the gas cooler acts as a gas cooler, and the evaporator acts as an evaporator in both cooling mode and heating mode, which simplifies system design and eliminates the use of valves. Two gas coolers and two evaporators are used to treat fresh air and exhaust air separately, which helps to reduce the energy loss in the mixing of air streams.

Figure 9.52 shows the schematic diagram of air reversing CO_2 air conditioning with ejector in the heating and cooling mode (Andreassen 2010). Based on the heating and cooling demands for the five cities, the performance of CO_2 train air conditioning was simulated. The simulation results show that COP is affected by the heating and cooling demand and occupancy rate of the city. The COP of cooling and heating decreases with the occupancy rate decreasing. The COP ranges from 3.1 to 6.0 in the cooling mode, while the COP ranges from 2.8 to 8.2 in the heating mode, whose variation is greater than that in the cooling mode (Andreassen 2010).

9.5.2 CO_2 Bus Air-Conditioning

9.5.2.1 Research Progress

Konvekta has developed CO_2 bus air-conditioning for more than 20 years. Since summer 1996, a public-transport bus MB O305 (Fig. 9.53a) with a CO_2 air conditioning unit has been in daily use (JK, SM and FS 2003). At the beginning of 1997 the second bus MB O305 N (Fig. 9.53b) was put into operation (JK, SM and FS 2003). These buses are equipped with mobile measurement to conduct a comparison with an R134a unit in parallel. The operational experience of more than 10 years and the evaluation of the field test show the good applicability of CO_2 as a refrigerant.

Due to this experience, Konvekta plans on achieving the series-production readiness of CO_2 units. Since August 2002 the first transport refrigeration unit with a

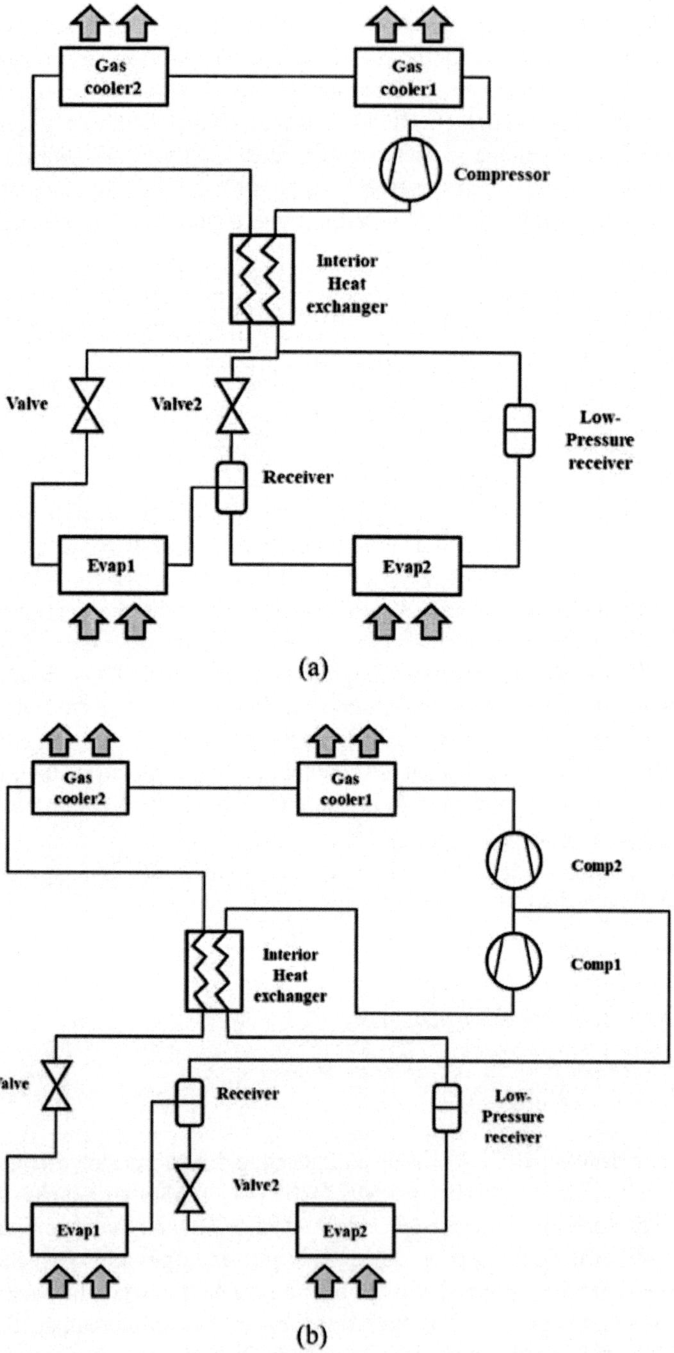

Fig. 9.50 Simulation schematic diagram of two CO_2 HVAC units (Christensen 2009)

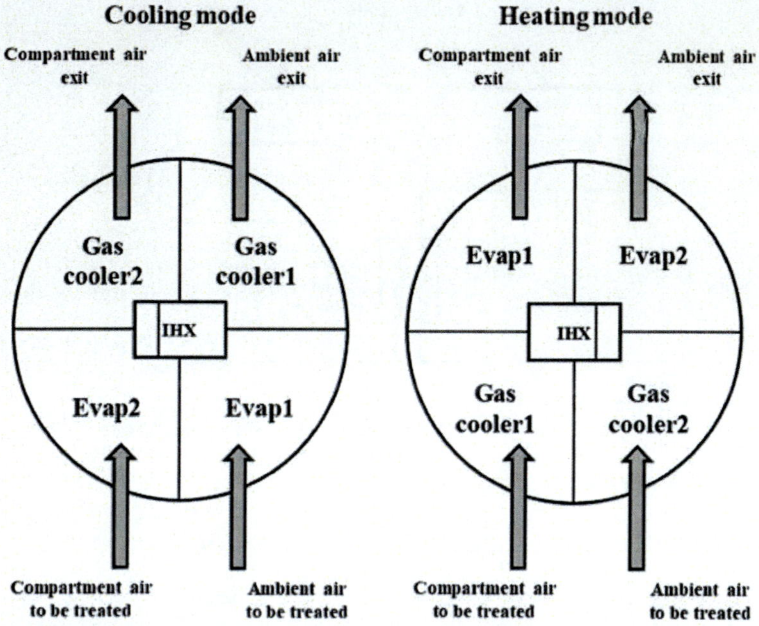

Fig. 9.51 Layout of the turntable prototype in cooling and heating mode (Andreassen 2010)

CO_2 compression circulation has been used by ALDI. In cooperation with the Volkswagen, a CO_2 air conditioning unit for a midibus was developed and tested in detail. In 2004, one of the first field tests with a CO_2 bus air conditioning unit was started by Konvekta with the company Saar-Pfalz-Bus GmbH (Konvekta 2015).

Besides, Konvekta has presented two generation CO_2 heat pumps in 2015. The first field test with the second generation of the Konvekta CO_2 heat pump has already commenced. After conducting a series of internal tests successfully, the unit is undergoing trial runs at the municipal utilities of Klagenfurt (Konvekta 2015). Figure 9.54 shows the heating test result of Solaris E 8,9 Klagenfurt (Konvekta 2017). After about 2500 s, the system will be stable and have excellent heating capacity.

With further research of CO_2 bus air conditioning system, until 2017, a total of 90 A/C systems and 10 heat pump systems are tested on operational vehicles in different areas such as Berlin, Bad Hersfeld, NVV and so on (Konvekta 2017). Two kinds of CO_2 conditioning systems on the Konvekta official website are shown in Fig. 9.55: UltraLight 700 CO_2 (Dimensions: L = 2655 mm/W = 2083 mm/H = 207 mm; Weight: 159 kg) and UltraLight 500 CO_2 Heat pump 2.0 (Konvekta 2018). The cooling capacity and heating capacity of the two types is presented in the following Table 9.4 (Konvekta 2018).

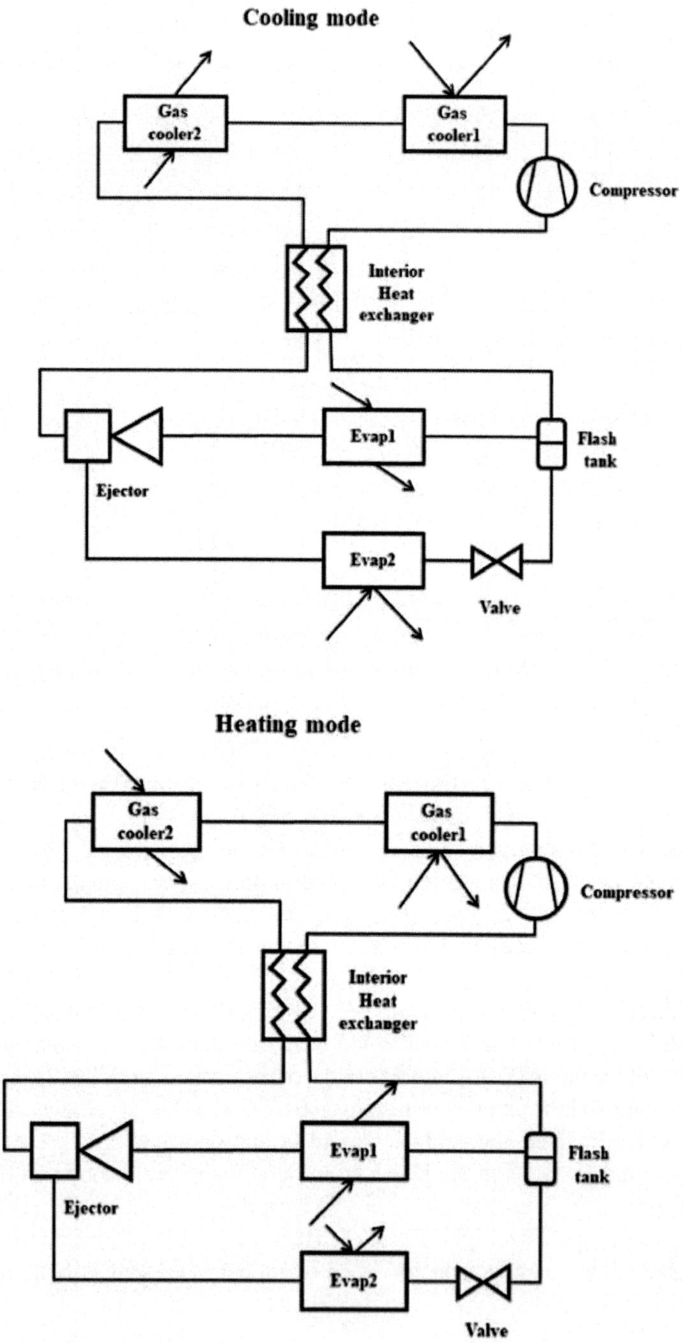

Fig. 9.52 The schematic diagram of air reversing CO_2 HVAC unit with ejector in heating and cooling mode (Andreassen 2010)

(a) (b)

Fig. 9.53 Two buses of Konvekta equipped with CO_2 air conditioning system (JK, SM and FS 2003)

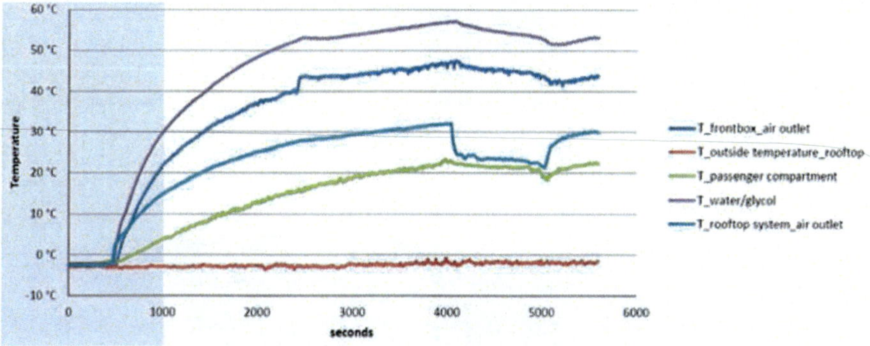

Fig. 9.54 Heating test result of Solaris E 8,9 Klagenfurt (Konvekta 2017)

Fig. 9.55 UltraLight 700 CO_2 and UltraLight 500 CO_2 Heat Pump 2.0, developed by Konvekta (2018)

Table 9.4 The capacity of UltraLight 700 CO_2 and UltraLight 500 CO_2 Heat Pump 2.0

Types	Cooling capacity (kW)	Heating capacity (kW)
UltraLight 700 CO_2	33	20
UltraLight 500 CO_2	38	18

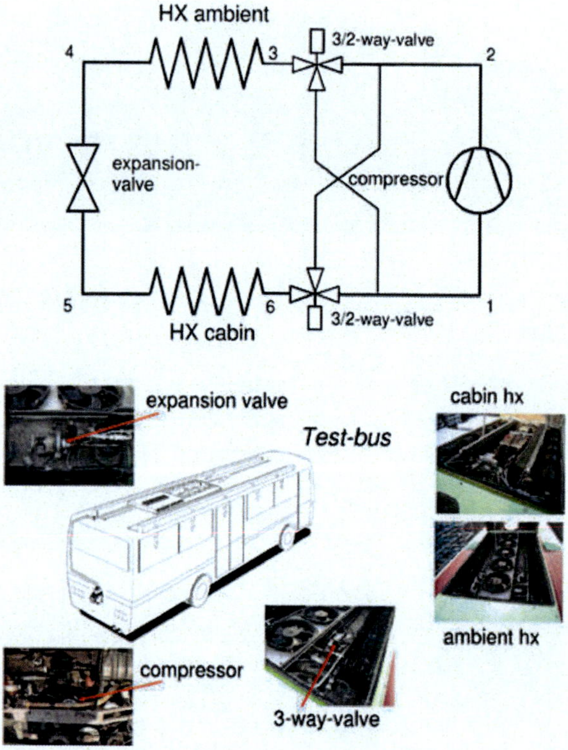

Fig. 9.56 Test rig of CO_2 bus air conditioning unit (Főrsterling and Sonnenkalb 2009)

9.5.2.2 Simulation of Bus CO_2 Air Conditioning Unit

Konvekta is the first thermo system supplier to install a CO_2 HVAC unit on bus. In 2009, Konvekta made a simulation model to evaluate the performance of CO_2 bus air conditioning. Figure 9.56 shows a CO_2 heat pump test rig for CO_2 bus air conditioning (Főrsterling and Sonnenkalb 2009). Figure 9.57 presents the simulation results of the CO_2 bus air conditioning (Főrsterling and Sonnenkalb 2009). At 715 min^{-1} n_{comp}, -10 °C ambient temperature and 50% fresh air rate, the COP is 3.8. As the ambient temperature increased from -10 to 15 °C, the COP increases slowly (Főrsterling and Sonnenkalb 2009).

9.5.3 *CO_2 Container Air-Conditioning*

9.5.3.1 Research Progress

Carrier Transicold launched the world's first container NATURALINE® in 2013 using CO_2 as refrigerant for the marine shipping industry. The Fig. 9.58 shows the

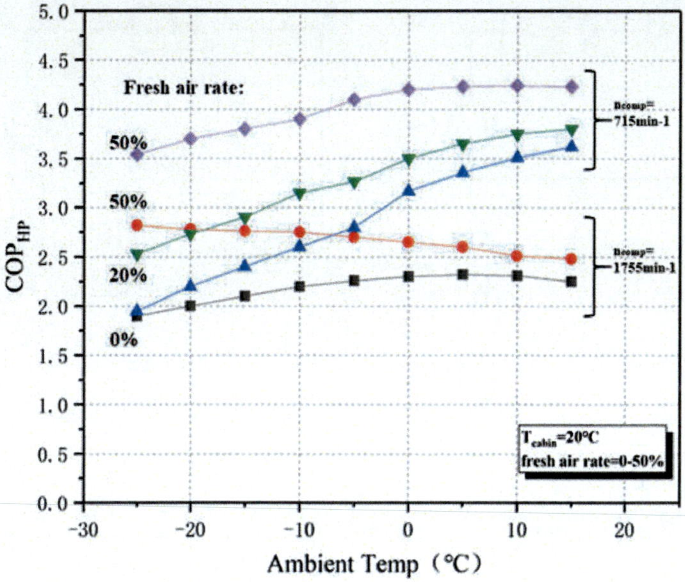

Fig. 9.57 Simulation results of CO_2 bus air conditioning unit (Fõrsterling and Sonnenkalb 2009)

NATURALINE front view. Cooling capacity is 9.4 kW (2 °C), 6 kW (−18 °C) and 4.4 kW (−29 °C), respectively (Transicold 2013).

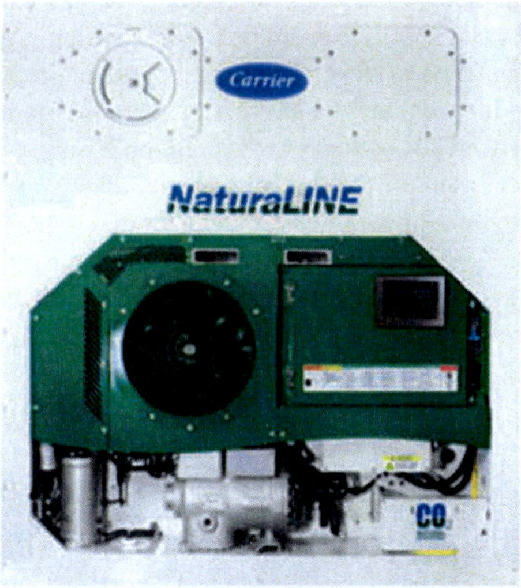

Fig. 9.58 The Naturaline front view (Transicold 2013)

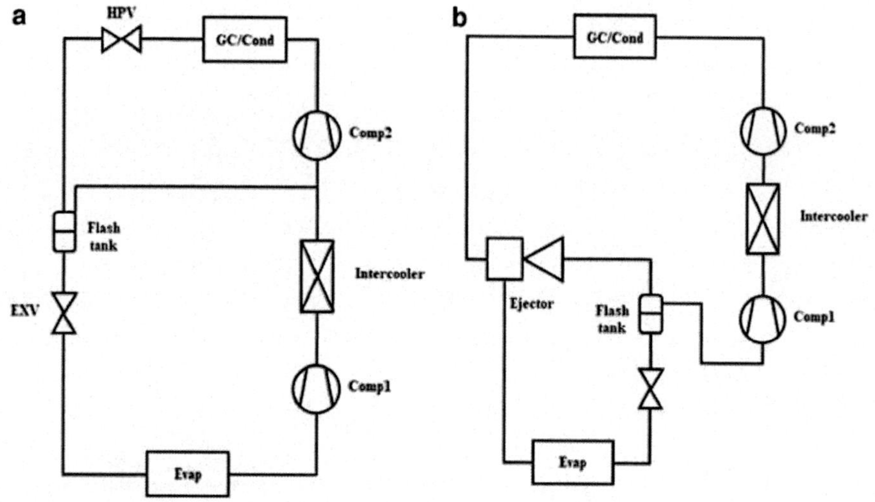

Fig. 9.59 **a** CO_2 container refrigeration system **b** CO_2 container refrigeration system with ejector schematic diagram (Poolman 2018)

9.5.3.2 CO_2 Container Refrigeration System with a Modulating Two Phase Ejector

R-134a and R-404A are the most common refrigerants in the container industry, with R-134a used in the vast majority of applications (Poolman 2018). For a CO_2 container, different from train or bus unit, a majority of refrigeration systems suitable for truck, trailer or marine reefer use an intercooled two-stage refrigeration circuit. This kind of design is to control the entire operating range (typically −30 to 50 °C) to meet capacity and efficiency requirements. Figure 9.59 shows the typical design of the CO_2 container refrigeration system (Poolman 2018).

Based on the typical CO_2 container refrigeration system, C. Poolman applies a modulating two phase ejector in the system and finishes a model-based analysis for it. Figure 9.60 presents the result of the analysis (Poolman 2018). Compared with the baseline system (Fig. 9.59), the container system with a modulating two phase ejector shows that the highest increase of COP is 15% (2 °C box @38 °C Amb) (Poolman 2018).

9.5.3.3 Two Stage CO_2 Container Refrigeration System

Oliver Finckh gives an introduction on a 2-stage refrigeration container system in Fig. 9.61, using an economizer for vapor injection (Finckh 2016). Initial testing at conditions of −20/30 °C and 0/30 °C resulted in a COP gap of only 10–20%, compared with the hydrocarbons (R290) container system (Finckh 2016).

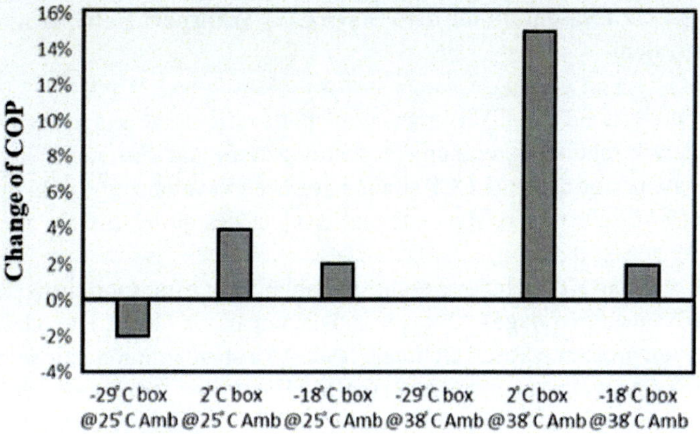

Fig. 9.60 CO_2 container refrigeration system with ejector schematic diagram (Poolman 2018)

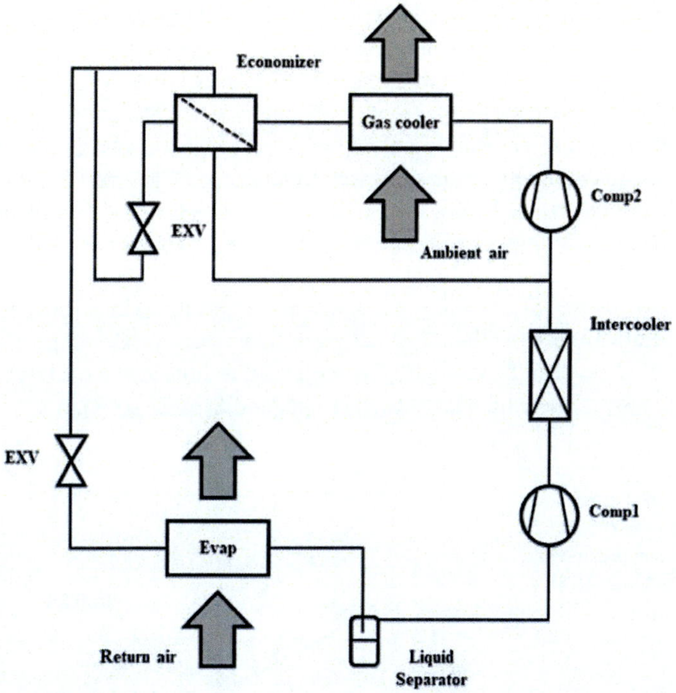

Fig. 9.61 Laboratory system set-up for trailer unit (Finckh 2016)

9.5.3.4 LCCP Evaluation for Two Stage CO_2 Transport Refrigeration System

Trying to understand the GWP impacts of both refrigerant and system design is a challenge for regulators, equipment manufacturers and customers. Specific life cycle climate performance (LCCP) tools have been developed to try to understand the tradeoffs in selecting refrigerants and product design by climate zone (Kujak et al. 2014).

Steve Kujak and Ken Schultz present an evaluation of a transport refrigeration unit using an advanced two stage CO_2 cycle and a comparison of its GHG (Green House Gas) emission profiles across 18 climate zones to a standard production electrically driven R452A unit (Kujak and Schultz 2018). The reason for choosing R452A unit as a baseline is that it is now a common lower GWP refrigerant used today in Europe to replace R404A transport trailer products (Kujak et al. 2014).

At first, Steve Kujak and Ken Schultz provide a summary of the performance (Fig. 9.62) obtained for the CO_2 prototype unit and for a standard production electrically powered R452A unit (Kujak and Schultz 2018). The CO_2 unit was run over a capacity range of 30 to 60 Hz using the adjustable frequency drive, refrigerated box temperatures of −20 to 10 °C and ambient temperatures of 0 to 40 °C (Kujak and Schultz 2018).

Next, the relationships from Fig. 9.62 for each refrigerant system are utilized to calculate the indirect energy emissions for climate zone. The direct emissions from refrigerants are determined from the LCCP data of Kujak study. Figure 9.63 is the result of LCCP Evaluation of CO_2 and R452A by the 18 Climate Zones (Kujak and Schultz 2018).

In terms of the result, the CO_2 total greenhouse gas emissions (both direct and indirect) are always greater than the total greenhouse gas emission for R452A. And in many climate zones, the amount of energy used at high ambients can be a very large percentage of the total energy used (Kujak and Schultz 2018).

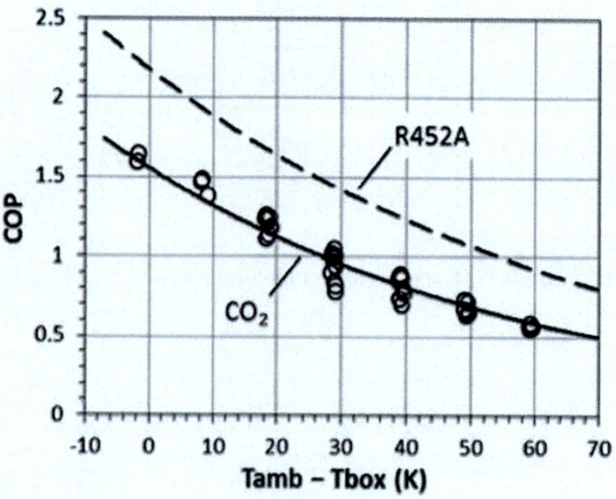

Fig. 9.62 COP Performance of CO_2 prototype and R452A electric drive production unit (Kujak and Schultz 2018)

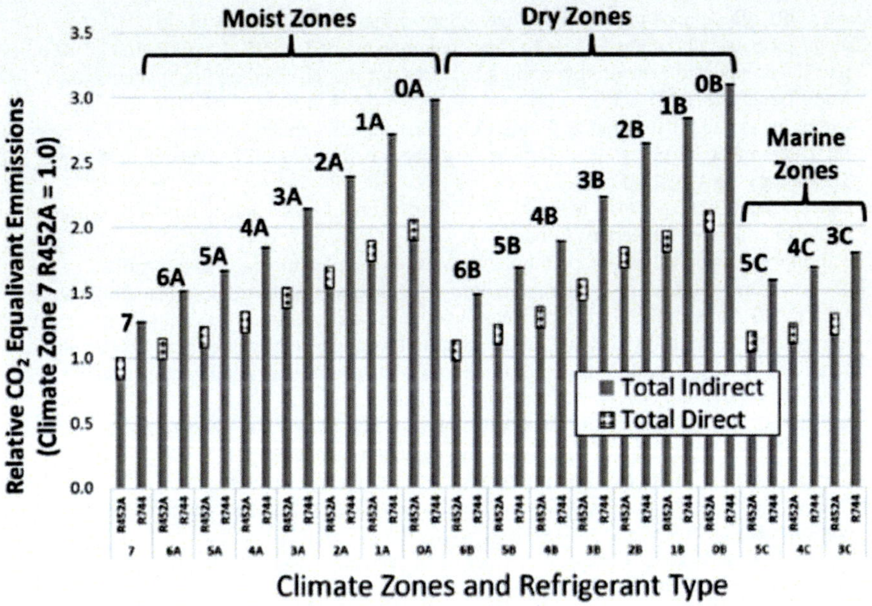

Fig. 9.63 LCCP evaluation of CO_2 and R452A by the 18 climate zones (Kujak and Schultz 2018)

References

Andersen S, et al (2017) Comparative manufacturing and ownership cost estimates for secondary loop mobile air conditioning systems (SL-MACs). In: WCX™ 17: SAE world congress experience. Detroit

Andreassen HEB (2010) Air reversing CO_2 heat pumps. In: Department of energy and process engineering. Norwegian University of Science and Technology, pp 26–58

Armin H, Petter N (2005) R744 (CO_2) technology maturity and availability. Workshop on technology cooperation for next-generation mobile air conditioning. http://www.ammonia21.com/files/pdf_118.pdf.

Armin HSEr (2009) CO_2/R744 mobile air-conditioning (MAC). In: Reulens W (ed) Natural refrigerant CO_2. Leonardo Da Vinci agency

Ayad F, Benelmir R, Souayed A (2012) CO_2 evaporators design for vehicle HVAC operation. Appl Therm Eng 36:330–344

Baek C et al (2013) Optimal control of the gas-cooler pressure of a CO_2 heat pump using EEV opening and outdoor fan speed in the cooling mode. Int J Refrig 36(4):1276–1284

Boewe DE et al (1999) Comparative experimental study of subcritical R134a and transcritical R744 refrigeration systems for mobile applications. University of Illinois, Air Conditioning and Refrigeration Center

Boewe D, et al (1999) The role of suction line heat exchanger in transcritical R744 mobile A/C systems. In: International congress & exposition. Detroit

Brown JS, Yana-Motta SF, Domanski PA (2002) Comparitive analysis of an automotive air conditioning systems operating with CO and R134a. Int J Refrig 25(1):19–32

Bullard CW, Yin JM, Hrnjak PS (2000) Transcritical CO_2 mobile heat pump and A/C system experimental and model results. https://www.researchgate.net/publication/242406810_TRANSCRITICAL_CO2_MOBILE_HEAT_PUMP_AND_AC_SYSTEM_EXPERIMENTAL_AND_MODEL_RESULTS

Center DRD. Developed product. http://www.doowonrnd.com/eng/?page_id=15
Cheng L, Ribatski G, Thome JR (2008) New prediction methods for CO_2 evaporation inside tubes: part ii—an updated general flow boiling heat transfer model based on flow patterns. Int J Heat Mass Transf 51(1):125–135
Christensen Ø (2009) Reversible R744 (CO_2) heat pumps applied in public trains in Norway. In: Department of Energy and Process Engineering, Norwegian University of Science and Technology, pp 4–5; 21–51
I Dewa Made Cipta Santosa MS (2015) Optimisation of gas coolers for CO_2 refrigeration application. Brunel University, London
Electronic expansion valves—ExV. https://www.carel.com/electronic-expansion-valve
EmJ—electronic modulating ejector. https://www.carel.com/emj-electronic-modulating-ejector
Finckh O (2016) Manufacturers experiences with developing transport applications using natural refrigerants. In: 12th IIR Gustav Lorentzen natural working fluids conference. Edinburgh
Fõrsterling, Tegethoff, and Sonnenkalb (2009) CO_2 air conditioning and heating applied to city buses (COACH). In: VDA-Wintermeeting. Saalfelden
GmbH KE (2016) Retrieved from kiepe knorr-bremse. http://www.kiepe.knorr-bremse.com/news/press-releases/new-CO2-air-conditioning-system2014a-milestone
Hafner A, Neksa P (2006) R&D on carbon dioxide refrigeration technology
Hafner A (2016) R744 train HVAC unit. In: 12th IIR Gustav Lorentzen natural working fluids conference. Edinburgh
Hammer H, Audi A (2002) Comparative study of AC-and HP-systems using the refrigerants R134a and R744. VDA alternate refrigerant winter meeting
Higuchi Y, et al (2017) Efficient heat pump system for PHEV/BEV. SAE International
Hu B (2011) Design and research on CO_2 mobile air conditioning system. Shanghai Jiaotong University, Shanghai
Chen JJ, C Z (2011) Development and validation of a microchannel evaporator model for a CO_2 air-conditioning system. Appl Therm Eng, 24(1):137–146
J K, S M, F S (2003) 7 years of on-the-road experience with the refrigerant CO_2. Technical University Braunschweig, Germany
Jian MY, Bullard CW, Hrnjak PS (2001) R-744 gas cooler model development and validation. Int J Refrig 24(7):692–701
Jin J (2011) Research on a carbon dioxide automobile air conditioning system using microchannel heat exchangers. PhD thesis, Shanghai Jiaotong University, Shanghai
Kim MH, Pettersen J, Bullard CW (2004) Fundamental process and system design issues in CO_2 vapor compression systems. Prog Energy Combust Sci 30(2):119–174
Kim SC, Won JP, Min SK (2009a) Effects of operating parameters on the performance of a CO_2 air conditioning system for vehicles. Appl Therm Eng 29(11):2408–2416
Kim SC et al (2009b) Performance evaluation of a stack cooling system using CO_2 air conditioner in fuel cell vehicles. Int J Refrig 32(1):70–77
Konvekta (2017) 20 years of CO_2 systems operated in city buses. Berlin, pp 2–10
Konvekta (2015) Press release busworld 2015: air conditioning technology that opens up new d. http://www.konvekta.de/press.html
Konvekta (2018). http://www.konvekta.de/en/bus-airconditioning/products/ultralight-500-CO2-heat-pump.html
Kujak S, Schultz K (2018) Analysis of an advanced CO_2 cycle and its impact on green house gas emissions by climate zone. In: 13th IIR Gustav Lorentzen conference. Valencia
Kujak S, Srichai R, Schultz K (2014) Assessment of life cycle climate performance (LCCP) tools for HVACR applications with the latest next generation refrigerant technology. In: The international refrigeration and air conditioning conference. Purdue
Kwon YC et al (2009) Experimental study on heat transfer characteristics of initial heat exchangers for CO_2 system under cooling condition. Mech Sci Technol 23:698–706
Kwon C et al (2017) Performance evaluation of a vapor injection heat pump system for electric vehicles. Int J Refrig 74:138–150

Lee D (2015) Experimental study on the heat pump system using R134a refrigerant for zero-emission vehicles. Int J Automot Technol 16(6):923–928

Lee GH, Kim HS (2008) High efficiency R744 compressor development for vehicle. In: VDA alternative refrigerant winter meeting. Saalfelden

Lemke N, Stulgies N (2005) R744 accumulator measurement and simulation. TLK-Thermo GmbH, Braunschweig

Li X et al (2004) A new method for controlling refrigerant flow in automobile air conditioning. Appl Therm Eng 24(7):1073–1085

Liebherr (2017). https://www.liebherr.com/en/deu/products/aerospace-and-transportation-systems/transportation-systems/products-and-solutions/air-conditioning-systems/air-conditioning-systems.html#!/air-conditioning-acc-start-module=air-cycle-acc-item-start-module+CO2-cycle-acc-item-start-module

Liu J (2003) Researches on throttling devices characteristics and related optimisation of transcritical CO_2 automobile air-conditioning system. PhD thesis. Shanghai Jiaotong University, Shanghai

Liu H, Chen J, Chen Z (2005) Experimental investigation of a CO_2 automotive air conditioner. Int J Refrig 28(8):1293–1301

Liu H (2005) Research on the characteristics of the carbon dioxide automotive air conditioning system. PhD thesis, Shanghai Jiaotong University, Shanghai

Lorentzen G, Pettersen J (1993) A new efficient and environmentally benign system for car air-conditioning. Int J Refrig 16(1):4–12

M B, T W, L N (2008) Transient modelling of a controllable low pressure accumulator in CO_2 refrigeration cycles. In: Proceedings of 6th International Modelica Conference. Bielefeld

Malvicino C, et al (2009) B-cool project—Ford Ka and Fiat Panda R-744 mac systems. SAE World Congress & Exhibition. http://citeseerx.ist.psu.edu/viewdoc/download?doi=10.1.1.585.1433&rep=rep1&type=pdf

Martin K, Lang G, Rieberer R (2005) Mobile HVAC-system with CO_2 as refrigerant—simulations and measurements. In: Vehicle thermal management systems conference and exhibition. Toronto

Mercedes-Benz to equip first vehicle models with CO_2 air conditioning systems. https://www.daimler.com/sustainability/product/further-environmental-technologies/CO2-air-conditioning-system.html

Mu J (2003) Development of autotive air-conditioning system using carbon dioxide as refrigerant. PhD thesis, Shanghai Jiaotong University, Shanghai

Mu J, Chen J, Chen Z (2003) System design and analysis of the trans-critical carbon-dioxide automotive air-conditioning system. J Zhejiang Univ-Sci B (Biomed & Biotechnol), 4(3):305–308

Parsch W, Hotz U (2008) Key success factors of R744 compressors for the next decade. http://www.ammonia21.com/assets/link/ixetic_specification_displacement.pdf

Pettersen J et al (1998) Development of compact heat exchangers for CO_2 air-conditioning systems. Int J Refrig 21(3):180–193

Pettersen J, Nekså P (2003) Consequences of the newest improvements in R744 systems. In: Automotive alternate refrigeration systems symposium. Scottsdale

Pitla SS, Groll EA, Ramadhyani S (2002) New correlation to predict the heat transfer coefficient during in-tube cooling of turbulent supercritical CO. Int J Refrig 25(7):887–895

Poolman C (2018) Modulating two phase ejectors in transport refrigeration. In: Gustav Lorentzen natural working fluid conference. Valencia

Preparation of a DIN specification series for CO_2 air-conditioning systems. https://www.vda.de/en/topics/safety-and-standards/standardization/preparation-of-a-din-specification-series-for-CO2-air-conditioning-systems.html

Raiser H, et al (2006) Transient behavior of R744 vehicle refrigeration cycles and the influence of the suction side accumulator design. In: SAE 2006 world congress & exhibition. Detroit

Raiser H, et al (2006) Transient behavior of R744 vehicle refrigeration cycles and the influence of the suction side accumulator design. doi:https://doi.org/10.4271/2006-01-0162

S G, M R, MA W (2013) Flow pattern map, heat transfer and pressure drops during evaporation of R-1234ze(E) and R134a in a horizontal, circular smooth tube: experiments and assessment of predictive methods. Int J Refrig, 36(2):478–491

Süβ J, Kruse H (1998) Efficiency of the indicated process of CO_2 compressors. Int J Refrig 21(3):194–201

Tamura T, Yakumaru Y, Nishiwaki F (2005) Experimental study on automotive cooling and heating air conditioning system using CO_2 as a refrigerant. Int J Refrig 28(8):1302–1307

Carrier Transicold (2013). https://www.carrier.com/container-refrigeration/en/worldwide/products/Container-Units/NaturaLINE/

Trygstad E (2017) R744 HVAC unit for NSB Flirt trains. Norwegian University of Science and Technology Department of Energy and Process Engineering, pp 58–93

Wang D (2018a) Heating performance evaluation of a CO_2 heat pump system for an electrical vehicle at cold ambient temperature. Appl Therm Eng 142:656–664

Wang D et al (2018) Heating performance characteristics of CO_2 heat pump system for electrical vehicle in a cold climate. Int J Refrig 85:27–41

Wang D (2018b) Experimental and theoretical study on the cooling performance of a CO_2 mobile air conditioning system. Energies, 11(8)

Wieschollek F, Heckt DR (2007) Improved efficiency for small cars with R744. VDA alternate refrigerant winter meeting. http://www.r744.com/files/pdf_109.pdf

Wolf F (2008) R744 system update. http://www.ammonia21.com/assets/link/P_20081020_Shanghai_OE_Nov.pdf

Wolf F, et al (2007) Emission reductions. http://www.r744.com/files/pdf_328.pdf

Yang T (2006) Dynamic simulation of the transcritical carbon dioxide automotive air conditioning system and research on the lubricant oil. Shanghai Jiaotong University, Shanghai

Yang WW, Fartaj A, Ting SK (2005) CO_2 automotive A/C system optimum high pressure control. SAE technical paper

Chapter 10
Industrial Cooling Systems

Lei Wang and Xin-Rong Zhang

Nomenclature

q_0	Specific refrigerating effect, kJ
h	Enthalpy value, kJ/kg
w_0	Specific compression work, kJ
COP	Coefficient of performance
t_0	Evaporating temperature, °C
t_k	Cooler outlet temperature, °C
p_k	Cooling pressure, bar
T_{atm}	Ambient temperature, °C
Δe	Specific exergy loss, kJ
s	Entropy value, kJ/(kg °C)
η_{ex}	Exergy efficiency

Subscripts

a, b, c, d, e, f	Status points
comp	Compressor

L. Wang · X.-R. Zhang (✉)
Department of Energy and Resources Engineering, College of Engineering, Peking University, Beijing 100871, China
e-mail: xrzhang@pku.edu.cn

L. Wang
Department of Refrigeration and Cryogenic Engineering, Faculty Environment and Life, Beijing University of Technology, Beijing 100124, China

X.-R. Zhang
Beijing Engineering Research Center of City Heat, Beijing 100871, China

X.-R. Zhang and T. M. Eikevik (eds.), *CO_2 Refrigeration Cycle and Systems*,
Lecture Notes in Energy 96, https://doi.org/10.1007/978-3-031-22512-3_10

gc	Gas cooler
val	Valve
evap	Evaporator

The development of industry has promoted the progress of civilization. The process of refrigeration and cooling in the industry is an indispensable part. With the development and progress of industrial technology, refrigeration and cooling technology accompanied by it is also advancing with the times. This chapter introduces the characteristics of refrigeration and cooling needs in the industry, and the thermodynamic characteristics of refrigeration systems in the industry. The chapter also presents examples of CO_2 cooling thermodynamic cycles and system applications in four application areas: food industry cooling, ice making in ice rink, power generation with refrigeration, and transport refrigeration.

10.1 Introduction

Industrial cooling generally refers to the use of physical methods for cooling or precise temperature regulation of industrial processes, both process requirements for temperature maintenance and the production process safety take the range of sizes as 10 kW to 10 MW of refrigerating effect, at cooling temperatures ranging from −50 to 20 °C. Industry as a pillar of national economic development is a multidisciplinary, multi-species, long history of the complex. Industrial cooling system and agriculture, light industry, textiles, food, materials, construction, transportation, energy and national defense and other sectors have close links, the service object has been and will continue to penetrate into all areas of the national and international economy.

Industrial cooling systems have some special characteristics compared to residential or commercial, for several reasons.

(1) The industrial production process involves a lot of dangerous materials.
Most of the raw materials and semi-finished products used in the production process in industrial production are flammable, explosive, toxic and corrosive chemically dangerous products. In some underground mineral mining industries, the mining process will produce many complex compositions of toxic, flammable, explosive gases. And so on and so forth industrial production processes will be more demanding requirements for industrial cooling systems.

(2) Industrial production requires strict process conditions
Some processes in industry require high temperature and pressure, some are to be carried out at low temperature, high vacuum. For example, the quenching process of steelmaking need to be heated to a certain temperature above the critical point (Grade 45 steel quenching temperature of 840−860 °C, carbon tool steel quenching temperature of 760−780 °C), maintain a certain time, and then cooled at an appropriate rate in water or oil to obtain martensite or bainite

organization. During the production of polyethylene in the refining process, the light diesel fuel is first cracked at 800 °C, and the cracked gas is separated under deep cooling conditions at −96 °C, and the ethylene gas obtained from the separation is polymerized under a pressure of 294 MPa.

(3) Large-scale production
With the double growth of productivity and consumption power in the international community, industrial processes have also evolved over the past decades to become larger and more concentrated. In China, for example, the crude steel production of the Chinese steel industry grew from 158,000 tons in 1949 to 928 million tons in 2018, and today China has 425 blast furnaces of 1,000 m^3 and above (of which 22 are of 3,000 m^3 class, 16 of 4,000 m^3 class, and over 7 of 5,000 m^3) and 403 converters of 100 tons and above (of which 300 tons and above 14). The use of large installations can significantly reduce the construction investment and production costs per unit of product, which is conducive to improving labor productivity. Therefore, various countries are actively developing large production units in industrial production.

(4) Production methods are becoming increasingly complex and advanced
Modern industrial production methods have been changed from the past manual operation, intermittent production to a high degree of automation, continuous production; production equipment from open to closed; production equipment from indoor to open air; production operations from decentralized control to centralized control, but also from manual operation to the development of computer AI control.

Due to these characteristics of industrial systems, the cooling systems serving industrial processes are subject to additional requirements, including precise temperature control, safety and environmental restrictions, and a greater degree of adjustability and multifunctional compatibility while meeting the diverse cooling needs of high loads. In this chapter, CO_2 cooling systems used in industrial systems are introduced, including not only single refrigeration systems but also complex systems with multifunctional cooling and power generation coupled with CO_2 cooling system cycles to maximize energy utilization. Theoretical approaches such as thermodynamics are used to analyze the CO_2 industrial refrigeration system cycle. The chapter also presents examples of CO_2 cooling system applications in four application areas: food industry cooling, ice making in ice rink, power generation with refrigeration, and transport refrigeration, as well as the role of the CO_2 refrigeration cycle in avoiding fossil fuel energy consumption and protecting the environment.

10.2 Thermodynamic Analysis

The CO_2 refrigeration system in industrial cooling is mainly divided into direct and indirect according to the form of cooling. The direct type refers to the effect of CO_2 in the evaporator directly absorbing the heat released in the industrial process to achieve cooling, as shown in Fig. 10.1. The indirect type refers to the CO_2 refrigeration system through secondary refrigerant system, CO_2 in the evaporator to absorb the heat of the secondary refrigerant. The low temperature secondary refrigerant enters the industrial production system for cooling, as shown in Fig. 10.2. The condensation method of the CO_2 refrigeration system in the heat sink can be air-cooled, water-cooled, evaporative cooling, or other mechanically assisted cooling methods.

In terms of refrigeration principles, CO_2 refrigeration systems in industrial cooling systems are no different from CO_2 refrigeration systems used in other fields. The energy performance indicators for CO_2 refrigeration units in industrial cooling systems can still use COP. The values of COP represent the amount of cold produced per unit of energy consumed by a CO_2 refrigeration unit in a given operating condition. However, these two metrics are based on the conservation of energy as a starting point, equating different qualities and quantities of energy. The energy use and output of CO_2 refrigeration units in industrial processes are more complex, and considering the energy match of energy use, we can also analyze the process of CO_2 refrigeration cycle by using the exergy analysis method, considering both quantity and quality, so as to reflect the real state of energy consumption, and closely correlate the quality and quantity of energy with industrial profits.

The subcritical refrigeration cycle of CO_2 is exactly the same as the normal vapor compression refrigeration cycle and will not be discussed here. The P-h diagram of the transcritical refrigeration cycle of CO_2 is shown in Fig. 10.3.

Unit cooling capacity of the refrigeration system is

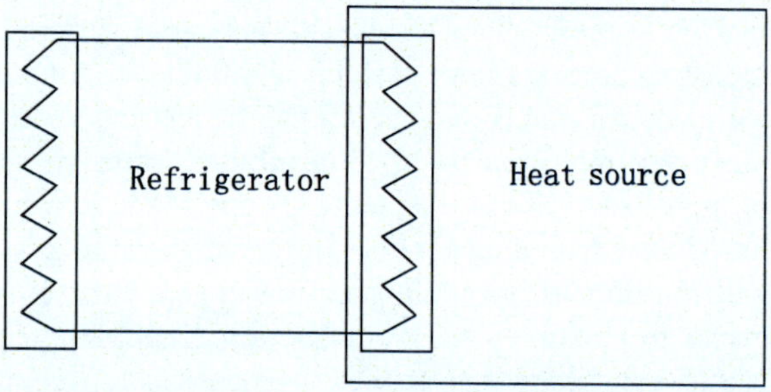

Fig. 10.1 Schematic diagram of direct cooling system

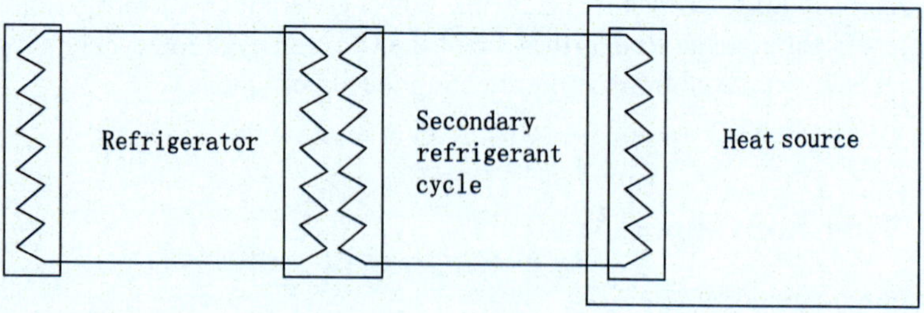

Fig. 10.2 Schematic diagram of indirect cooling system

Fig. 10.3 P-h diagram of a transcritical refrigeration cycle for CO_2

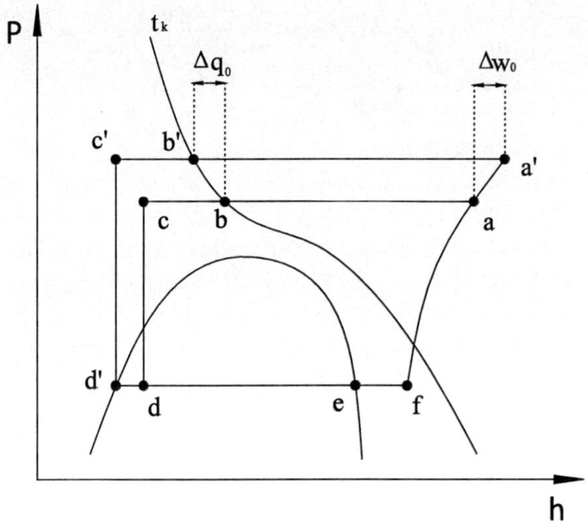

$$q_0 = h_e - h_d = h_e - h_c \tag{10.1}$$

If the heat regenerative cycle is used, the

$$h_f - h_e = h_b - h_c \tag{10.2}$$

Then

$$q_0 = h_f - h_b \tag{10.3}$$

The compression work per unit mass is

$$w_0 = h_a - h_f \tag{10.4}$$

System coefficient of performance is

$$COP = \frac{q_0}{w_0} = \frac{h_f - h_b}{h_a - h_f} \tag{10.5}$$

Assuming that the compression process is an ideal reversible adiabatic process, when the ambient conditions are constant, i.e., the evaporating temperature t_0 and the condenser outlet temperature t_k are constant, if the condensing pressure p_k changes, the point a moves on the isentropic line and the point b is certain on the isothermal line, the unit cooling capacity and the compression work per unit mass of the cycle change, and when the condensing pressure increases Δp_k, the cooling capacity increases Δq_0 and the compression work increases Δw_0. The increment of the performance coefficient is

$$\Delta COP = \frac{q_0 + \Delta q_0}{w_0 + \Delta w_0} - \frac{q_0}{w_0} = \frac{w_0 \Delta q_0 - q_0 \Delta w_0}{\Delta w_0 (w_0 + \Delta w_0)} \tag{10.6}$$

When $w_0 \Delta q_0 - q_0 \Delta w_0 > 0$, i.e., $\frac{\Delta q_0}{\Delta w_0} > \frac{q_0}{w_0}$, $\Delta COP > 0$ and the performance factor increases.

Greenfield et al. showed that for a transcritical CO_2 refrigeration cycle, increasing the evaporation temperature of the cycle can effectively improve the COP, especially at condensation pressure above a certain value, and using the heat regenerative method to increase the suction superheat will significantly improve the system performance at low cooling pressures p_k, while using the reheat method at high condensation pressures p_k has little significance in improving the performance coefficient of the system. is not significant. In addition, there exists an optimal condensation pressure p_k for a certain condenser outlet temperature t_k to maximize the COP (Greenfield et al. 1999).

The temperature and exergy diagram of the transcritical refrigeration cycle of CO_2 is shown in Fig. 10.4, where the cyclic process of CO_2 is in the clockwise direction. Assuming that there is no heat transfer temperature difference between the heat transfer processes in the heat exchanger and the heat sink of condenser/gas cooler at the ambient temperature T_{atm}, it is easy to calculate the exergy loss of each process of the cycle (Srinivasan et al. 2003).

Exergy loss in compression process is

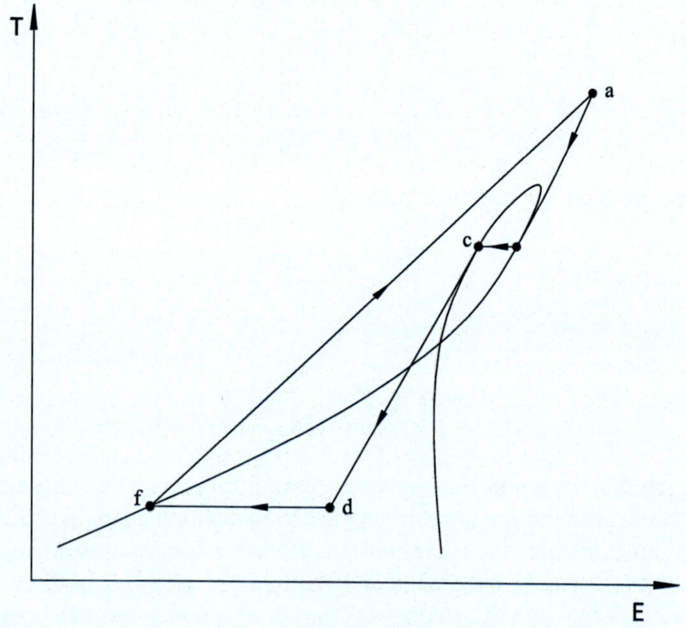

Fig. 10.4 T-E diagram of the transcritical refrigeration cycle of CO_2

$$\Delta e_{comp} = T_{atm}(s_a - s_f) \tag{10.7}$$

Exergy loss in condenser/gas cooler is

$$\Delta e_{gc} = h_a - h_c - T_{atm}(s_a - s_c) \tag{10.8}$$

Exergy loss in throttling is

$$\Delta e_{val} = T_{atm}(s_c - s_d) \tag{10.9}$$

Exergy loss in evaporator is

$$\Delta e_{evap} = \frac{T_{atm}(h_d - h_f)}{T_f} - T_{atm}(s_d - s_f) \tag{10.10}$$

Total system exergy loss is

$$\Delta e = \Delta e_{comp} + \Delta e_{gc} + \Delta e_{val} + \Delta e_{evap} \tag{10.11}$$

Refrigeration cycle exergy efficiency is

$$\eta_{ex} = 1 - \frac{\Delta e}{h_a - h_f - T_{atm}(s_a - s_f)} \tag{10.12}$$

Studies have shown that the exergy loss in the compressor, gas cooler, evaporator and expansion valve is relatively high, while the exergy loss in the internal heat exchanger is small. The exergy loss in the compression process, the throttling pressure loss in the inlet and outlet valves, and the heat flow loss due to the ambient heat transfer are the basic causes of the exergy loss in the compressor, although the latter two can be neglected by making the heat flow loss less pronounced through proper selection of the temperature difference in the surrounding environment. The results of the study on the relationship between compressor isentropic efficiency and exergy loss show that a 10% increase in the isentropic efficiency of the compression process can improve the efficiency of the exergy by about 3% (Sarkar et al. 2005). The exergy loss in the gas cooler and evaporator comes from the fluid temperature difference, pressure loss and flow imbalance in the exchanger, as well as the heat transfer with the surrounding environment. Appropriate increase in heat exchanger size can improve COP and exergy efficiency. In addition, the use of an expander instead of a throttle valve can effectively reduce the exergy loss, using an expander with 60% isentropic efficiency can make the exergy loss of 35% of the entire thermal cycle, the same conditions, the use of an expander with 85% isentropic efficiency can improve the system COP and exergy efficiency by about 22%, which is particularly suitable for industrial cooling systems where large installations are commonly used.

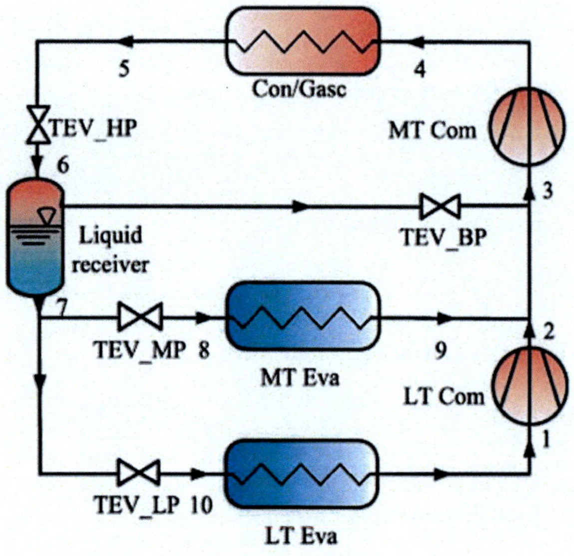

Fig. 10.5 Flow chart of a CO_2 transcritical two-stage compression refrigeration system (Sun et al. 2020)

CO_2 refrigeration compression cycle in industrial processes may face problems of large span of evaporation and condensation temperatures, large compression ratio, large throttling losses, high power consumption in the compression process, and multiple evaporation temperature requirements. To meet these challenges, a two-stage compression refrigeration cycle can reduce system throttling losses, and improve system performance coefficients. Figure 10.5 shows a CO_2 transcritical two-stage compression refrigeration system flow chart.

Replacing the throttle valve with an expander in the refrigeration cycle is the fundamental way to reduce the throttling loss of the system, recover the expansion work, and improve the COP and exergy efficiency. The system flow of CO_2 transcritical single-stage compression with expander is shown in Fig. 10.6.

The ejector is a throttling mechanism that converts the expansion energy of CO_2 into kinetic energy, and then converts the kinetic energy into the pressure energy of CO_2. The use of the expander effectively reduces the capacity requirements of the compressor and lowers the compression ratio and compression energy consumption. Figure 10.7 shows the flow diagram of a CO_2 transcritical refrigeration cycle with an ejector.

In order to solve the problem of multiple evaporation temperature demand, Zhang et al. proposed to adopt a new multi-temperature complex system as shown in Fig. 10.8, which effectively utilizes the heat transfer law and temperature gradient change in the circulation process to achieve a higher thermal level and thermal economy while meeting multiple evaporation temperature demands (Zhu et al. 2021).

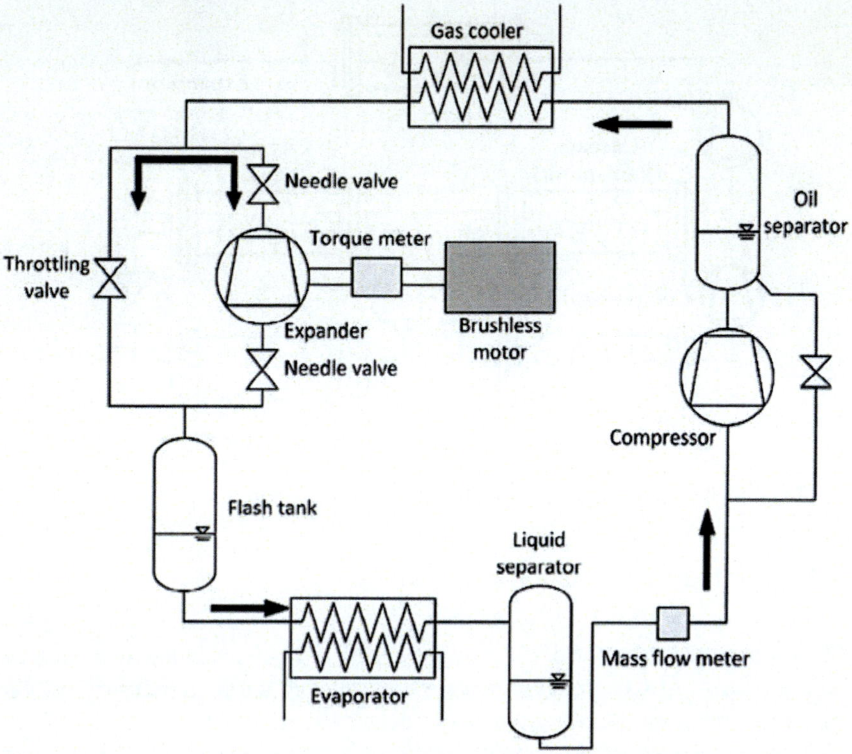

Fig. 10.6 System flow diagram of CO_2 transcritical single-stage compression with expander (Ferrara et al. 2016)

Fig. 10.7 Flow diagram of a CO_2 transcritical refrigeration cycle with an ejector (Fangtian and Yitai 2011)

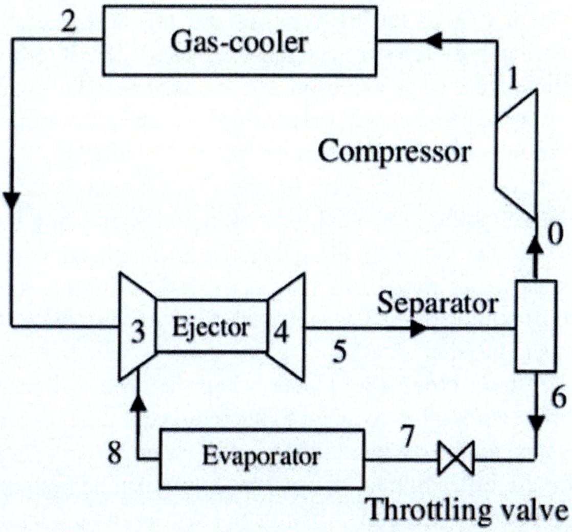

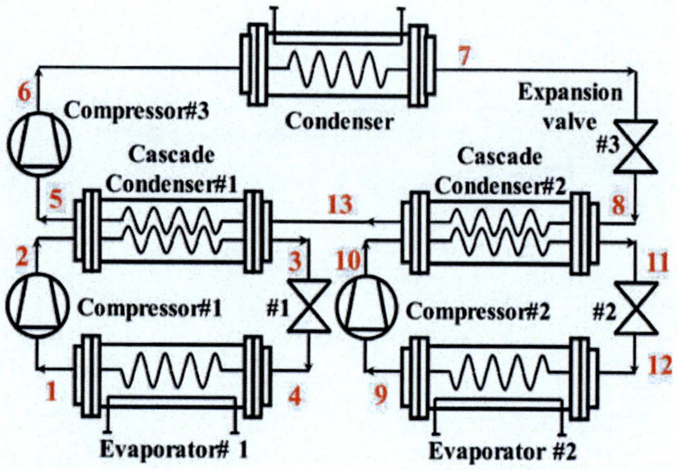

Fig. 10.8 A multi-evaporation temperature CO_2 refrigeration system

10.3 Food Industry Cooling

CO_2 is considered to be one of the most promising refrigerants in the food industry. The basic forms of CO_2 refrigeration system cycles in the food industry include: cascade system, transcritical system, and ejector system.

The basic CO_2 cascade refrigeration system structure is shown in Fig. 10.9 (Bansal 2012). Most commonly used in the food industry, especially in some warm climate regions, this type of system is suitable for more conventional components and is very stable in operation. Two different types of refrigerants are used in the cascade system, CO_2 as the low temperature stage refrigerant, CO_2 cycle is responsible for cooling, high temperature stage can use R717, R134a, R404, R1270, etc. The high temperature stage is responsible for absorbing the condensation heat of CO_2 cycle.

It is well known that the cooling problem of the motor in vapor compression refrigeration systems has been one of the factors affecting the performance of refrigeration systems. In order to solve the motor cooling problem of R717/R744 cascade refrigeration system, Liu et al. proposed a method using the principle of thermosyphon to cool the motor by the refrigerant in the reservoir of the high temperature stage, as shown in Fig. 10.10 (Liu et al. 2021). By this method, the compressor power consumption of the CO_2 refrigeration cycle is reduced by 5.46% and the exergy loss was reduced by 3.51%.

In food processing plants, when the need to freeze a variety of food products, different food has to adapt to the temperature requirements, so that the refrigeration system for the refrigeration temperature proposed a variety of complex requirements, the use of traditional design means of a single system under only one evaporation temperature, which is difficult to adapt to the production requirements. Saini et al. designed a new cascade system as shown in Fig. 10.11 (Saini et al. 2021). This system has multiple evaporators at both high and low temperature stages, which

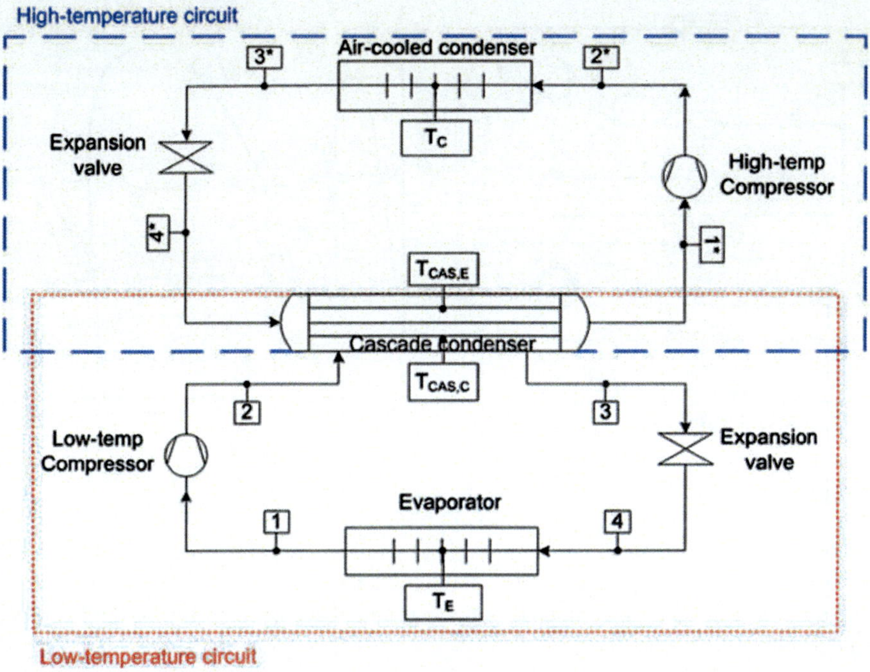

Fig. 10.9 Schematic of a two-stage cascade refrigeration system

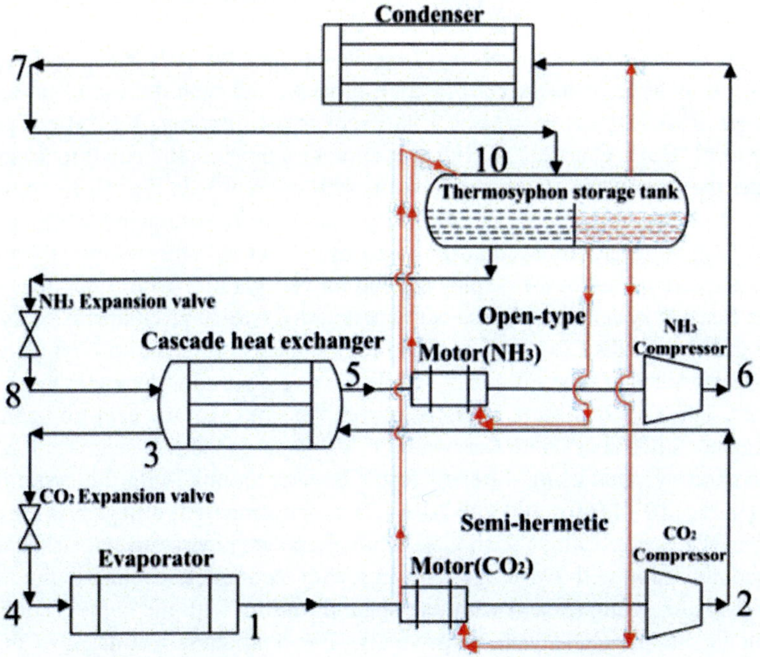

Fig. 10.10 R717/R744 refrigeration system with thermosiphon motor cooling

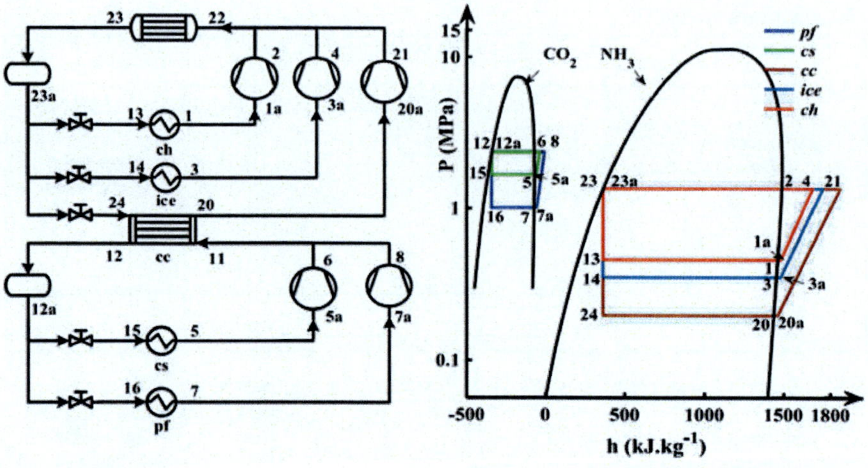

Fig. 10.11 R717/R744 refrigeration system with multiple evaporation temperatures

ensures the diversity of evaporation temperatures. The maximum average annual energy consumption was reduced by 8.3% when processing surimi in the Mumbai (India) climate.

The required refrigeration temperature in the dairy industry processing is not particularly low (above −10 °C), and requires multi-temperature cooling and simultaneous heating, Dasi et al. designed a CO_2 ejector refrigeration system with a dual evaporator, as shown in Fig. 10.12 (Dasi et al. 2020). The use of a transcritical cycle is acceptable when the temperature difference between the heat source and the heat sink is not as large as in the refrigeration process, and with the use of an ejector, a portion of the expansion work can be recovered to improve the system performance level (Gullo et al. 2017). Higher evaporation temperature conditions can also consider the use of two-stage compression with intercooling. Fig. 10.13 shows the two-stage compression two-stage throttling intermediate incomplete cooling transcritical CO_2 refrigeration system proposed by Sun et al. This system is generally used in supermarkets for the display of food for refrigeration (Sun et al. 2021).

The cascade system can also be combined with a two-stage compression system, which can change the CO_2 cycle in the low temperature stage into subcriticality and improve the refrigeration efficiency, while lower evaporation temperatures can be obtained. This type of system is relatively complex, and Mosaffa et al. compared the performance differences between two R717/R744 cascaded two-stage compression refrigeration systems using different forms of intercooling, with the system form shown in Fig. 10.14 (Mosaffa et al. 2016). The high temperature stage was the same, with the R717 cycle using a flash tank, while the low temperature stage used a two-stage compression with two-stage throttling with incomplete cooling in the middle and a two-stage compression with one-stage throttling with incomplete cooling in the middle for the CO_2 cycle, respectively. The final study results show that the performance level and the efficiency of the exergy are almost the same.

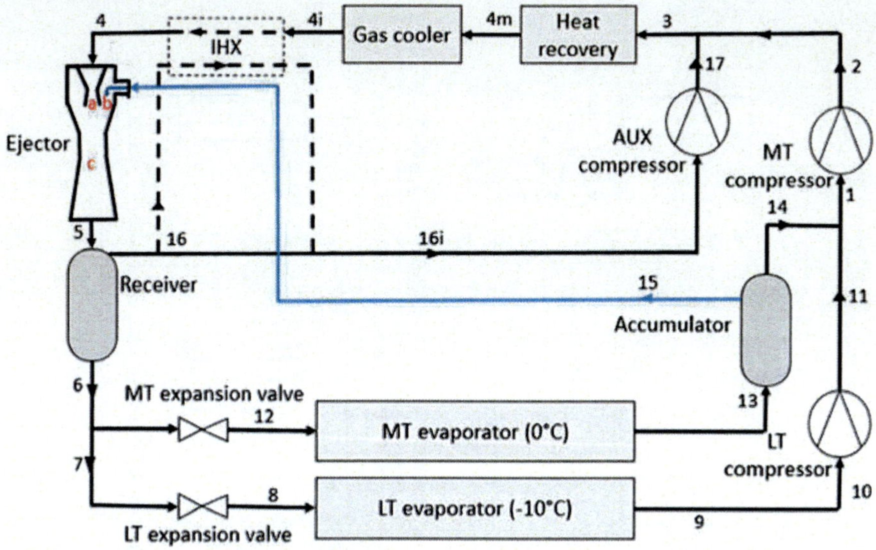

Fig. 10.12 Refrigeration system of R744 with injector

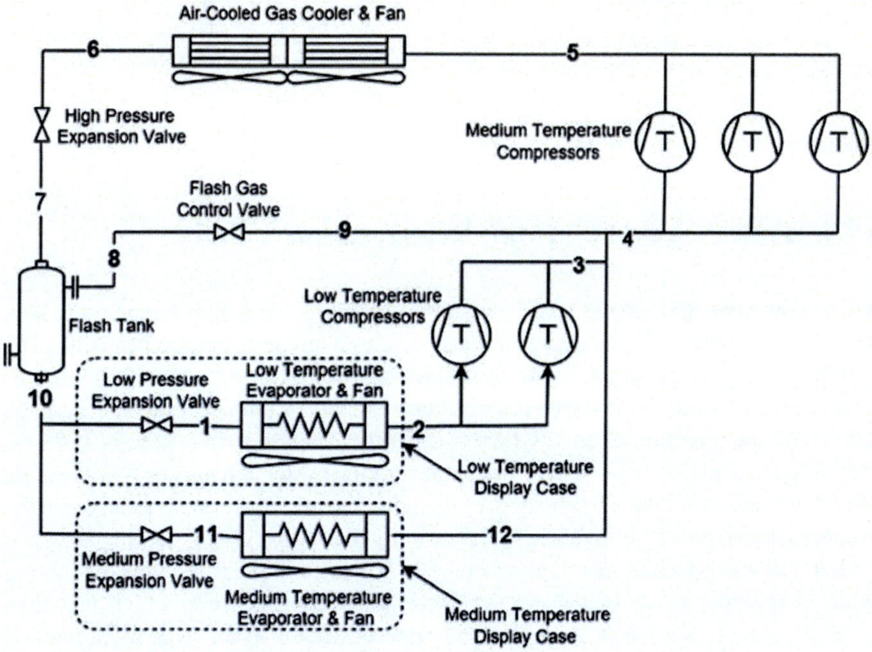

Fig. 10.13 Two-stage compression transcritical CO_2 refrigeration system

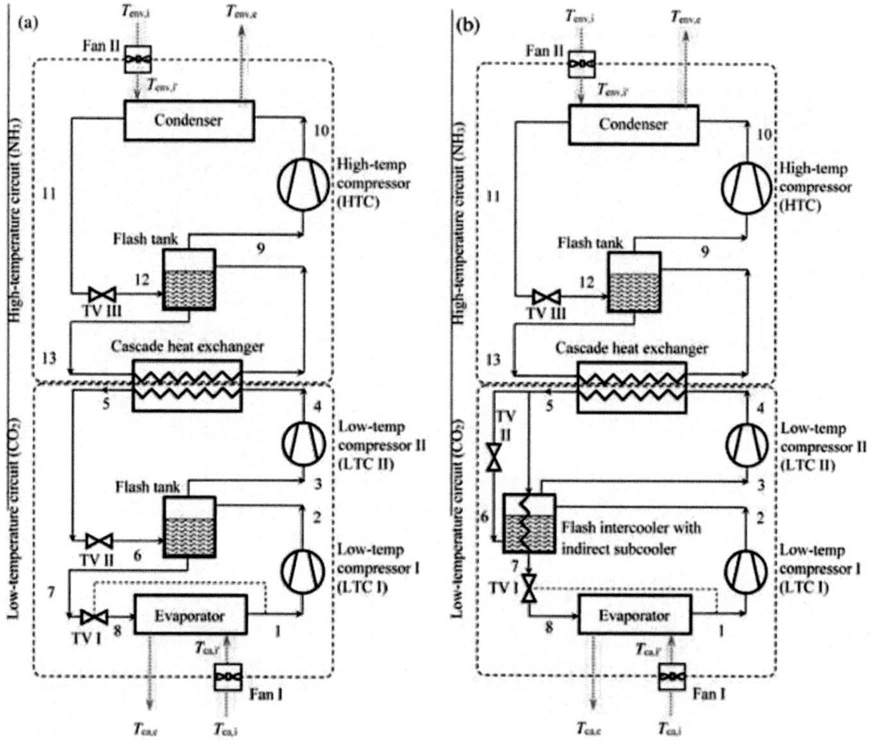

Fig. 10.14 Two different R717/R744 cascade two-stage compression refrigeration systems

10.4 Power Generation with Refrigeration

The power generation cycle is the absorption of heat from a high temperature heat source to do work and the release of waste heat to a low temperature heat sink. The refrigeration cycle consumes work to absorb heat from the heat source and releases heat to the heat sink. So it seems that the key elements of both cycle systems are the same only the direction of energy flow is different. Therefore, in industrial systems, the refrigeration system serves the power generation system on the one hand, and on the other hand, it can be a by-product of the power generation system to improve the comprehensive energy utilization of the power generation system.

Early power generation systems used the CO_2 vapor compression cycle for cooling while recovering waste heat from the power generation process, which was then used to generate hot water at higher temperatures. Zhang et al. proposed a method to recover low temperature heat from the exhaust water vapor of fossil fuel-fired thermal power plants using the CO_2 vapor compression thermodynamic cycle which is shown in Fig. 10.15, and experimental results showed that the COP of this system form could reach 5.0 (Zhang and Zhang 2013).

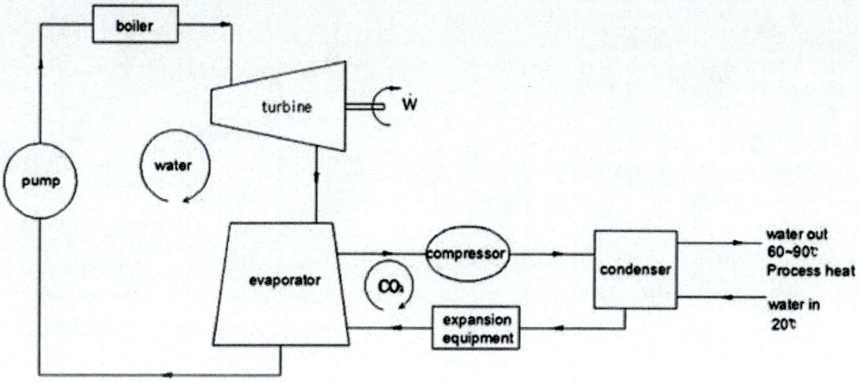

Fig. 10.15 Flow diagram of fossil fuel thermal power plant with heat recovery by CO2 vapor compression heat pump cycle

It has been proposed that the difference in energy grade between the power generation cycle and refrigeration cycle can be cleverly exploited to constitute an integrated system that has both power output and can provide refrigeration, as shown in Fig. 10.16 (Sun et al. 2021). SUN et al. used a CO_2 mixture as the work mass to combine the power cycle and the refrigeration cycle through a condenser, and the two cycles can exchange heat and mass in the condenser, and the combined system relies on heat drive to generate electricity. The CO_2 mixture of the work mass after power generation and the discharge of the compressor of the refrigeration cycle can realize the condensation process in the condenser relying on the ambient temperature and the temperature difference between the two fluids, and the driving force of the refrigeration compressor relies entirely on the power generation of the power cycle. It is shown that this type of system can improve the performance of the system by adjusting the ratio of the CO_2 in the power cycle and the refrigeration cycle by performing CO_2 splitting in the condenser, and the power generation capacity of the system after splitting adjustment is improved by 5.18% compared to the conventional system.

To meet the diversified needs of power and cooling, Zhang et al. investigated a CO_2-based multi-mode combined cooling and power cycle as shown in Fig. 10.17, which can achieve full power mode, simultaneous power and cooling mode, and full cooling mode. The improved system shows a significant improvement in performance and potential for diversified energy supply through energy and exergy analysis. In particular, the improved single-stage compression system has a 4.9% increase in power output and a 21.7% increase in cooling output compared to the base system under refrigeration conditions (Zhang et al. 2020).

The temperature of the SCO_2 power system is still high after power generation, and direct cooling causes a waste of heat. From the perspective of energy utilization, Mishra et al. proposed a combined solar SCO_2 power generation absorption refrigeration system as shown in Fig. 10.18 (Mishra and Singh 2018). This system is based on the supercritical CO_2 solar power generation cycle and uses the waste

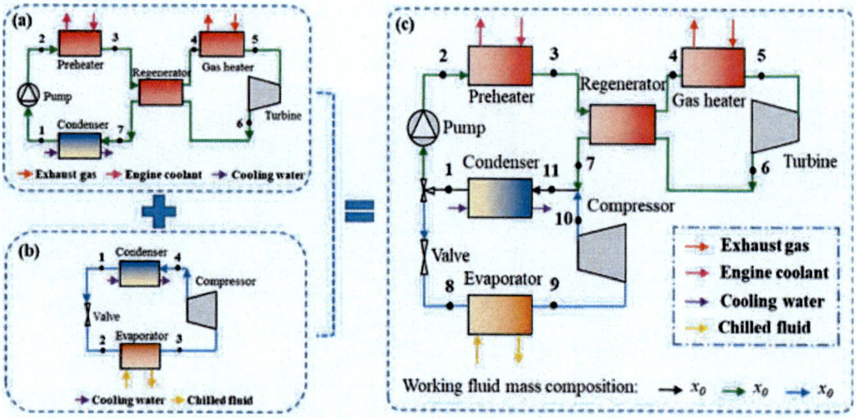

Fig. 10.16 Combined cooling and power cycle

heat from the supercritical CO_2 power generation to drive an absorption refrigeration system, and the cooling capacity generated by the absorption refrigeration system is in turn used to cool the CO_2 in the power generation system, which greatly reduces the capacity requirement of the cooling system in the cooling process of the conventional SCO_2 power generation system and improves the system's efficiency in terms of the exergy. Wang et al. also proposed a supercritical CO_2 cooling, heating and power triple-supply system based on the same idea (Wang et al. 2020).

Another form of system for solar electric cooling and heating co-generation, where the system is entirely CO_2-based, is shown in Fig. 10.19 (Lykas et al. 2022). At the same time, the system generates cooling capacity to meet the cooling needs of production, useful heat to meet the heating needs, and electricity to meet the consumption of the refrigeration cycle, which is sold to the national grid. The CO_2 in the system participates in the Brayton cycle for power generation, and the CO_2 after power generation participates in the throttling cooling, and the heat dissipation in the intermediate process uses waste heat recovery, which perfectly solves the problem of cooling and waste heat utilization in the power cycle. The annual energy efficiency of the system is 67.8% and the annual exergy efficiency is 10.1%. It is very suitable for use in energy and power systems.

10.5 Transport Refrigeration

Transport refrigeration is the process of creating a low temperature environment during the transportation of goods so as to ensure that the quality of the goods is changed as little as possible during the transportation process. Long-term refrigerated transport in the industry is reflected in the ocean-going fishing boats for keeping the freshness of the seafood, fishing boats operate for a few days to several months, and the preservation of caught seafood is entirely dependent on the refrigeration

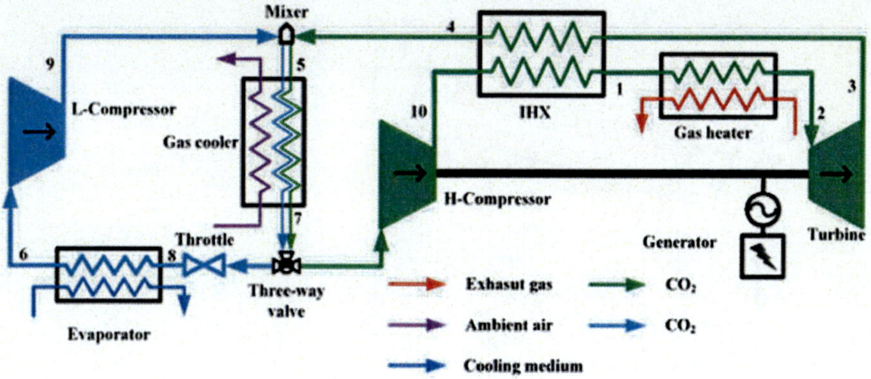

(a) Schematic diagram of the cycle

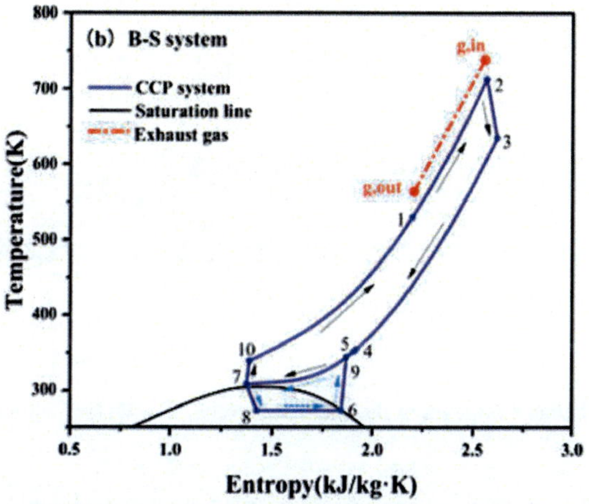

(b) System T-S diagram

Fig. 10.17 CO_2-based multi-mode combined cooling and power cycle

equipment on board. Figure 10.20 shows the frozen tuna out of the warehouse in the fishing boat.

The refrigerated transportation of seafood by fishing boats using CO_2 refrigeration technology can adopt indirect refrigeration and direct refrigeration methods. Indirect refrigeration can use CO_2 ice making equipment to produce a large amount of ice before going to sea and keep it on board, generally, one ton of seafood needs to store one ton of ice for refrigeration, or use flake ice making machine to make ice ready to use. Another method is to use the system as in the principle of Fig. 10.21, using CO_2 refrigeration device to freeze seawater, and then the seafood will be continuously immersed in the frozen seawater to achieve the purpose of storage. Direct refrigeration generally used plate freezer, the temperature of −30 °C~−35 °C, and large fish with shelf type freezer, as shown in Fig. 10.22.

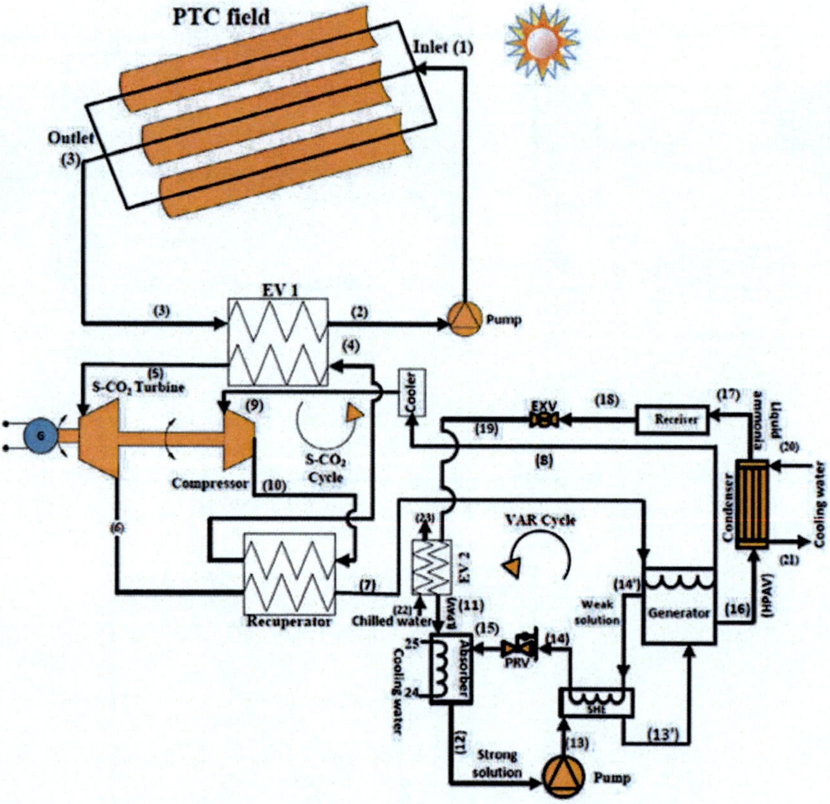

Fig. 10.18 Schematic diagram of the solar-powered SCO_2-VAR combined cycle

In the land to adapt to a long-time refrigerated transport method to take refrigerated container transport. Refrigerated container structure is the CO_2 mechanical refrigeration device installed in one end of the standard container, chillers are installed inside the container. During transportation, the power supply of the vehicle is connected to the container refrigeration unit, and the container can be refrigerated for a long time. Figure 10.23 shows a refrigerated container.

In refrigerated transport, in addition to the use of mechanical refrigeration mode, another important way of modern refrigerated transport is to use consumable phase change materials. Dry ice is solid CO_2, dry ice under atmospheric pressure will be directly sublimated to gas, while absorbing a large amount of heat, and can achieve a constant low temperature environment of −70 °C. There is no residue after sublimation of dry ice into gas, non-toxic, odorless, and certain sterilization effect. In refrigerated transport, especially in low temperature air transport, dry ice refrigeration not only saves space, but also eliminates the trouble of replacing the materials. It is commonly used for storage and transportation of quick-frozen foods, biological drugs, living tissue cells, biological enzymes, and pharmaceuticals, etc. The working environment of mechanical refrigeration during transportation is harsh, the technical

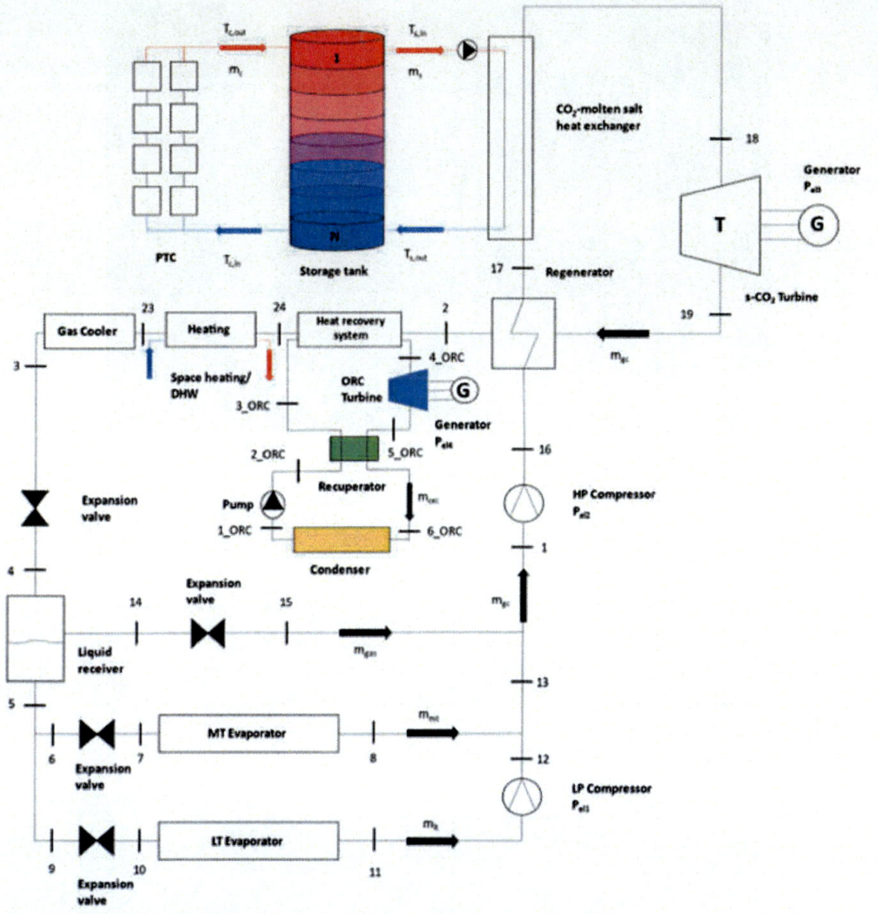

Fig. 10.19 CO_2 multi-generation system

requirements for the refrigerator are high, the refrigeration effect is unstable, and it is difficult to achieve a lower refrigeration temperature. And the cost of dry ice compared to liquid nitrogen is lower, so the use of dry ice for cryogenic transport materials is the most ideal choice.

Figure 10.24 shows a dry ice refrigerated truck. When the dry ice refrigeration truck works, the air is exchanged with dry ice first, and then the cooled air is circulated and cooled in the cabin with the help of a ventilator, and the CO_2 after heat absorption and sublimation is discharged out of the car through the exhaust pipe. In the 1960s, dry ice refrigerated trains were used in Japan, with dry ice scaffolding installed on the upper side wall of the refrigerated cargo, which can be filled with 500–800 kg of dry ice for refrigerated heat absorption. For some smaller-scale refrigerated transport, needs can be carried out using a dry ice holding tank as shown in Fig. 10.25.

Fig. 10.20 Refrigerated transport of fishing boat

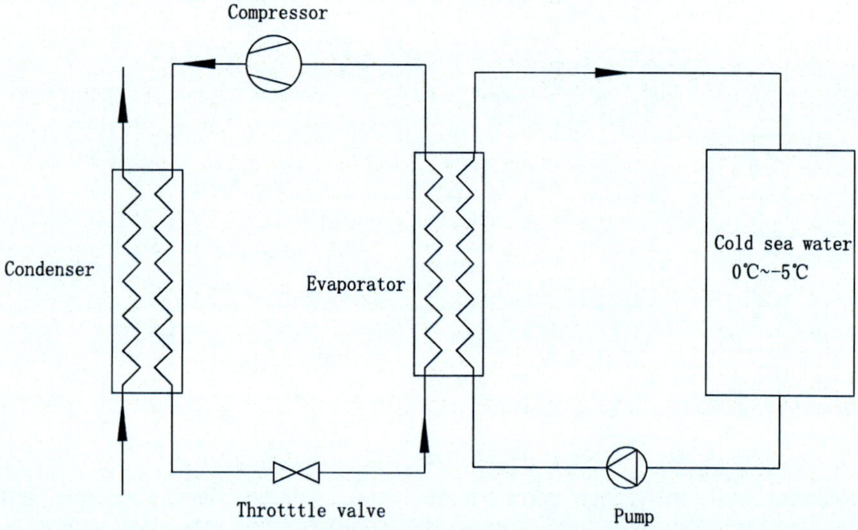

Fig. 10.21 Frozen seawater preservation

Then the use of liquid CO_2 instead of dry ice can also be used to achieve the refrigeration effect during transportation. The American Frozen Food Institute and the International Cold Storage Association Refrigeration Foundation jointly organized a research group in 1981, the research group used insulated car shipment of frozen food, each time before the start according to the amount of cold needed, filled with the appropriate amount of liquid CO_2. After more than one year of six consecutive tests, the use of liquid CO_2 as a cold source is considered better than liquid nitrogen because of the low investment and better versatility of liquid CO_2. In 1982, the U.S. company GE began to produce a new type of liquid CO_2 reefer, with a

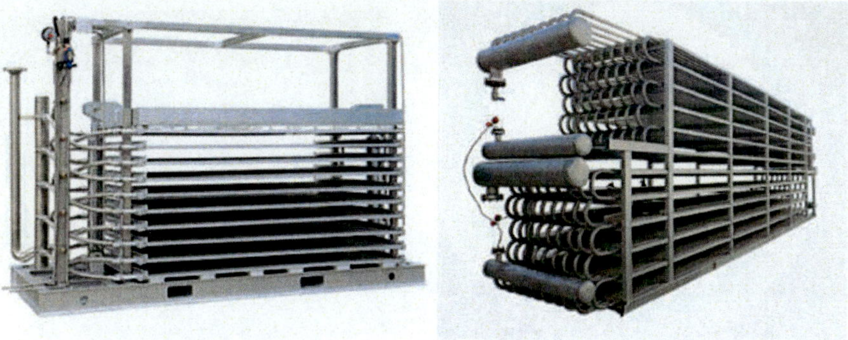

Fig. 10.22 Plate and shelf freezer

Fig. 10.23 Refrigerated container

Fig. 10.24 Dry ice refrigerated truck

Fig. 10.25 Dry ice incubator

storage container under the floor of the car. In addition to CO_2 injection on the cargo before departure, CO_2 injection can be added to the cargo during transport, triggered by a thermostatically controlled sensor.

10.6 Conclusions

This chapter introduces the principles and methods of applying CO_2 refrigeration technology in industrial cooling processes. In industrial cooling, CO_2 has emerged as a refrigerant in the food industry, industrial power generation, and transport refrigeration. The food industry is relatively more mature and widespread in the application of CO_2 refrigeration, but it is still possible to expand the use of CO_2 in the refrigeration process, and the combination of industrial power generation and refrigeration technology will be of greater value in the next new energy revolution. In the field of refrigerated transportation, as the quality of life continues to improve, the short time required for cold storage and preservation will lead to more economical and convenient CO_2 cooling becoming more common.

Since G. Lorentzen, the former president of the International Institute of Refrigeration, proposed the use of CO_2 as an environmentally friendly refrigerant and the theory of transcritical cycle, the advantages of CO_2 in terms of environmental protection and performance have increasingly attracted the attention of scholars around the world. Although CO_2 as a refrigerant is constrained by its thermodynamic properties, and the construction of thermal systems and thermal devices is more difficult than traditional refrigerants, with the development of new technologies and theories that break through the performance and use boundaries of CO_2 refrigeration systems, there will be more and more scenarios in which CO_2 refrigeration technology will be used in industrial processes. This is very important for the protection of the global environment and the sustainable development of energy systems.

In the global process of carbon neutrality, the reduction of CO_2 emissions from industrial processes and the resource utilization of CO_2 will become the direction of attention, which will lead to new opportunities for the expansion of CO_2 refrigeration methods and applications, and how more maturely apply CO_2 refrigeration technology in more segments of industrial cooling and improve the cost effectiveness of CO_2 refrigeration will also be of greater interest.

Acknowledgment The support of the National Key Research and Development Program of China (2018YFD0901002) is gratefully acknowledged.

References

Bansal P (2012) A review–Status of CO2 as a low temperature refrigerant: Fundamentals and R&D opportunities. Appl Therm Eng 41:18–29

Dasi K, Singh S, Guruchethan AM, Maiya MP, Hafner A, Banasiak K, Neksa P (2020) Performance evaluation of ejector based CO_2 system for simultaneous heating and cooling application in an Indian dairy industry. Therm Sci Eng Prog 20:100626

Fangtian S, Yitai M (2011) Thermodynamic analysis of transcritical CO_2 refrigeration cycle with an ejector. Appl Therm Eng 31(6–7):1184–1189

Ferrara G, Ferrari L, Fiaschi D, Galoppi G, Karellas S, Secchi R, Tempesti D (2016) Energy recovery by means of a radial piston expander in a CO_2 refrigeration system. Int J Refrig 72:147–155

Greenfield ML, Mozurkewich G, Schneider WF, Bramos GD, Zietlow DC (1999) Thermodynamic and cycle models for a low-pressure Cycle CO_2 refrigeration cycle. SAE Trans. 1622-1631

Gullo P, Tsamos K, Hafner A, Ge Y, Tassou SA (2017) State-of-the-art technologies for transcritical R744 refrigeration systems–a theoretical assessment of energy advantages for European food retail industry. Energy Procedia 123:46–53

Liu L, Yang Q, Wu J, Li L, Zhang Y, Zhang X (2021) Thermodynamic analysis of NH3/CO_2 cascade refrigeration system with thermosyphon refrigerant cooling screw compressor motor. Int J Refrig 130:1–13

Lykas P, Georgousis N, Bellos E, Tzivanidis C (2022) Investigation and optimization of a CO_2-based polygeneration unit for supermarkets. Appl Energyss 311:118717

Mishra RS, Singh H (2018) Detailed parametric analysis of solar driven supercritical CO_2 based combined cycle for power generation, cooling and heating effect by vapor absorption refrigeration as a bottoming cycle. Therm Sci Eng Prog 8:397–410

Mosaffa AH, Farshi LG, Ferreira CI, Rosen MA (2016) Exergoeconomic and environmental analyses of CO_2/NH3 cascade refrigeration systems equipped with different types of flash tank intercoolers. Energy Convers Manag 117:442–453

Saini SK, Dasgupta MS, Widell KN, Bhattacharyya S (2021) Comparative analysis of a few novel multi-evaporator CO_2-NH3 cascade refrigeration system for seafood processing & storage. Int J Refrig 131:817–825

Sarkar J, Bhattacharyya S, Gopal MR (2005) Transcritical CO_2 heat pump systems: exergy analysis including heat transfer and fluid flow effects. Energy Convers Manag 46(13–14):2053–2067

Srinivasan K, Lim YK, Ho JC, Wijeysundera NE (2003) Exergetic analysis of carbon dioxide vapour compression refrigeration cycle using the new fundamental equation of state. Energy Convers Manag 44(20):3267–3278

Sun J, Kuruganti T, Munk J, Dong J, Cui B (2021) Low global warming potential (GWP) refrigerant supermarket refrigeration system modeling and its application. Int J Refrig 126:195–209

Sun X, Shi L, Tian H, Wang X, Zhang Y, Shu G (2021) A novel composition tunable combined cooling and power cycle using CO_2-based binary zeotropic mixture. Energy Convers Manag 244:114419

Sun Z, Wang C, Liang Y, Sun H, Liu S, Dai B (2020) Theoretical study on a novel CO_2 Two-stage compression refrigeration system with parallel compression and solar absorption partial cascade refrigeration system. Energy Convers Manag 204:112278

Wang S, Liu C, Li J, Sun Z, Chen X, Wang X (2020) Exergoeconomic analysis of a novel trigeneration system containing supercritical CO_2 Brayton cycle, organic Rankine cycle and absorption refrigeration cycle for gas turbine waste heat recovery. Energy Convers Manag 221:113064

Zhang XR, Zhang Y (2013) Experimental investigation on heat recovery from condensation of thermal power plant exhaust steam by a CO_2 vapor compression cycle. Int J Energy Res 37(14):1908–1916

Zhang Y, Shu G, Tian H, Shi L, Sun X (2020) Modified CO_2-based combined cooling and power cycle with multi-mode and adjustable ability. Energy Convers Manag 226:113485

Zhu YD, Peng ZR, Wang GB, Zhang XR (2021) Thermodynamic analysis of a novel multi-target-temperature cascade cycle for refrigeration. Energy Convers Manag 243:114380

Chapter 11
CO_2 Trans-Triple-Point Refrigeration Method

Qiu-Yun Zheng and Xin-Rong Zhang

This Chapter presents a new cryogenic refrigeration method using CO_2. This refrigeration is achieved by micro CO_2 solid particle sublimation, not by CO_2 liquid evaporation. In this chapter, new cryogenic CO_2 refrigeration cycles will be introduced here, which achieves below −56.6 °C refrigeration by using CO_2 solid-gas sublimation process. Basic flow dynamic and heat transfer of micro CO_2 particle sublimation are also presented.

11.1 Introduction

CO_2 is an old and natural refrigerant during the first decades of the twentieth century, it was widely used in marine systems. But this natural refrigerant was replaced by 'safe refrigerants', CFCs and HCFCs, from the 1930s in most applications. With the CFCs and HCFCs problems becoming pressing issues in the late 1980s, CO_2 as a refrigerant/working fluid was increased considerably throughout the 1990s (Kim et al. 2004). CO_2 is a non-flammable natural fluid with no Ozone Depletion Potential (ODP = 0) and a negligible Global Warming Potential (GWP = 1). On the other hand, CO_2 is also mainly responsible for the greenhouse effect. So if CO_2 is recycled and used as a refrigerant, this will become a good way for relieving the greenhouse effect.

The CO_2 thermodynamic and transport properties seem to be good at heat transfer and pressure drop, compared to other refrigerants (Kim et al. 2004; Zhang and

X.-R. Zhang (✉)
Department of Energy and Resources Engineering, College of Engineering, Peking University, Beijing 100871, China
e-mail: xrzhang@pku.edu.cn

Q.-Y. Zheng · X.-R. Zhang
Beijing Engineering Research Center of City Heat, Beijing 100871, China

X.-R. Zhang and T. M. Eikevik (eds.), *CO_2 Refrigeration Cycle and Systems*, Lecture Notes in Energy 96, https://doi.org/10.1007/978-3-031-22512-3_11

Yamaguchi, 2007), where the critical pressure and temperature are 7.38 MPa and 31.1 °C, respectively. Because of the advantages above, CO_2 fluid has received much attention in the trans-critical compression refrigeration thermodynamics cycle of air conditioners and heat pumps. In general, in these processes, using CO_2 as a working fluid, the refrigeration temperature range is about from −30 to 0 °C. From 2005, Yamaguchi and Zhang et al. found a new refrigeration method by using CO_2 as the working fluid (Yamaguchi and Zhang 2009; Yamaguchi et al. 2008; Yamaguchi et al. 2005; Zhang and Yamaguchi 2011). This new refrigeration method using CO_2 as a working fluid can achieve a cryogenic temperature below CO_2 triple point temperature −56.6 °C. In the next sections, CO_2 as a working fluid will be discussed with trans-triple-point refrigeration methods, the micro-particle sublimation flow dynamics, refrigeration thermodynamic cycles, and below −56.6 °C refrigeration methods.

11.2 Trans-Triple-Point Refrigeration Method

The temperature of the CO_2 triple point is −56.6 °C, the pressure is 0.518 MPa. If CO_2 wants to achieve tans-triple-point refrigeration, a solid–gas two-phase flow of CO_2 is needed. As shown in Fig. 11.1 (Ilchi-Ghazaani and Parvin 2011), phase change along and below the sublimation line (between the sublimation point and triple point) will directly happen from solid states to gas states. If CO_2 is in liquid state below −20 °C, the solid–gas two-phase flow will be generated if properly controlled below the triple-phase point by fast-expanding liquid CO_2. This feature of CO_2 fluid allows the possibility of a new cryogenic operation or refrigeration below the triple-phase point. Figure 11.2 shows a schematic diagram of this CO_2 refrigeration principle, in which the refrigeration is achieved by liquid CO_2 expanding into a solid–gas two-phase fluid. The process of a-b represents the liquid CO_2 expansion into the two-phase flow, the dry-ice-produced region, shown in Fig. 11.2a. In this process, CO_2 goes down through the triple point in the P−h diagram. By the CO_2 expansion process, the CO_2 solid–gas two-phase fluid is obtained, and the temperature is below −56.6 °C. In the b-c process in the CO_2 P–h diagram in Fig. 11.2a, CO_2 solid particles obtained from the a-b process sublimate absorb heat quantity when flowing through a pipe. In this process, CO_2 solid–gas two-phase fluid flows through a pipe, and solid CO_2 sublimates, the temperature can sustain to stay below −56.6 °C, then trans-triple-point refrigeration is obtained. Figure 11.2b is a schematic diagram of CO_2 trans-triple-point refrigeration.

Obviously, heat transfer of CO_2 solid–gas two-phase flow in a channel is very important for the efficiency of the new refrigeration technologies. Especially, heat transfer of CO_2 solid–gas flow with CO_2 particles sublimation is a very important scientific problem in this refrigeration method, and the next section will discuss it.

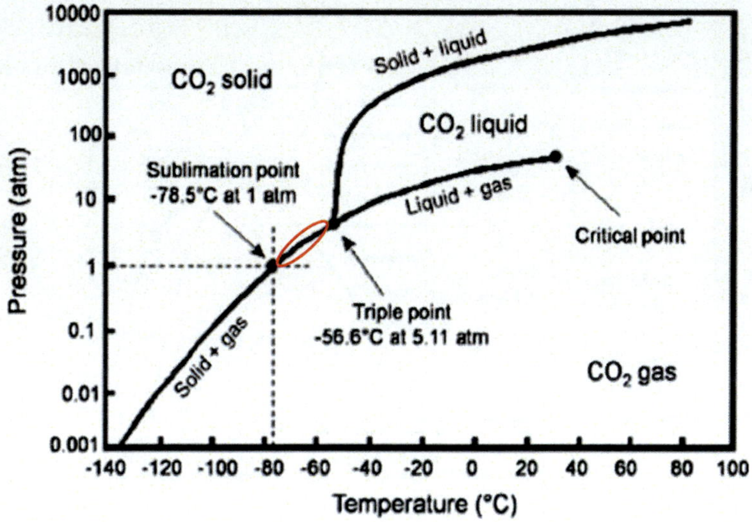

Fig. 11.1 Pressure–temperature phase diagram for CO_2 (Ilchi-Ghazaani and Parvin 2011)

11.3 CO_2 Micro Particle Sublimation Flow Dynamic

The basic dry ice sublimation model can reference Aoki's study (Aoki et al. 2002), where the vapor film and transition boundary are noted. As shown in Fig. 11.3, two different states of dry ice sublimation are shown when dry ice is immersed in different liquid tanks. It can be clearly shown that local non-equilibrium of sublimation particle interface may affect the sublimation rate a lot for different locations in Fig. 11.4a, and detailed radial distribution of parameters can be founded in Fig. 11.4b. In recent years, some studies have considered the modeling of the transient sublimation process of solids, proving a general analytical method for both the sublimation and melting process (Sahin and Dincer 2000). These studies can be advantageous for researching dry ice sublimation flow dynamics.

For the flow dynamics study of dry ice sublimation flow, models are generally based on Navier–Stokes equations and phase coupling methods. At the early stage, studies mainly focused on equilibrium or near-equilibrium conditions and focused on the thermodynamic evolution of the solid–gas interface (Charwat 1965). In recent years, major particle sublimation flow dynamics studies are based on the flow dynamics of multi-phase flow research studies (Bi et al. 2000; Ding et al. 2008; Dong et al. 2008a, 2008b; Subramaniam 2013; Tsuji 2007; Wang et al. 2010). Dry ice particle sublimation flows should include particle deformation and mass transfer, momentum, and energy transportation between phases. Therefore the development of sublimation flow should be dependent on the basic methods about solid–gas two-phase flows. Wang's review article (Wang et al. 2010) shows that multi-scale methods are also indicative for the future development of multi-scale sublimation

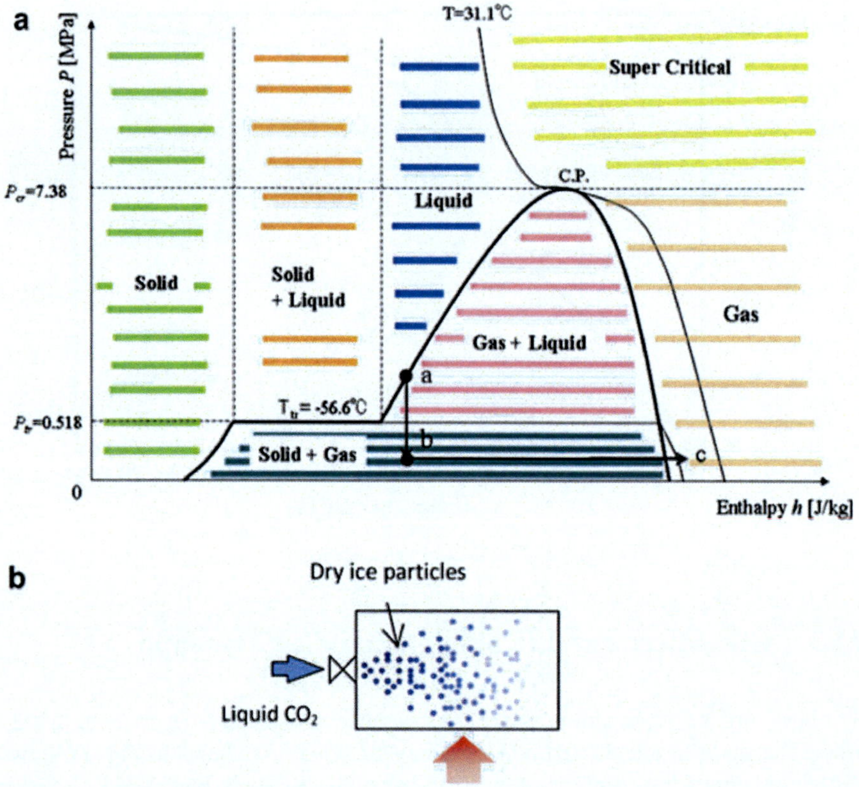

Fig. 11.2 Principle schematic of the new refrigeration method using CO_2 solid–gas two-phase flow. **a** CO_2 P–h diagram. a-b process represents liquid CO_2 expanding into the solid–gas two-phase fluid flow, which goes below the CO_2 triple point of −56.6 C. b-c is the sublimation process of the CO_2 solid–gas fluid. **b** image of the evaporator using CO_2 solid–gas two-phase fluid flow, in which dry ice particles sublimate, absorb heat quantity from the refrigeration objective and therefore achieve refrigeration below −56.6 C (Zhang and Yamaguchi, 2011)

flow analysis. And here we may focus on some benchmark studies in the microparticle sublimation flow field.

In Michaelides's group's work (Michaelides and Lasek 1987, 1991; Michaelides et al. 1992), they set up mathematical models for solid–gas sublimation flows. In their model, they considered the non-equilibrium status of the thermal and mechanical processes in sublimation, respectively. The numerical results showed that the sublimation rate will increase with the increased wall temperature or decrease in particle size. They also found that turbulence plays an important role in the phase change process and that the evaporation time is reduced by 20% when the turbulence level is doubled from the normal, that was to say, turbulence enhancement elements may be advantageously used for the improvement of phase change processes. Gan et al. also research one theoretical model for the analysis of single-particle or multiple-particle sublimation evolution under channel gas flow conditions (Gan et al., 2003a;

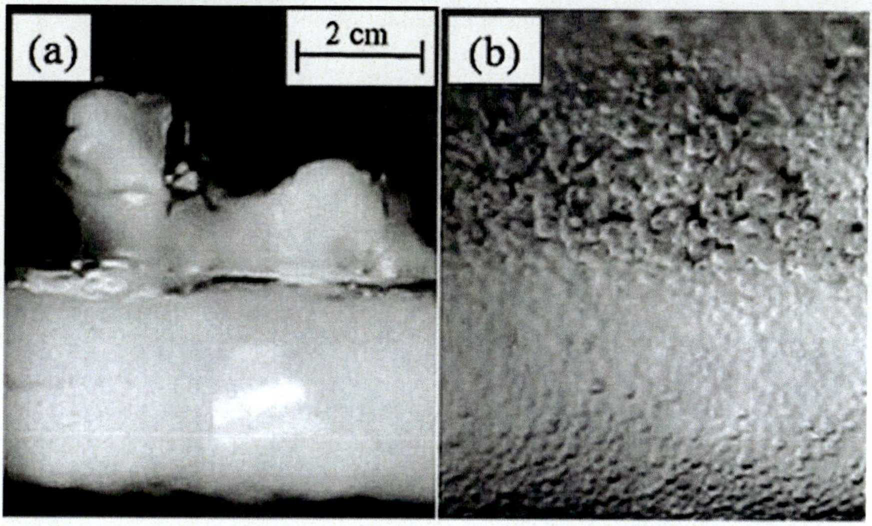

Fig. 11.3 Sublimation patterns of dry ice in liquid. **a** film-state sublimation (water of 25 °C); **b** nucleate-state sublimation (ethanol of 17 °C) (Aoki et al. 2002)

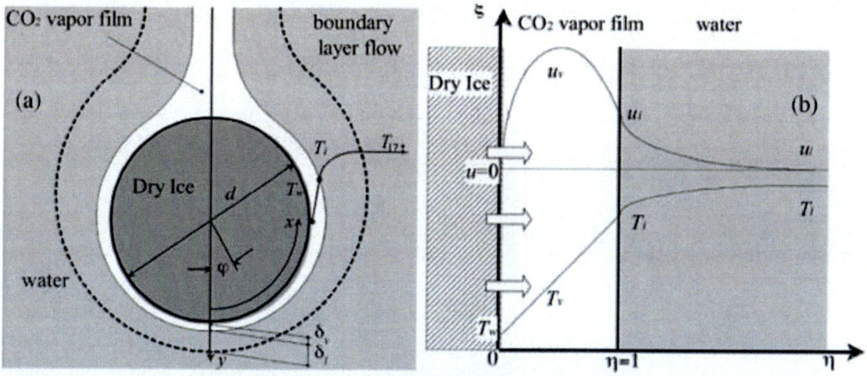

Fig. 11.4 Analytical model for dry ice sublimation: **a** Schematic diagram of dry ice sublimation model; **b** Dry ice sublimation interface (Aoki et al. 2002)

Gan et al., 2003b). The model for typical particle sublimation inside the vertical channel is shown in Fig. 11.5a. Figure 11.5b and a show the streamlines of a particle sublimation and velocity of two particles inside a vertical channel at some parameters, respectively. That study systematically presented the effect of the channel, particle geometry, and Reynolds number, Grashof number effects, and shown that two particles separate from each other at low Grashof number but attract each other at higher Grashof number. In Martynove's works (Martynov et al., 2016), a compressible flow Computational Fluid Dynamics (CFD) model was developed to predict the formation of dry ice during a transient decompression of CO_2 pipelines. Based on the Homogeneous Equilibrium Mixture (HEM) assumption and utilized an extended Peng-Robinson equation of state, the model predicted the physical properties of CO_2 in vapor, liquid, and solid states during the transient phase change at the CO_2 triple point.

Yamaguchi and Yamasaki et al. researched CO_2 micro particle sublimation by experimentation (Yamaguchi et al. 2011; Yamaguchi et al. 2008; Yamasaki et al. 2017; Zhang and Yamaguchi 2011). Figure 11.6 shows a schematic diagram of the experimental set-up designed, constructed, and tested for dry ice sublimation in CO_2 solid–gas two-phase flow. In this experiment, a needle valve is used as an expansion valve. By the expansion valve, the CO_2 liquid fluid expands, and solid-phase CO_2 is produced by the Joule–Thomson effect. Thus in the visualization section, dry ice particle sublimation can be visualized in solid–gas two-phase flow. In Fig. 11.1.7 (Yamaguchi et al., 2008), the photographs of solid–gas two-phase fluids are shown. By using a high-speed video camera (Photron Firstcam), the particle density, particle velocity, and particle behaviors are measured. In Fig. 11.7a, the particle size is approximately the same, the diameters of most particles are about 1.0 mm, and the mean particle size is measured to be 1.023 mm. But when the flow velocity decreases, it is observed that there is a sedimentation phenomenon, and also large particles are seen in Fig. 11.7b. Zhang and Yamaguchi also experimentally study the heat transfer of CO_2 solid–gas two-phase flow with dry ice particle sublimation (Zhang

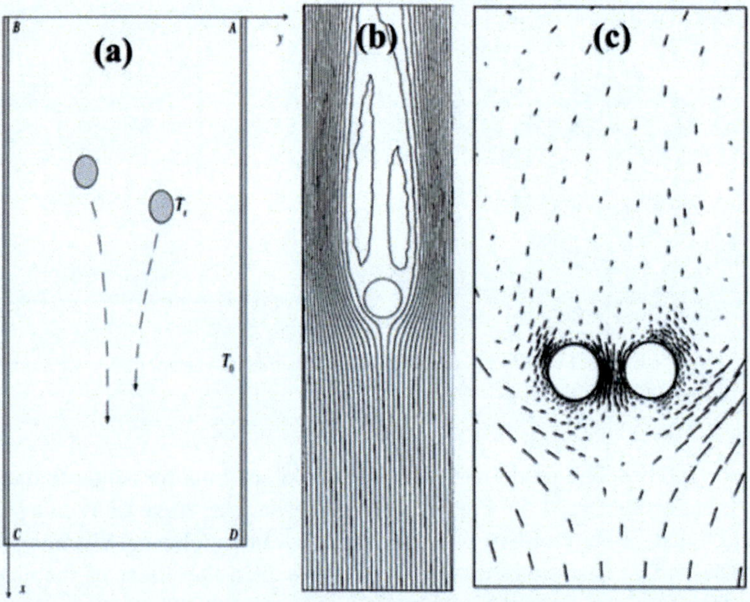

Fig. 11.5 Schematic model of particle sublimation flow configurations. **a** Basic model design; **b** One cold particle streamlines; and **c** Two particles magnified views of the velocity field (Gan et al. 2003a; Gan et al. 2003b)

and Yamaguchi, 2011), in this study, the two-phase fluid temperature can be continuously below the triple point temperature (−56.6 °C) by the particle sublimation. Yamasaki et al. give another method to visual CO_2 dry-ice behavior in a sublimator, just like Fig. 11.8. For the sake of solving the blockage problem of dry-ice particles and promoting good heat transfer characteristics for dry-ice solid–gas two-phase flow, they compared two channels with the swirl promoter and one without the swirl promoter. Since the swirl flow induced by the swirl promoter increases the momentum of dry-ice particles toward the radial direction of the pipe, the volume concentration of dry-ice particles at the inlet of the visualization section should be increased after the expanding channel, compared with the case without the swirl promoter. As the volume concentration of dry-ice particles increases, the heat absorbed is strongly increased by particle sublimation. They also find that the generation process of dry-ice balls induces the decrease of the heat absorbed of CO_2 solid–gas two-phase flow in the channel.

However, from general dry ice particle sublimation to heat transfer calculations, the size effect, the interaction between the solid particle and solid wall, between solid particle and gas, among solid particles, and flow instabilities during transient mass diffusion and transportation from particle to gas, has raised new critical problems in the dry ice particle sublimation flows. The method of numerical calculation and experimental study these problems are under discussion.

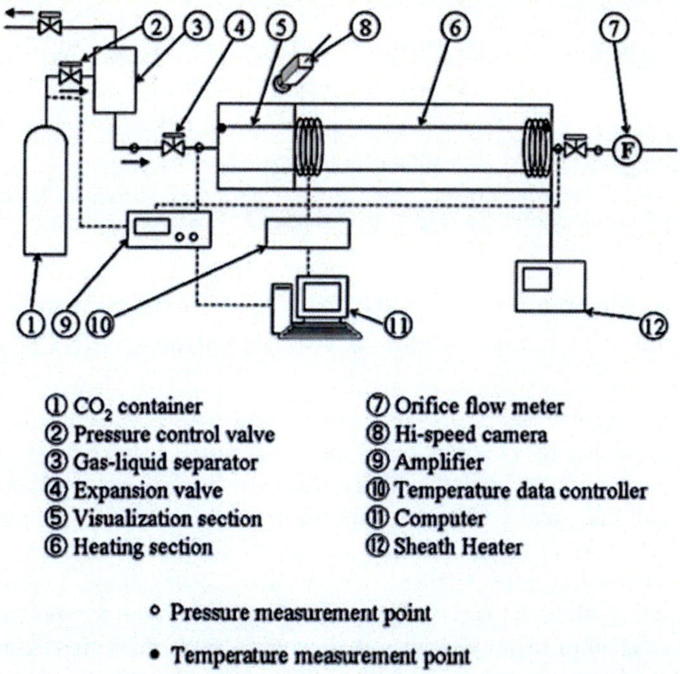

Fig. 11.6 Schematic of the experimental set-up made to investigate the feasibility of liquid CO_2 expanding into solid–gas two-phase fluid flow by an expansion valve (Yamaguchi et al. 2008)

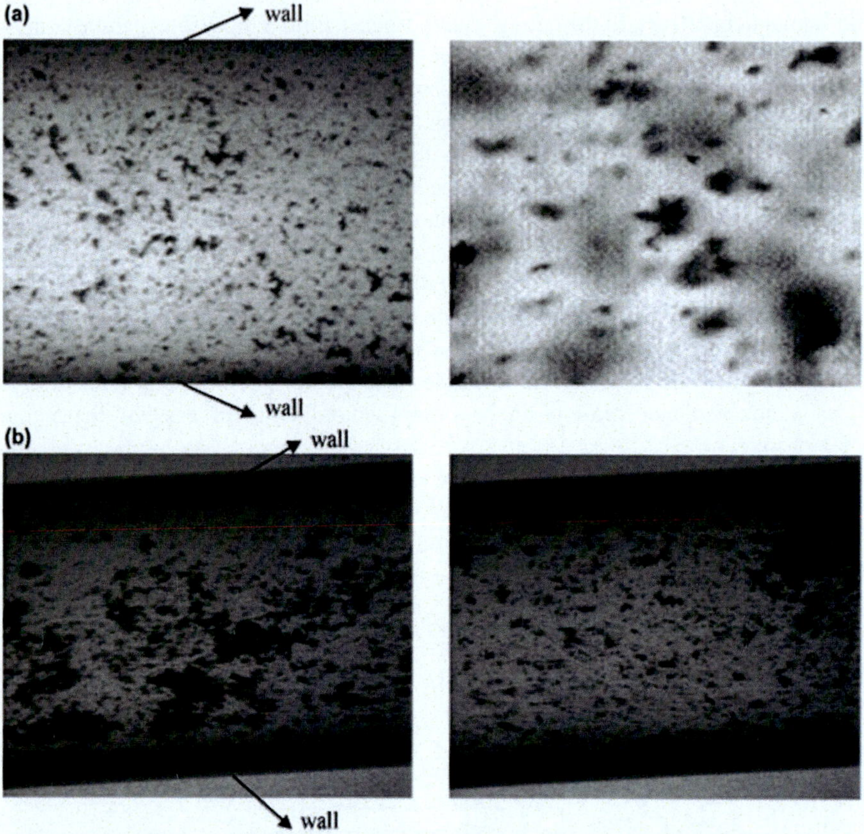

Fig. 11.7 Photographs of CO_2 solid–gas two-phase fluid flow in the visualization section. CO_2 solid–gas flow at a CO_2 flow velocity measured at **a** 2.50 m/s and **b** 0.78 m/s. The left photograph is given at 4500 fps and the right photograph is at 13,500 fps (Yamaguchi et al. 2008)

11.4 New Cryogenic CO_2 Refrigeration Thermodynamic Cycle and Under −56.6 °C Refrigeration Using CO_2

In the new cryogenic CO_2 refrigeration system, the cryogenic temperature can be below −56.6 °C. In Fig. 11.2, it is found that if the system can continuously obtain CO_2 solid–gas two-phase fluid, the dry ice particles will continue to sublimation in this two-phase fluid, and cryogenic temperature below −56.6 °C will also obtain continuously. Figure 11.9 is a schematic diagram of the basic thermodynamic cycle for new cryogenic CO_2 refrigeration (Yamaguchi and Zhang 2009). In this thermodynamic cycle system, the high-pressure cycle (HPC) and the low-pressure cycle (LPC) are coupled to each other for CO_2 refrigeration. In this cascaded refrigeration cycle, the high-pressure cycle is like to the traditional CO_2 heat pump cycle, which is based on the CO_2 trans-critical compression refrigeration cycle. The gas

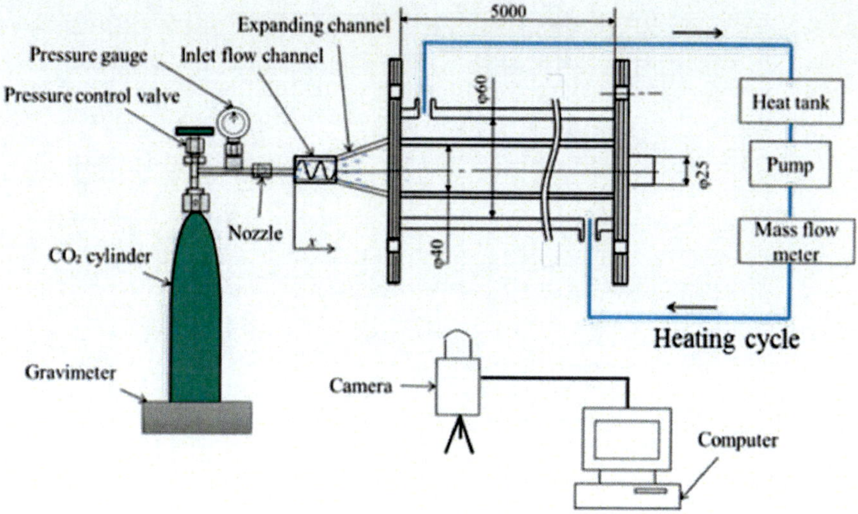

Fig. 11.8 Schematic diagram of visualization test (Yamasaki et al. 2017)

cooling process of g-h and evaporating process of e–f in the high-pressure CO_2 heat pump cycle are respectively above and below the CO_2 critical point. In the compression process of f-g, the low-temperature CO_2 gas is compressed to high-temperature super-critical status, and then in the cooling process of g-h, the high-temperature super-critical CO_2 is cooled to low-temperature super-critical status. In the cooling process, thermal energy can supply to the user, such as hot water reaching above 60 °C. In the isenthalpic expansion process of h-e, supercritical CO_2 is expanded and entered into the liquid–gas two-phase zone, and liquid–gas phase CO_2 is vaporized by absorbing heat in the e–f process, the temperature of about −30 °C can be obtained continuously. In the low-pressure CO_2 refrigeration cycle, the processes of a-b and b-c are based on the refrigeration method described in Fig. 11.2a. Because the temperature of point a, before the liquid CO_2 expanding process, is low, the low-pressure CO_2 refrigeration cycle is coupled with a high-pressure CO_2 heat pump cycle, which is also shown in Fig. 11.9. That is to say, in this cascade CO_2 refrigeration cycle, the low-pressure condensing process d-a is coupled with the high-pressure evaporation process of e–f, which can be obtained by liquid CO_2 about −20 °C continuously in LPC and also provided heat to HPC. In the proposed cascade cycle, In the low-pressure CO_2 refrigeration cycle, the condensing process of d-a and the evaporation process of b-c are respectively above and below the CO_2 triple point, so it is called the trans-triple point refrigeration cycle. In this cycle, the a-b process can obtain solid–gas two-phase flow by expanding, and the dry ice particle is sublimated in the b-c process, so it is expected to have a potential of achieving the refrigeration below −56.6 °C continuously in LPC. Therefore, the proposed CO_2 cascade refrigeration technology, which is described in Fig. 11.9, not only can achieve the refrigeration below −56.6 °C but also supply thermal energy above 60 °C. It is a high-efficient technology for refrigeration and to supply heat at the same time.

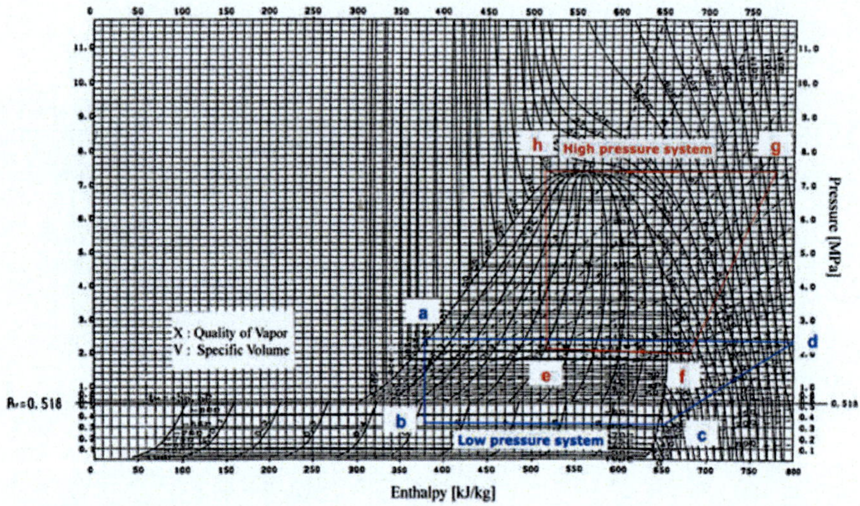

Fig. 11.9 Thermodynamic cycle of the proposed CO_2 cascade refrigeration system, which is shown in CO_2 P–h diagram (Yamaguchi and Zhang 2009)

Figure 11.10 is a typical schematic of the CO_2 cascade refrigeration system, which matches with the refrigeration cycle in Fig. 11.9. Comparing with Fig. 11.9, the high-temperature side is consistent with the high-pressure cycle and the low-temperature side is consistent with the low-pressure cycle, respectively, in Fig. 11.10. In this refrigeration system, a needle valve is used as an expansion valve. Other expansion valves can also be used to expand liquid CO_2 in this refrigeration system. On the high-temperature side, CO_2 is compressed into a high-temperature and high-pressure supercritical fluid. The first condenser is cooled by 50 °C or higher temperature hot water from natural gas cogeneration. The second condenser is cooled with 30 °C cool water from the cooling tower. After these two condensers, the CO_2 fluid is cooled into the liquid state. After that, the CO_2 fluid is expanded by an expansion valve, becomes a relatively low temperature and low-pressure state, and flows into an evaporator. In an evaporator, low-temperature CO_2 fluid can cool brine, the latter can be cooled below −20 °C. On the low-temperature side, relatively high temperature and pressure CO_2 gas, which is obtained by the compressor, is cooled by three condensers. After the gas cooler by brine, the CO_2 fluid becomes a liquid state and the temperature is about −20 before entering the expansion valve. Through the expansion process, the dry ice-gas two-phase fluid is achieved. And in the test section in Fig. 11.10, dry ice particles are sublimated by a heater and changed to CO_2 gas before entering into the compressor. These CO_2 cascade refrigeration systems can finish HPC and LPC in Fig. 11.9 and achieve cooling capacity continuously at the temperature below −56.6 °C.

According to the research results by Yamaguchi and Zhang et al. (Yamaguchi et al., 2011; Yamaguchi and Zhang 2009; Yamaguchi et al. 2008; Zhang and Yamaguchi 2011), COP (Coefficient of Performance) is defined. The low-temperature system COP is defined as a useful power output, cooling capacity here, compared with the compressor work:

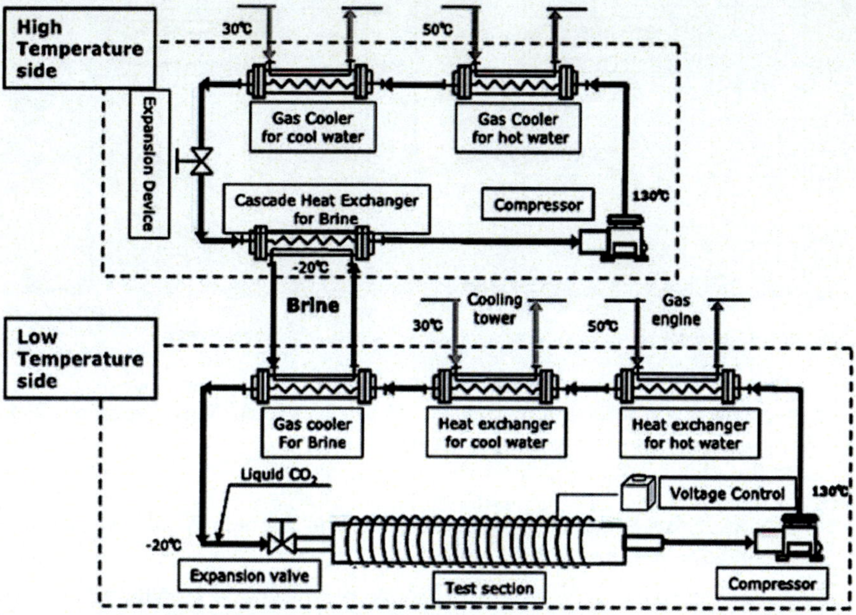

Fig. 11.10 Schematic of the CO_2 cascade refrigeration system (Yamaguchi et al. 2011)

$$COP_{LT} = \frac{\Delta h_{lcool}}{W_l} \tag{11.1}$$

Here COP_{LT} is the coefficient of performance for the low-temperature side in Fig. 11.10, $\Delta h_{lcool,}$ and A_{wl} are the cooling capacity and compressor work on the low-temperature side. For the whole cascade system, if only the cooling capacity on the low-temperature side is considered as a useful output, the COP value is defined as:

$$COP_{SL} = \frac{\Delta h_{lcool}}{W_l + W_h} \tag{11.2}$$

Here *why* is compressor work in the high-temperature side. Yamaguchi and Zhang's research results show that the average COP_{LT} value is about 2.45 and the average COP_{SL} value is about 0.85, the condensation temperature has an obvious influence on system performance, the COP value increases with a decrease of the condensation temperature (Yamaguchi and Zhang 2009). In this new cryogenic CO_2 refrigeration thermodynamic cycle, the refrigeration temperature is not only related to the condensation temperature but also the heating input. In Fig. 11.11, it is shown that the refrigeration temperature decreases with a decrease of condensation temperature and heating input, and all fluid temperature is below −56.6 °C, especially the lowest temperature is about −61.2 °C. In this CO_2-solid–gas ultra-low temperature cascade refrigeration cycle system based on CO_2-solid–gas sublimation, dry ice blockage

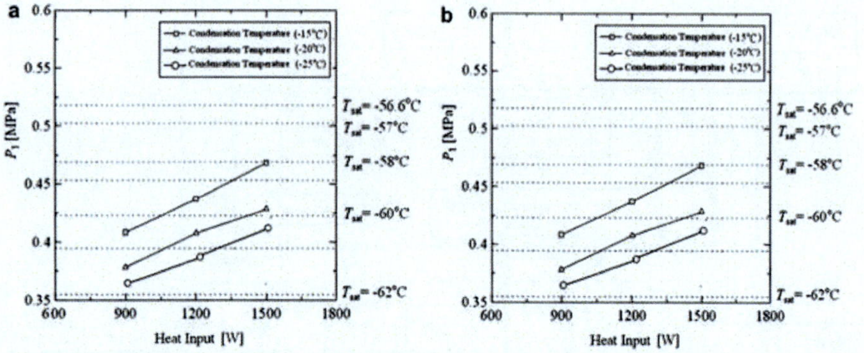

Fig. 11.11 variations of the experimental data of P1 with the heating input under different condensation temperatures: **a** expansion valve opening of 15 mm; **b** expansion valve opening of 10 mm (Yamaguchi et al. 2011; Zhang and Yamaguchi 2011)

is easy to occur. But it can be avoided by increasing the input heat fluxes and the opening condition of the expansion valve.

There is another new refrigeration cycle system using CO_2 vapor–solid as a refrigerant, it also can be a cooling capacity for temperatures lower than triple point temperature of −56.6 °C. Figure 11.12 (Huang et al., 2008) is the schematic diagram of this refrigeration cycle system and its T-s diagram. This refrigeration cycle is based on Rozhentsev et al.'s research studies, where a three-phase mixture of CO_2 is contained in a big container and exchanges heat with the heat source and the heat transfer rate is very small, at only about 0.5–16 W (Dvornitsyn et al. 2006; Rozhentsev 1995). According to Fig. 11.12a, in this refrigeration cycle, the compressor sucks CO_2 gas and compresses it to high pressure firstly. After that, the compressed CO_2 flows into the gas cooler and gets cooled. And then the cooled CO_2 is throttled by the throttling valve, where gas–liquid CO_2 under medium pressure but slightly higher than the triple point pressure is got. Gas–liquid CO_2 is separated by a liquid separator, and saturation CO_2 liquid from the bottom of the separator flows into the adjustable nozzle, where it mixes with saturated CO_2 gas from the compressor suction side and the high-pressure regulating valve outlet from the top of the liquid separator. After the adjustable nozzle, the mixture becomes solid–gas two-phase fluid by the Joule–Thomson effect. The solid–gas two-phase fluid flows into sublimation and the dry ice particle is sublimated, so the temperature under the triple point temperature of −56.6 °Ccan be achieved. In this refrigeration system, the size of the solid particles at the sublimator inlet can be controlled by controlling the adjustable nozzle and it can avoid the blockage of the sublimation pipe. In Fig. 11.12b, it is shown that the CO_2 solid–gas refrigeration system could be 50% higher than that of a conventional CO_2 liquid–gas refrigeration system. The reason is dry ice sublimation latent heat is about 540 kJ/kg which is larger than 350 kJ/kg which is the CO_2 vaporization latent heat at the triple point pressure. But more research studies are needed for the efficiency between this refrigeration cycle and the cascade cycle, which both can achieve the refrigeration temperature under −56.6 °C by using CO_2.

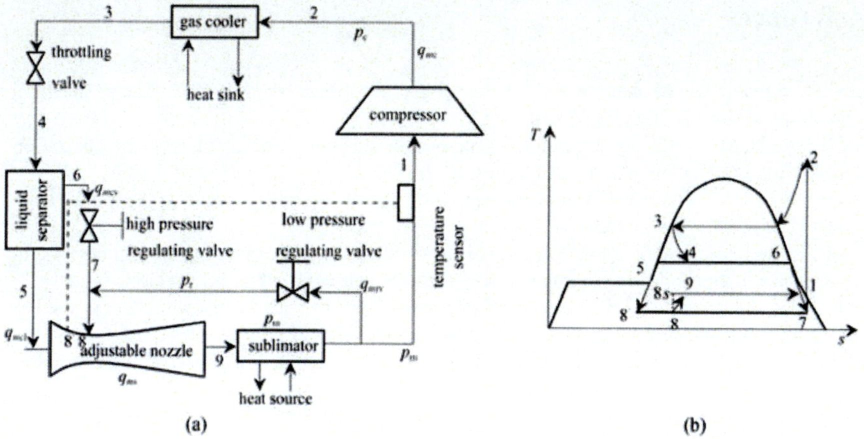

Fig. 11.12 CO_2 vapor–solid refrigeration cycle system and its refrigeration process in T-s diagram **a** CO_2 vapor–solid refrigeration cycle system; **b** refrigeration process in T-s diagram (Huang et al., 2008)

11.5 Conclusions

In this chapter, a new cryogenic refrigeration method using CO_2 is studied, which is achieved by micro CO_2 solid particle sublimation, not by CO_2 liquid evaporation. This new method can achieve a temperature below −56.6 °C. Because of CO_2 solid–gas sublimation process in this new method, basic flow dynamic and heat transfer of micro CO_2 particle sublimation are also presented. With the help of modern techniques, visual experiments and numerical simulation techniques, more detailed information on the dry ice sublimation process can be revealed. Both numerical and experimental progress show that the dry ice particle sublimation system is quite different from the liquid–gas phase change process in the conventional CO_2 refrigeration system. And two typical CO_2 trans-triple-point refrigeration cycle systems are presented in detail. But the multi-scale and multi-disciplinary investigation of flow dynamic and heat transfer for CO_2 solid–gas sublimation process need more research in the future. The related thermo-physical characteristics and various effects on the CO_2 solid–gas sublimation process also need to be considered deeply. More CO_2 trans-triple-point refrigeration cycle systems for ultra-low temperature refrigeration will be designed in the future and their various effects analysis, structural features, applied range, and so on are all very interesting problems in this field for the future.

References

Aoki K, Sawada M, Akahori M (2002) Freezing due to direct contact heat transfer including sublimation. Int J Refrig 25:235–242

Bi HT, Ellis N, Abba IA, Grace JR (2000) A state-of-the-art review of gas-solid turbulent fluidization. Chem Eng Sci 55:4789–4825

Charwat AF (1965) The effect of surface-evaporation kinetics on sublimation near the leading edge. Int J Heat Mass Transf 8:383–394

Ding YL, He YR, Cong NT, Yang W, Chen HS (2008) Hydrodynamics and heat transfer of gas-solid two-phase mixtures flowing through packed beds - a review. Prog Nat Sci 18:1185–1196

Dong WG, Wang W, Li JH (2008a) A multiscale mass transfer model for gas-solid riser flows: Part 1 - Sub-grid model and simple tests. Chem Eng Sci 63:2798–2810

Dong WG, Wang W, Li JH (2008b) A multiscale mass transfer model for gas-solid riser flows: Part II - Sub-grid simulation of ozone decomposition. Chem Eng Sci 63:2811–2823

Dvornitsyn A, Naer V, Rozhentsev A (2006) Experimental investigation of a bottle-sublimation cooler. Int J Refrig 29:101–109

Gan H, Chang JZ, Feng JJ, Hu HH (2003a) Direct numerical simulation of the sedimentation of solid particles with thermal convection. J Fluid Mech 481:385–411

Gan H, Feng JJ, Hu HH (2003b) Simulation of the sedimentation of melting solid particles. Int J Multiphas Flow 29:751–769

Huang D, Ding G, Quack H (2008) New refrigeration system using CO_2 vapor-solid as refrigerant. Front Energy Power Eng Chin 2:494–498

Ilchi-Ghazaani M, Parvin P (2011) Characterization of a dry ice heat exchanger. Int J Refrig 34:1085–1097

Kim MH, Pettersen J, Bullard CW (2004) Fundamental process and system design issues in CO_2 vapor compression systems. Prog Energ Combust 30:119–174

Martynov S, Zheng W, Brown S, Mahgerefteh H (2016) Numerical simulation of CO_2 flows in pipes with phase transition across the triple point. In: International Conference on Heat Transfer, Fluid Mechanics and Thermodynamics

Michaelides EE, Lasek A (1987) Fluid solids flow with thermal and hydrodynamic non-equilibrium. Int J Heat Mass Transf 30:2663–2669

Michaelides EE, Lasek A (1991) Particulate flow with sublimation or evaporation and with thermal and hydrodynamic non-equilibrium. Int J Heat Mass Transf 34:601–609

Michaelides EE, Liang L, Lasek A (1992) The effect of turbulence on the phase change of droplets and particles under non-equilibrium conditions. Int J Heat Mass Transf 35:2069–2076

Rozhentsev AV (1995) A throttle-sublimation refrigerating machine using the solid coolant with changeable porous structure. In: Proceedings of 19th International congress of refrigeration. International Institute of Refrigeration, Paris, pp. 176–183

Sahin AZ, Dincer I (2000) Analytical modelling of transient phase-change problems. Int J Energ Res 24:1029–1039

Subramaniam S (2013) Lagrangian-Eulerian methods for multiphase flows. Prog Energ Combust 39:215–245

Tsuji Y (2007) Multi-scale modeling of dense phase gas-particle flow. Chem Eng Sci 62:3410–3418

Wang W, Lu BN, Zhang N, Shi ZS, Li JH (2010) A review of multiscale CFD for gas-solid CFB modeling. Int J Multiphas Flow 36:109–118

Yamaguchi H, Niu XD, Sekimoto K, Neksa P (2011) Investigation of dry ice blockage in an ultra-low temperature cascade refrigeration system using CO_2 as a working fluid. Int J Refrig 34:466–475

Yamaguchi H, Zhang XR (2009) A novel CO_2 refrigeration system achieved by CO_2 solid-gas two-phase fluid and its basic study on system performance. Int J Refrig 32:1683–1693

Yamaguchi H, Zhang XR, Fujima K (2008) Basic study on new cryogenic refrigeration using CO_2 solid-gas two phase flow. Int J Refrig 31:404–410

Yamaguchi H, Zhang XR, Kitai T, Fujima K (2005) Heat transfer characteristics of solid-gas two phase flow with sublimation of carbon dioxide. In: The 2nd International Exergy, Energy and Environment Symposium (IEEES2), July 2–7. KOS, Greece

Yamasaki H, Yamaguchi H, Hattori K, Neksa P (2017) Experimental observation of CO_2 dry-ice behavior in an evaporator/sublimator. Energy Procedia 143:375–380

Zhang XR, Yamaguchi H (2007) Forced convection heat transfer of supercritical CO_2 in a horizontal circular tube. J Supercrit Fluid 41:412–420

Zhang XR, Yamaguchi H (2011) An experimental study on heat transfer of CO_2 solid-gas two phase flow with dry ice sublimation. Int J Therm Sci 50:2228–2234